Flow Cytometry in Microbiology

Technology and Applications

Edited by

Martin G. Wilkinson

Department of Life Sciences
University of Limerick
Limerick
Ireland

Caister Academic Press

Caister Academic Press
Norfolk, UK

www.caister.com

British Library Cataloguing-in-Publication Data
A catalogue record for this book is available from the British Library

ISBN: 978-1-910190-11-1 (hardback)
ISBN: 978-1-910190-12-8 (ebook)

Cover design adapted from Figures 2.22, 7.1, 8.5 and 9.8.

Contents

Current Books of Interest

Shigella: Molecular and Cellular Biology	2016
Aquatic Biofilms: Ecology, Water Quality and Wastewater Treatment	2016
Alphaviruses: Current Biology	2016
Thermophilic Microorganisms	2015
Probiotics and Prebiotics: Current Research and Future Trends	2015
Epigenetics: Current Research and Emerging Trends	2015
Corynebacterium glutamicum: From Systems Biology to Biotechnological Applications	2015
Advanced Vaccine Research Methods for the Decade of Vaccines	2015
Antifungals: From Genomics to Resistance and the Development of Novel Agents	2015
Bacteria-Plant Interactions: Advanced Research and Future Trends	2015
Aeromonas	2015
Antibiotics: Current Innovations and Future Trends	2015
Leishmania: Current Biology and Control	2015
Acanthamoeba: Biology and Pathogenesis (2nd edition)	2015
Microarrays: Current Technology, Innovations and Applications	2014
Metagenomics of the Microbial Nitrogen Cycle: Theory, Methods and Applications	2014
Pathogenic *Neisseria*: Genomics, Molecular Biology and Disease Intervention	2014
Proteomics: Targeted Technology, Innovations and Applications	2014
Biofuels: From Microbes to Molecules	2014
Human Pathogenic Fungi: Molecular Biology and Pathogenic Mechanisms	2014
Applied RNAi: From Fundamental Research to Therapeutic Applications	2014
Halophiles: Genetics and Genomes	2014
Molecular Diagnostics: Current Research and Applications	2014
Phage Therapy: Current Research and Applications	2014
Bioinformatics and Data Analysis in Microbiology	2014
The Cell Biology of Cyanobacteria	2014
Pathogenic *Escherichia coli*: Molecular and Cellular Microbiology	2014
Campylobacter Ecology and Evolution	2014
Burkholderia: From Genomes to Function	2014
Myxobacteria: Genomics, Cellular and Molecular Biology	2014
Next-generation Sequencing: Current Technologies and Applications	2014
Omics in Soil Science	2014
Applications of Molecular Microbiological Methods	2014
Mollicutes: Molecular Biology and Pathogenesis	2014

Full details at www.caister.com

Contributors

Ultan Cronin
Flow Cytometry Unit
Biotechnology Programme
Spanish National Cancer Research Centre
Madrid
Spain

upcronin@cnio.es

Mario Díaz
Department of Chemical Engineering and Environmental Technology
Faculty of Chemistry
University of Oviedo
Oviedo
Spain

mariodiaz@uniovi.es

Imelda A. Doolan
Department of Life Sciences
University of Limerick
Limerick
Ireland

imelda.doolan@ul.ie

Thomas Egli
Department of Environmental Microbiology
Swiss Federal Institute for Aquatic Science and Technology (Eawag)
Dübendorf; and
MICROBES-IN-WATER GmbH
Feldmeilen
Switzerland

egli@eawag.ch
egli@microbes-in-water.ch

Mónica Herrero
Department of Chemical Engineering and Environmental Technology
Faculty of Chemistry
University of Oviedo
Oviedo
Spain

herreromonica@uniovi.es

Dara K. Hickey
Department of Life Sciences
University of Limerick
Limerick
Ireland

dhickey@cpingredients.com

Deirdre Kennedy
Food Chemistry and Technology Department
Teagasc Food Research Centre
Moorepark
Fermoy
Ireland

deirdre.kennedy@teagasc.ie

Stefan Kötzsch
Department of Environmental Microbiology
Swiss Federal Institute for Aquatic Science and Technology (Eawag)
Dübendorf
Switzerland

stefan.koetzsch@eawag.ch

Susann Müller
Department of Environmental Microbiology
Helmholtz Centre for Environmental Research
Leipzig
Germany

susann.mueller@ufz.de

Tim W. Overton
Bioengineering
School of Chemical Engineering
University of Birmingham
Birmingham
UK

t.w.overton@bham.ac.uk

Kathleen Schleinitz
Department of Environmental Microbiology
Helmholtz Centre for Environmental Research
Leipzig
Germany

kathleen.schleintz@ufz.de

Howard M. Shapiro
The Center for Microbial Cytometry
West Newton, MA
USA

hms@shapirolab.com

Martin G. Wilkinson
Department of Life Sciences
University of Limerick
Limerick
Ireland

martin.wilkinson@ul.ie

Kenji Yasuda
Institute of Biomaterials and Bioengineering
Tokyo Medical and Dental University
Tokyo
Japan

yasuda.bmi@tmd.ac.jp

Introduction

Martin G. Wilkinson

Microbiologists continually strive to develop methods which can enumerate bacteria, yeasts and mould populations and to distinguish the various physiological states of key microorganisms in industrial fermentations, waste treatment plants, water supply systems, during bioremediations or in clinical applications. Traditional methods have largely relied on various non-selective and selective media followed by more sophisticated analysis using PCR or immunological serotyping. The advent of genetic methodologies such as PCR and its variants has enabled a huge step forward in terms of both rapid and highly specific analysis of cells in various matrices. When combined with bioinformatics and phylogenetic analysis the transformation of these data has enabled new and exciting relationships to be drawn regarding species and strain evolution. However, while these technologies are highly useful, a key aspect that has not been clarified is the viable properties of these cells at the point of analysis. Indeed a key question for microbiologists is whether we can replace viability counting on agar plates with something faster but which can also generate a range of data enabling a better understanding of all bacterial cells within a given population. One such technology which promises to deliver further advances towards these objectives is flow cytometry abbreviated as FCM or FC. Flow cytometry is indeed a multi-parametric technique and a flow cytometer brings together various physical, biological, chemical and engineering disciplines to deliver a unique analytical package for the study of cell biology.

While FCM holds primacy in human clinical applications especially for immunological analysis it has now begun to make serious inroads into a new and challenging area; microbiology, in all its manifestations and complexities. In this text a number of authors have brought together a wealth of experience demonstrating the power and limitations of FCM as it currently stands within microbiology today. The book commences with a contribution in Chapter 1 from probably the most eminent scientist in the area of flow cytometry, not just microbial flow cytometry, Professor Howard Shapiro. This scientist who has already written a seminal text in the area elegantly describes the evolution of cytometry from the early microscopes of Antonie van Leeuwenhoek to the staining techniques of Paul Ehrlich and the development of the first true flow cytometers and casts an eye forward to its future position within microbiology based on technological advances and the ability to provide equipment which is simple to operate and is cost-effective. Thereafter the various chapters deal with individual areas of microbiology where FCM applications have advanced the basic understanding of the science through analytical innovation driven by FCM and/or its more advanced aspect, cell sorting. To the microbiologist the ability to use FCM to visualize, enumerate and analyse a population of cells into subpopulations of varying physiological status followed by the use of cell sorting to then actually physically isolate and sort specific cells for further analysis is truly amazing! Traditional microbiology has relied on separate techniques to examine growth (plate counting), physiology (biochemical testing) and morphology (light or fluorescence microscopy) of microorganisms of interest. These techniques while well accepted are quite slow and heavily labour-intensive with quite a considerable lag time before a complete data profile can be generated. A new and exciting development, which is only recently reaching the field of microbial cytometry, is imaging flow cytometry. This technique combines flow cytometry with advanced cell imaging enabling rapid differentiation of individual cells within a flow stream on the basis of shape/integrity/morphology. Using this type of flow cytometry, imaging analysis of morphology and the degree of uptake of fluorescent metabolic stains and/or expression of GFP constructs to enable monitoring of cellular responses at genetic level can be undertaken. This type of flow cytometer such as the Image Stream X Mark II from Amnis Corp, (Seattle, WA, USA) acquires a range of images for an individual cell in a flow stream using various imaging modes. The image quality is enhanced by using 40–60× microscopy objectives and it can acquire six images of each cell. Most reported applications of imaging flow cytometers involve mammalian cells. However a number of papers reporting the progression of bacterial infection of mammalian cells have provided interesting data on the possibilities for other non-medical microbiology applications. These cytometers have a reported flow through of ~300 cells per second. While the imaging system is purported to generate 60,000 images of 10,000 cells in about 30 seconds and 600,000 images of 100,000 cells approximately 5 minutes (Basiji *et al.*, 2007). In Chapter 2, Professor Kenji Yasuda outlines the development of on-chip imaging flow cytometry systems for analysis of bacterial behaviour in terms of single cell movement and growth kinetics. It is clear from this work that imaging flow cytometry may not be far behind in its implementation at on-chip level with numerous potential applications at basic and applied levels. A key feature of FCM is the ability to gain a range of data based on multi-parametric analysis, this feature is especially of interest when applied to understanding bacterial population heterogeneity. The traditional 'plate and wait' approach to bacterial population enumeration is already recognized as being too limited in its scope and merely allows certain bacterial cells which satisfy particular growth criteria to be enumerated. The use of enrichment and/or recovery media combined with more sophisticated techniques such as PCR enables a better composite profile of a bacterial population. However it is clear that multi-parametric FCM is superior in many aspects to many of the traditional techniques. In a number of Chapters the application of FCM to understanding population dynamics is discussed from a range of applications. Firstly, from the point of view of Environmental Bioprocessing where a new insight into understanding the key populations of the microbial consortia within biotreatment plants is discussed in Chapter 3 by Mónica Herrero and Mario Díaz. These authors outline the major advances that have been made when FCM and cell sorting have been combined with genetic analysis to isolate key bacterial species responsible for biodegradation reactions. The ability of FCM and cell sorting to isolate putative or candidate bacteria with potentially important metabolic activities which have even yet to be grown or identified by conventional technologies is also outlined. As these authors state, the numbers of unidentified or unculturable bacteria species in soil or water is immense but it is clear that cell sorting will advance our understanding of the environmental microflora. This theme of using FCM to gain a deeper understanding or fingerprinting of complex microbial communities has been developed by Susann Müller and Kathleen Schleinitz in Chapter 4. In this chapter an experimental case study is described in which the complexities involved in degradation of m-xylene by successive microbial populations are revealed using FCM. These and other authors outline a recurring theme related to analysis of microorganisms in complex samples (soil, water, food) by FCM, namely the requirement to release maximum number of bacterial cells without affecting their physiological status.

All sample preparation steps for microbial FCM analysis must generate an analyte solution free of interfering particles, with minimum auto-fluorescence and contain populations of cells capable of being analysed by the particular cytometer in use. To date, no optimal sample preparation method has been established for general use in analysis of matrices such as soil, food or environmental sources. Publications generally describe in-house sample preparation methodologies which in many cases includes filtration and centrifugation (including density gradient centrifugation). However, antibody based systems such as Immunomagnetic Separation (IMS) show promise when looking to concentrate/isolate specific target populations in complex matrices. While this approach has been developed for some commercialized PCR and FCM based technologies it is still a very difficult technical proposition to clean up samples, release the total microbial population and capture sufficient cells for FCM analysis. An evaluation of commercial systems which recovered *E. coli* 0157:H7 from spinach using IMS capture found that efficiencies ranged from ~1 to 63%. This variation was dependent on extraction medium, sample volume, presence of background microflora and the use of spiking (Chen *et al.*, 2014). Thereafter a 4- to 6-hour enrichment was required for detection either by PCR or FCM (Chen *et al.*, 2014). Buzatu *et al.*, (2013, 2014); Wilkes *et al.*, (2012) describe the evolution of a commercial flow cytometric system for the rapid detection of *E. coli* 0175H7 in foods such as spinach and outline in great detail the technical issues involved including overcoming matrix interference and cell recovery steps. It is evident that for direct rapid FCM analysis for a range of pathogens in foods much work still remains to provide simple, cost-effective, high efficiency methods for capture of cells from complex matrices and which eliminate or minimize a cell enrichment step.

While FCM can analyse samples rapidly the data are only as good as the sample presented for analysis. In Chapter 5 the detection of pathogenic bacteria by FCM in a range of samples including food, environmental and clinical origins are discussed. The themes both Deirdre Kennedy and myself as authors have sought to highlight are the methodologies required for isolation of target bacteria from diverse samples and the necessity to set up the cytometer properly by running an extensive series of control cell populations. For all scientists or practitioners in the field of microbial cytometry it is vital to have first ascertained where live, dead and damaged target cells occur on a typical cytograph for the strain of interest. This can be carried out initially in media and thereafter in samples which have been cleaned up by various methodologies as outlined in this chapter. It is also necessary to understand that species and strain-related differences may well occur in the uptake of many common stains used in FCM analysis and some stains may prove toxic to cells thus producing erroneous data. Hence, a series of experiments should be undertaken to ensure that cell viability or metabolic activities are not affected by the staining procedures. In the case of cell sorting, the sorting procedure may also affect viability and should also be checked prior to embarking on extensive microbial studies.

A key issue for microbial FCM is the necessity to selectively stain and/or label target cells within a complex microflora. To date, the stains commercially available do not allow for any major species and strain selectivity of the extent required for routine pathogen analysis. In contrast to mammalian FCM, the range of satisfactory commercial antibodies suitable for FCM analysis is still quite limited. In part, this may be due to market forces which are still not driving the development of commercial antibodies probably due to the limited commercial application of microbial FCM. A major technical obstacle is the necessity to determine unique antigenic properties of a range of various species and strains and also to find epitopes which can also be used classify live, dead and damaged cells. Especially when these cells are present at very low levels in complex samples and preferably without an enrichment step. In Chapter 6, the application of FCM to detection and study of spore formers such as *Bacillus* or *Clostridia* species is outlined in detail by Ultan Cronin. Currently spore formers in foods are a major issue as they can survive almost all current conventional heating and dehydration technologies and subsequently germinate on storage of the product causing various illnesses, some fatal. The issue of rapid methods to detect spore formers is of major interest to the Dairy and Infant Formula industries who are currently using slow, laborious and non-selective quality control methods. However as the author explains there are a number of morphological and physiological states in the spore to vegetative cell life cycle that may provide avenues which can be exploited for FCM detection. Hence, in this chapter, the potential for FCM – when combined with immuno-labelling, staining and/or nucleic acid probes – to offer the possibility to generate rapid and selective assays for spore formers and vegetative cells is highlighted. The range of applications of FCM and cell sorting to the study of yeasts and fungi is explored extensively in Chapter 7 by Tim Overton. Yeasts and fungi play a major role in industrial production of high value products such as beers, wines and industrial enzymes. Hence, bioprocess control using on/at line FCM technology would clearly be of economic value. In the medical field, the use of FCM for rapid detection of yeast or fungal infections and the identification of optimal drug based therapeutic strategies is already providing new insights into drug–host cell interactions. The use of yeast surface display (YSD) in combination with FCM and cell sorting is a major biotechnological application for this technology for recombinant protein production and is outlined in detail.

Another important group of industrial microorganisms are lactic acid bacteria (LAB) principally involved in the production of a range of fermented dairy foods including cheese, yogurt and fermented milks. Indeed, LAB also play a key role in the generation of particular flavours in wines through the malo-lactic fermentation process. In Chapter 8, authors, Doolan, Hickey and Wilkinson outline the development of FCM and cell sorting to the study of LAB during cheese production and ripening along with some other food applications. LAB contribute to acidification of cheese products during their viable state thus ensuring their safety, while their enzymes generate flavour compounds during the non-viable state which proceeds to extensive cell autolysis and release of intracellular enzymes which hydrolyse the casein and lipid content of the cheese to generate typical flavours. In the selection of LAB cultures for cheese production a key factor is the ability of cells to release enzymes through early autolysis. The traditional methodology to establish strain related autolytic profiles relies heavily on the use of viable cell counts on solid media with a three day incubation step. This is usually combined with data obtained from measurement of released intracellular enzymes such as lactate dehydrogenase (LDH) or peptidases such as Pep X into aqueous cheese extracts which act as markers of autolysis. However, while these methods are sufficiently useful to allow determination of variations in autolysis it was not clear until recently whether release of enzymes was possible or even influenced by differences in cell permeabilization properties. In this chapter the important contribution of FCM towards the establishment of rapid methods for estimation of cell viability and the establishment of the importance of strain related permeabilization properties in intracellular enzyme release is outlined. It is also clear that the application of FCM and cell sorting will provide new methodologies for selection of LAB cultures for particular industrial processes based on the ability to select strains or even strain subpopulations which perform key reactions within a food system. In addition, the potential for FCM to unravel and provide a deeper insight into the potential interactions between a human host cell in the gastrointestinal tract and a probiotic LAB culture are only just becoming to be realized.

In the final chapter, Thomas Egli and Stefan Kotzsch outline a fascinating journey from laboratory-scale work on water quality using FCM to replace or enhance traditional methodologies for drinking water quality to the eventual full scale acceptance and implementation of newly developed FCM based standards of quality control. Their work has been so successful that FCM is now used as a key quality control technology for drinking water quality in major cities of Switzerland such as Zurich. In this Chapter the authors have outlined the obstacles they faced both technical, regulatory and attitudinal towards the use of FCM for water monitoring. However, they also show how a 'bottom up' approach works best when laboratory analysts take on board and are first convinced by the new technology and then drive the regulators towards adoption of the technology, provided all validation steps are satisfactorily achieved. Currently, FCM-based drinking water quality testing is written into Swiss Federal regulations and the technology is progressing towards on-line quality testing. This is a situation that the pharmaceutical, food and allied healthcare industries should strive for, and in my view, this will become a reality in certain of the industrial processes provided they are amenable to rapid sample preparation, have access to species/strain related discriminatory staining or antibody/DNA based tagging procedures and the availability of cheap, simple and perhaps specifically modified microbial flow cytometers.

A key advantage to flow cytometry for the microbiologist is the instant output of data which can quantify various cell subpopulations. Increasingly cytometers are providing us with the ability to manipulate these data and develop statistically valid outputs which will certainly enhance acceptability for publications and for uptake by the regulatory community. However, what may be of benefit is the integration of modelling and predictive growth responses, already available in the literature or as accessible databases from regulatory authorities, for a range of pathogens into the data analytics package of a cytometer. This must of course be based on using samples which allow the capture and analysis of a particular strain or species, for example by immuno-magnetic separation (IMS). For example, in Food Microbiology there is quite a body of work on modelling of growth patterns of pathogens, e.g. *Listeria monocytogenes* under different conditions of temperature, pH and salt concentrations. These data could also be supplemented with FCM profiles using differential staining to provide an enhanced database for major bacterial species of interest. These data could then be available on a flow cytometer analytics package such that, when working with specific cultures obtained from samples with particular compositional/storage parameters, the analyst may be able to relate certain FCM profiles with the statistical probability of the preponderance of a particular physiological status, e.g. the higher or lower likelihood of the presence of cell subpopulations in a VBNC state. These decisions could be based on certain pre-programmed criteria from a combination of predictive modelling, FCM profiling and other physiological data available within the cytometer's analytics package. Another factor critical to the uptake of microbial flow cytometry is development of specially adapted cytometers for microbial detection. Fortunately, this is also becoming a commercial reality with a range of engineering and electronic modifications and pre-selected gates such as those described for the RAPID-B model 9013 flow cytometer for detection of microbial pathogens (Wilkes *et al.*, 2012).

Finally, I wish to make the reader aware of the limitations of this book which for various logistical reasons has not dealt with other areas of microbial flow cytometry such as the exciting work on monitoring marine microbiology conducted by Dr Gerald Gergori and colleagues in Marseille, the work on use of cytometry for monitoring microbial toxins by Francis Ligler and colleagues in the USA. It would be totally remiss of me to not acknowledge the pioneering and ongoing research undertaken into microbial flow cytometry undertaken by Chris Hewitt and Gerhard Nebe-von-Caron, whose work has inspired many traditional microbiologists, including myself, to take the step into microbial cytometry. These workers and others, including J. Paul Robinson at Purdue University, have laid the foundations and hopefully all others can build on them to develop a unified, robust methodology which provides a unique insight into the microbial cell and its interactions with humans in all its manifestations both good and bad.

References

Basiji, D.A., Ortyn, W.E., Liang, L., Venkatachalam, V., and Morrissey, P. (2007). Cellular image analysis and imaging by flow cytometry. Clin. Lab. Med. *27*, 653–670, viii.

Buzatu, D.A., Cooper, W.M., Summage-West, C., Sutherland, J.B., Williams, A.J., Bass, D.A., Smith, L., Woodruff, R.S., Christman, J.M., Reid, S., *et al.* (2013). Photobleaching with phloxine B sensitizer to reduce food matrix interference for detection of *Escherichia coli* serotype O157:H7 in fresh spinach by flow cytometry. Food Microbiol. *36*, 416–425.

Buzatu, D.A., Moskal, T.J., Williams, A.J., Cooper, W.M., Mattes, W.B., and Wilkes, J.G. (2014). An integrated flow cytometry-based system for real-time, high sensitivity bacterial detection and identification. PLoS ONE *9*, e94254.

Chen, J., Shi, X., Gehring, A.G., and Paoli, G.C. (2014). Automated immunomagnetic separation for the detection of *Escherichia coli* O157:H7 from spinach. Int. J. Food Microbiol. *179*, 33–37.

Wilkes, J.G., Tucker, R.K., Montgomery, J.A., Cooper, W.M., Sutherland, J.B., and Buzatu, D.A. (2012). Reduction of food matrix interference by a combination of sample preparation and multi-dimensional gating techniques to facilitate rapid, high sensitivity analysis for *Escherichia coli* serotype O157 by flow cytometry. Food Microbiol. *30*, 281–288.

Microbial Cytometry: What It Was, Is, and May Be

1

Howard M. Shapiro

Abstract

Microbes were discovered, counted, and sized in a microfluidic device by Antonie van Leeuwenhoek in the late 1600s, but their importance in nature and in health did not become apparent until the late 1800s, when many classical methods of microbial analysis were developed. It has only been since the mid-1900s that cytometry has made it possible to detect, identify, and characterize single microorganisms and other cells in terms of their structure, function, and genetics. Modern multiparameter cytometry typically makes multiple physical measurements of a cell, most commonly of light scattering and fluorescence, which may be emitted from cellular constituents themselves or from reagents or probes added to facilitate their detection and quantification. Flow cytometers, in which cells are analysed as they flow through the apparatus in a fluid stream, can be equipped with sorting capability to allow separation of cells with preselected characteristics; however, when sorting is not required, an increasing proportion of the measurements now made in flow cytometers can be made in simpler, less expensive imaging systems. By 2020, improvements along this line should allow the benefits of cytometric technology to be applied to problems and in places for and in which it was previously unaffordable. Microbiologists, who have been among the 'have-nots' with respect to cytometry, even in affluent countries, stand to benefit substantially from this.

Introduction

It has now been many years since the molecular biologists confirmed the evolutionary biologists' long-held suspicion that microbes are our distant, very much older relatives, and we continue to discover new species and the interesting, often useful, sometimes frightening things they have learned to do since we took different paths away from our common ancestry. We are just beginning to examine ourselves as microbial habitat; it has, after all, only been a few dozen human generation times since microscopy first made us aware of the existence of microorganisms and, some generations later, of our own somewhat more elaborate cellular structure. And it has only been in the last ten human generations, representing five million generations for some bacterial species, that, aided by greatly improved microscopy and by cytometry, we have come to appreciate the incredible phenotypical and functional heterogeneity of microorganisms and its implications for us and our fellow eukaryotes.

If the reader of this highly specialized volume has arrived at this point without some familiarity with modern microscopy and cytometry and is expecting a quick introduction, I would suggest he or she turns elsewhere. My lengthy book on flow cytometry (Shapiro, 2003), available free online, includes an introductory first chapter; there are also two extensive and excellent review articles dealing with microbial cytometry in particular (Davey and Kell, 1996; Müller and Nebe-von-Caron, 2010).

The word 'cytometry' puts the cell first and its measurement second. The very discovery of microorganisms became possible only after both the technology and the philosophy of science had reached a certain level of sophistication, and our perspective on microbiology continues to change as we develop and employ new techniques for the study of microorganisms, adding new terms to our vocabulary, and changing the meanings of older ones, as we go along. In the remainder of this chapter, I will consider how we have gotten where we are and how both old and new knowledge and technology may change the way we do things in the near future.

Tasks in microbial cytometry: motives and methods

The first microbial cytometry of which we are aware was done simply to satisfy the curiosity of a 17th-century Dutch fabric merchant whose original interest in microscopy probably reflected a desire to use it to assess the quality of his goods. Antonie van Leeuwenhoek had apparently seen the magnified images of various textiles in Robert Hooke's *Micrographia* (Hooke, 1665). In thinly sliced cork, at about 50× magnification, Hooke saw hollow structures he called cells, reminiscent of the similarly shaped larger spaces found in a honeycomb. His term was eventually adopted to describe what are now familiar to us as the basic elements of living systems (Harris, 1999). The 'cyte' in 'cytology' and 'cytometry,' words used since the 1800s, comes from a Greek term for a cell in an insect nest, used by Aristophanes in *The Wasps*.

Although lenses had been used to aid vision before 1300, it was not until the early 1600s that they were combined to bring distant objects closer, creating telescopes, and to permit visualization of objects too small to be seen with the unaided human eye, creating microscopes. Hooke's two-lens instrument fit the modern definition of a compound microscope, but, although his contemporaries and he eventually managed to observe a variety of animal and plant cells, no such instrument then had sufficient magnifying power to make microorganisms visible. Van Leeuwenhoek, whose microscopy hobby became an obsession, learned to make simple lenses with magnifying power 200× and higher, and eventually encountered not only blood cells but also 'little animalcules,' a designation he applied to human and animal sperm and a variety of protists and bacteria on the basis of their motility.

The means available to Hooke and van Leeuwenhoek for moving specimens into and out of the field of view were limited; Hooke adopted van Leeuwenhoek's technique for observing small objects in a glass tube or 'straw,' heated and drawn 'fine as a human hair.' Both were therefore, according to our modern definition, employing microfluidics in their instruments. Their only available high-brightness light source was the Sun; candles and lamps became progressively less useful as magnification increased. Specimens were often observed using oblique illumination rather than transmitted light; this provided an approximation to modern dark field and reflectance microscopy, both of which allow visualization of objects below the resolution limit of transmitted light.

Hooke, a skilled illustrator, made drawings of what he saw; van Leeuwenhoek described what he observed to an artist, persevering until what he deemed an acceptable drawing emerged. Both men counted objects under the microscope, and both attempted to measure their size. In England, which had established measures of length, Hooke could express dimensions in fractions of an inch. Lacking equivalent standards, van Leeuwenhoek compared his specimens to objects such as grains of sand or 'the eye of a large louse' (both were readily available at the time). Thus van Leeuwenhoek, at least, can be said to have practised cytometry, or the measurement of cells, and, more specifically, microbial cytometry. The words 'microbe' and 'cytometry' would, however, not be coined until about two hundred years later, by which time optics had improved and photography could provide more objective pictorial documentation; the limited resolution available to van Leeuwenhoek allowed him to fancy he saw his animalcules copulating and giving birth, and calculate the sizes their internal organs ought to be.

Detecting and counting microorganisms in a specimen remain the simplest tasks in microbial cytometry. Were we to land a spacecraft on Mars, equipped only with one of van Leeuwenhoek's microscopes and enough robotics, optics, and electronics to collect and image a soil specimen, video showing one or more small objects in motion could allow us to conclude that there was life on Mars, and count organisms defined as viable on the basis of their motility. Doing only this would not exclude the possibility that we had brought them with us. Without the video, we might be able to detect the presence of nucleotide sequences similar or even identical to those found on Earth; that, however, would not tell us whether any associated life forms were even viable, let alone establish their planet of origin.

Between 1700 and 1850, it became obvious that, although motility might be a sufficient criterion for establishing that a small particle was a living organism, the capacity for reproduction was a necessary one. Non-motile fungi, which had provided us with bread, beer, wine, and cheese for millennia, were initially dismissed by van Leeuwenhoek as mere 'globules,' but later work by others showed that both motile and non-motile microorganisms could reproduce by fission. Lazzaro Spallanzani, in particular, provided strong evidence that they could not arise by spontaneous generation; Louis Pasteur was to close the case for all but the conspiracy theorists.

The magnification of compound instruments made in the 1880s could exceed that of van Leeuwenhoek's lenses, while aberrations were greatly diminished. Detection and counting of small particles, living or otherwise, however, remained based on human perception, with the aid of microscopy, of their size, shape, and motility. Alphonse Laveran's 1880 identification of the malaria parasite in unstained blood, based on the motility of exflagellated gametocytes, could have been done using one of van Leeuwenhoek's microscopes, although optical discrimination between different types of microorganisms would soon after be facilitated by staining specimens with synthetic and natural dyes.

Pasteur's work in the 1850s and 1860s applied his training in chemistry, a science which matured greatly from the 1700s on, to establishing the role of yeasts and bacteria in fermentation and the dependence of industrial fermentation processes on maintaining the proper mix of organisms. He moved on to identify specific bacteria as animal and human pathogens, and to develop pasteurization and immunization to prevent and treat bacterial and viral disease. Beginning in the 1870s, Robert Koch further systematized the microbiology of infectious disease, identifying the bacterial causes of anthrax, tuberculosis, and cholera, refining techniques for microbial microscopy, and developing methods for pure culture of organisms on solid media, all the while maintaining an acrimonious rivalry with Pasteur.

It is fair to say that Paul Ehrlich was the prime mover in staining; he was introduced to the concept by his older cousin Karl Weigert, who worked with the great pathologist Rudolf Virchow.

Virchow had determined to put pathology on a cellular basis even as the last details of the cell theory were falling into place; it was difficult to do this when little of the internal structure of most cells could be visualized clearly by microscopy of untreated specimens. Ehrlich utilized synthetic textile dyes then newly available, demonstrating in his 1878 medical thesis that mixtures of acidic and basic dyes permitted different blood cell types to be distinguished by the coloration of their organelles. The stains used in haematology and malaria diagnosis today are but a few steps removed from Ehrlich's eosin–methylene blue mixtures (Clark and Kasten, 1983).

Koch had adapted Ehrlich's methylene blue technique to observe the anthrax bacillus, but found *Mycobacterium tuberculosis* far more difficult to stain; Ehrlich joined Koch's laboratory in 1882 and quickly developed a faster, more precise staining method based on the 'acid-fastness' conferred on mycobacteria by the lipid composition of their cell walls. Today's venerable Ziehl–Neelsen and auramine O staining procedures for TB are only slightly removed from Ehrlich's, and his work inspired Christian Gram to develop his own well-known stain for bacteria, which discriminates 'Gram-negative' and 'Gram-positive' organisms. We now know that the cell walls of the former contain lipopolysaccharides, whereas those of the latter do not.

Ehrlich's concept of chemical specificities in cells led him in several significant directions. Although his 1908 Nobel Prize reflected his contributions to immunology, it was his development of chemotherapy for syphilis, based on screening hundreds of dyes for activity against the pathogen, which brought him worldwide fame. Recognizing the affinity of methylene blue for malaria parasites, he introduced it as a drug; the compound is even now incorporated into combinations under test for treating the disease. More significantly, many of the dye companies evolved into drug companies, eventually yielding compounds such as ethidium bromide that are now far more useful in cytometry than in clinical medicine. Paul De Kruif's classic *Microbe Hunters* (De Kruif, 1926) still provides an absorbing account of the progress of microbiology to its time; when it was written, Ehrlich's arsenical compounds, Salvarsan and Neosalvarsan, active against the agent of syphilis, were the only known active antibacterial drugs.

There are now many variations on the basic cytometric theme of detection and counting. Organisms in pure culture may need to be identified and/or characterized as to their growth, metabolism, viability, and, in clinical situations, interaction with antimicrobial agents, and the most complex exercises in microbial cytometry typically require detection, identification, counting, and characterization of each of a number of organisms in a mixed population. This can involve removing cells other than those of interest from samples and/or detecting and measuring one or more identifying characteristics of the cellular target(s). Much microbial cytometry is still done by human observers using techniques of transmitted light microscopy available since the late 1800s. From then on, however, our characterization of microbes at both population and single cell levels became increasingly dependent on chemistry, as Table 1.1 indicates.

From microscopy to cytometry: more than meets the eye

The term 'microbe,' coined in 1878 to apply to the microorganisms visible under the microscope, sidestepped the question of whether they were animals or plants according to the 'two-kingdom' classification of life then used. By around 1900, after improved microscopy had facilitated the discovery of numerous bacterial and protozoan pathogens, it became clear that some other infectious diseases could be transmitted by agents which passed through filters that would retain the smallest known observable microorganisms; our nomenclature has eliminated the adjective 'filterable' and now refers to such agents simply as viruses.

The word 'cytometer,' like 'microbe,' dates from the 1870s; it initially denoted a chamber of defined volume in which an observer could determine the number of cells per unit volume of a specimen by microscopy, rather than an apparatus that could measure cells without constant human attention. The specimen was most often blood, and the device commonly used was referred to as a haemocytometer. In one of the more famous early publications on cytometry (Student, 1907), however, the cells involved were definitely microbial, of the species *Saccharomyces cerevisiae*. The author, who described what we would now call the Poisson statistics governing the precision of counting cells (and photons and many other things), adopted a pseudonym to conceal the fact that his work was of potential commercial importance to his employer, the Guinness brewery.

Getting from 'Student' to Table 1.1 is a long trip, involving many long stories. I will cut to the chase and refer you to my cytometry book (Shapiro, 2003) and a history chapter (Shapiro, 2010) I wrote for another cytometry book for the historic extras.

What we can measure in the microbes, which have been around a few billion years longer than we have, are mostly things they have carried on their journey for quite some time. Even the genes responsible for antimicrobial resistance, which have become considerably more widespread due to human influence, can be found in caves where syringes have never been dumped. We do have to take the rap for fluorescent proteins and other products of genes we have put into microorganisms, but that has only gone on for a few decades.

Almost all measurements now done in cells, microbial and otherwise, with flow or laser scanning cytometers utilize fluorescence. The most complicated flow cytometers can include at least nine laser beams (meaning that one will have more by the time you read this) and measure fluorescence in at least 24 spectral regions (see last set of parentheses), but very few instruments measure light scattering at more than one wavelength or more than two angles. I will discuss scattering in some detail and, of necessity, take a narrower approach to fluorescence, covering only a few selected topics.

Table 1.1 Microbial parameters measurable by cytometry

Parameter	Measurement method other than fluorescence
Intrinsic structural parameters (no probe added)	
Cell size	Small angle light scattering; electronic impedance [DC]; dye exclusion volume
Cytoplasmic granularity, vacuoles, etc.	Large angle light scattering; electronic impedance [AC]
Birefringence	Depolarized light scattering
Pigment content (e.g. photosynthetic pigments)	
Intrinsic functional parameters (no probe added)	
Redox state (endogenous pyridine and flavin nucleotides)	
Fluorescent protein expression	
Extrinsic structural parameters	
DNA content	
DNA base ratio	
Nucleic acid sequence (DNA and RNA)	
RNA content (double-stranded)	
Total double-stranded nucleic acid	
Total protein	Large-angle light scattering
Lipids	
Surface and internal antigens	
Surface sugars (lectin binding sites)	
Gram staining status	
Extrinsic functional parameters	
Surface receptors (including phage receptors)	
Surface charge	
Membrane fusion or turnover	
Cell division	
Membrane integrity ('viability')	
Membrane permeability (dye/drug/substrate uptake/efflux)	
Intracellular receptors	
Enzyme activity	Chromogenic substrates
Oxidative metabolism	Chromogenic substrates
Sulfhydryl groups/glutathione	
DNA synthesis	
Membrane potential	
Intracellular pH	

Intrinsic parameters

Light scattering and cell size

Scatter signals are traditionally divided into 'forward (FSC),' or 'small angle,' measuring light collected at 0.5 to perhaps 5° from an incident light beam, and 'side (SSC),' or 'large angle,' measuring light collected at angles above 15°. Most flow cytometers use vertically polarized laser beams for illumination, meaning the laser emits fluorescence polarized coplanar with the sample stream. Occasionally, additional optics, which can be as simple as pieces of a sunglass lens, are added to a detector to permit measurement of 'depolarized' SSC, in a horizontal plane perpendicular to the plane of polarization of the laser.

It is likely that van Leeuwenhoek was able to see some of the smaller objects he described only because they scattered light at relatively large angles. Two centuries later, John Tyndall established that particles too small to be detected easily by conventional microscopy, notably bacteria and fungi, could scatter measurable light, and dark field illumination came into use in microscopy. The dark-field 'ultramicroscope' developed by Richard Zsigmondy in the early 1900s arrived just in time to provide experimental confirmation of Albert Einstein's theory of Brownian motion; and it was used to observe viruses and large colloid molecules, helping Zsigmondy earn the 1925 Nobel Prize in chemistry. The apparatus itself was later modified to permit an aerosol sample to be flowed through it intermittently, allowing particles to be observed and counted in a chamber of defined volume; it was noted that, as 'Student' had found, the counts obeyed Poisson statistics.

'Forward scatter measures size' is the major urban legend of cytometry. Sometimes it does; more often, it does not, dependent on the details of the optical design of the apparatus, refractive index variations in the particles analysed and, probably, the horoscope for the day. It's actually somewhat better for bacteria, particularly spherical ones, than for larger cells. What forward scatter is good for is giving you a signal whether or not the cell or particle you are analysing contains anything fluorescent, which can be used to 'trigger' the instrument to recognize an 'event.' Although this works reliably with most eukaryotic cells, FSC signals from bacteria are often not detectable above scattered background, and, when they are, the instrument may not have enough electronic gain to get the measured values away from the bottom end of the scale. The low gain problem has been solved by replacing the relatively inexpensive silicon diodes used as FSC detectors in almost all older flow cytometers with more sensitive devices, such as photomultiplier tubes (PMTs).

Background problems are harder to deal with. One should, however, remember that, although particles scatter more light at small angles than at larger ones, background interference from laser light is typically substantially higher in FSC measurement channels than in channels used for SSC. Although SSC signals are more dependent on particle structure than are FSC signals, the information contents of the two signals become more similar as particles get smaller. SSC, which has been reported to track total protein content (which itself has been reported to track the 'dry mass' measurable by interference microscopy) in both microbes and eukaryotic cells, can often provide an acceptable indicator of the size of microorganisms. SSC is also typically a better trigger signal, since its background is lower.

The scattering behaviour of microorganisms contributes substantially to the bulk properties of microbial suspensions. In the early 1900s, it became common practice to estimate counts of bacteria in high concentrations (typically 10,000,000 or more per ml) from the turbidity they imparted to the medium; stable inorganic standards developed by Joseph McFarland in 1907 improved the precision of visual estimation. Such standards are still used, with photoelectric turbidimeters or nephelometers usually replacing the human observer.

Although photoelectric detectors were developed in 1904, they were not incorporated in scientific and medical instruments, cytometric or otherwise, until decades later. The first working flow cytometer, a direct descendant of the ultramicroscope, was built by Frank Gucker *et al.* (1947) during the Second World War; it measured light scattered at large angles from a beam derived from a Ford headlamp, using a PMT as a detector. The device, paid for by the US Army, was intended to detect and count biowarfare agents, e.g. anthrax spores, in aerosols. It incorporated a 'sheath' of filtered air to confine the air sample stream to the central portion of a flow chamber. The 'sheath flow' or 'hydrodynamic focusing' principle was adapted in the 1950s for blood cell counting. The terms 'flow cytometer' and 'flow cytometry' themselves did not come into use until the mid-1970s.

Electronic impedance-based cell counting and sizing technology was developed in the 1950s by Wallace Coulter, an electrical engineer who abandoned attempts to build an optical blood cell counter after learning that cells, by virtue of being membrane-bounded, are relatively poor conductors of electricity. This property had allowed the haematocrit, i.e. the percentage of blood volume accounted for by cells, to be determined by bulk measurement of the conductivity of blood, which might be viewed as exemplifying an electronic version of Archimedes' principle. If cells suspended in saline are passed through a small orifice, the displacement of conductive saline by a nonconductive cell increases the electrical impedance of the orifice in proportional to the volume of the cell, and the change can be detected and quantified by relatively simple electronics. If the orifice is made small enough, it is possible to detect single virus particles and large molecules using the Coulter principle; the sensitivity of light scattering-based cell counters can be increased to similar levels, but sample throughput and robustness and reliability of the apparatus is typically decreased for both methodologies as the particles of interest become smaller. Mack Fulwyler's first droplet cell sorter selected cells based on a Coulter volume signal; one can now buy flow cytometers, but not sorters, that measure Coulter volume as well as light scattering and fluorescence, but their orifice diameters, intended for analysis of a range of eukaryotic cells, are too large to permit measurements of microbial volume.

A recent paper (Tzur *et al.*, 2011) provides a method for determining the optical parameter(s) most representative of cell volume based on Coulter volume measurements of cell fractions sorted based on optical characteristics. In experiments with several eukaryotic cell lines, these authors found a different optical parameter best for each cell type, suggesting that their empirical approach might also be useful for microorganisms. Anyone who has a cell sorter should be able to find enough money for a low-end impedance-based cell counter.

Pulse width measurement is useful for sizing eukaryotic cells. If an illuminating laser beam is focused to an elliptical spot with its smaller axis in the direction of sample flow, objects with dimensions progressively larger than that axis dimension will produce signal pulses of progressively greater lengths. The area or integral of a fluorescence pulse will be proportional to the total amount of fluorescent material in the object; the height or peak value of the pulse will be representative of the brightness or fluorescence density of the object, and the duration above threshold or width of the pulse will be representative of volume. In flow cytometers that process pulses digitally, a 'width' is typically calculated from peak and integral values; in older instruments, hybrid circuitry was used to derive and store values of pulse widths, peaks, and integrals. Extinction signals were frequently used for pulse width based sizing, and measurement quality was good enough to be useful for red cell sizing in clinical instruments.

The beam axis dimension typically used for pulse width cell sizing is about 5 μm. This works well for eukaryotic cells; bacteria and fungi, however, are typically substantially smaller than that in any linear dimension, and relatively large changes in bacterial dimensions produce little or no change in pulse widths when that axis dimension is used. Although space precludes my including a detailed explanation, it is impractical to make the beam axis small enough for effective pulse width sizing of bacteria and other particles similar or smaller in size. It is also difficult to use extinction signals for sizing bacteria, whereas they work well for eukaryotic cells, just as it is much harder to detect a transit of Venus than a total solar eclipse.

Another recent paper (Lababidi *et al.*, 2011) demonstrates that precise volume measurements of microorganisms can be obtained using inexpensive imaging equipment and a reagent likely to be available at grocery stores, i.e. a food colouring dye. Such dyes are required to be nontoxic and therefore carefully checked to verify that they do not enter cells or bind excessively to their surfaces. If cells in a medium containing the dye are placed in a chamber of uniform thickness, each cell displaces a volume of dye equal to its own volume; the difference between the optical signal obtained from an image of a region containing the cell and the signal obtained from an image of a region the same size containing only dye provides an indication of cell volume.

The optical signal used by Lababidi *et al.*

(2011) was extinction; a much older paper (Gray *et al.*, 1983) describes a similarly Archimedean approach using a fluorescent dye and a flow cytometer. Cell volume is obtained by adding dye to samples and keeping the core stream diameter larger than the diameter of the cells of interest and the core flow rate constant, and monitoring the fluorescence signal from the core, which decreases in proportion to cell volume when cells pass through the observation region. Using the flow cytometric method with microorganisms might be tricky because of the need to maintain a very small core diameter in order to detect small volumes. It would, however, be relatively easy and cheap to implement for larger objects, requiring only a small change in detector electronics, and, unlike any other optical method available in flow cytometers, does provide a signal directionally proportional to volume.

Intrinsic fluorescence

The cyanobacteria have essentially donated their photosynthetic pigments to science; both phycoerythrins and allophycocyanin are used alone and in tandem combination with other fluorophores as fluorescent labels. The amounts of different pigments in a given species vary with the amount of light available in its environment and with various chemical factors. The presence of the pigments also provides challenges to investigators who want to use fluorescently labelled antibodies or other ligands to measure extrinsic parameters in cyanobacteria. In general, it is advisable to determine the intrinsic fluorescence of both organisms of interest and the media in which they are to be found before selecting dyes and labels, and this becomes more important as the expected intensity of signals from the target organisms decreases.

The reduced coenzyme NADH exhibits UV-excited blue autofluorescence; the oxidized form, NAD^+ does not. Laflamme *et al.* (2005), examining aerosols using a specially designed flow cytometer, found that *Bacillus* spores with higher autofluorescence were less likely to exhibit membrane damage and had higher plating efficiencies. Autofluorescence signatures in general also provide an approach to discriminating bacterial aerosol particles from other organic and inorganic particles.

Extrinsic parameters

Fluorescent dyes and labels

For at least the past decade, more sorting has been based on fluorescent protein expression than on fluorescent antibody binding. Cytometry has played a critical role in the development of both monoclonal antibodies and fluorescent proteins as useful tools, but matching the tools and apparatus to one another and to the biological questions being asked remains critical. Fluorescent proteins are remarkable in that they provide themselves with fluorescent labels by altering their amino acids; they can be engineered to report various physiological conditions in cells, but the more sophisticated applications demand skill and care on the part of experimenters.

Selection of fluorescent labels for antibodies and/or other ligands also demands increasing skill and care as the number of ligands used in an experiment increases. Although current compensation protocols can do a reasonable job of separating signals from five or more labels excited at a given wavelength, it is usually advisable to use the labels with the shortest emission wavelengths for the ligands expected to provide the weakest signals, as most labels have more pronounced emission spectral tails above, rather than below, their emission maxima. It is rarely possible to get good quantitative fluorescence measurements from more than two or three labels, and doing even this usually requires that a separate excitation beam be used for each label, minimizing side effects of compensation.

Large fluorescent particles, e.g. dye-impregnated plastic or silica beads, quantum dots, and doped metallic colloids, can produce very bright fluorescence signals when bound to cells, but are also much more likely to bind non-specifically than are ligands labelled with less bulky materials. It is discouraging to observe bright spots of fluorescence from a cell preparation stained with particles and then to find, after adding a permeant nucleic acid dye, that the bright fluorescence is coming from aggregates of labelled particles nowhere near a nucleus, while the label fluorescence in the immediate vicinity of the nuclei is barely detectable. A lot of smart people have been trying to solve this problem for a long time; most newcomers would benefit from greater familiarity with the literature.

Fluorogenic and chromogenic substrates

The activity of oxidative and hydrolytic enzymes, either in bulk or at the single cell level, can be measured using compounds that change their fluorescence or absorption when acted on by the enzyme. If one is looking at a tube full of organisms, it does not much matter whether the coloured or fluorescent products diffuse throughout the medium; for single cell analyses, however, it is preferable to use fluorogenic and chromogenic substrates that form products likely to be retained to at least some extent in or on the cells containing the enzyme.

Since the enzyme typically retains its catalytic property after interacting with a molecule of a fluorogenic or chromogenic substrate, the presence of a relatively small number of molecules in or on a cell can generate a much larger number of molecules of product. This makes selected enzymes, e.g. peroxidase, good labels for antibodies or ligands intended to detect small numbers of antigens or binding sites, since the enzymatic reaction can amplify signals considerably.

Binding peroxidase to an antibody molecule or oligonucleotide yields a substantially larger molecule; although cytoplasmic membranes can readily be permeabilized to admit the conjugates, it may be more difficult to get optimal permeabilization of bacterial cell walls. Microperoxidases, oligopeptides produced by hydrolysis of cytochromes, can yield much smaller conjugates but retain peroxidatic activity sufficient for some amplification protocols; they may be of interest for work with bacteria.

Nucleic acid dyes: quantification and sequence

By 1900, it was clear that the basic dyes in the blood stains developed by Ehrlich and his successors stained the nuclei and chromosomes of all cells and the cytoplasm of actively growing ones, including cancer cells. Ehrlich himself had verified that acid-fast staining of *M. tuberculosis* and other mycobacteria could be accomplished using any of a large variety of basic dyes. Nucleic acids had been identified, but their structure was not well understood.

In 1924, Robert Feulgen developed a procedure for binding dye covalently to DNA, allowing its quantification but destroying its structure. By the 1930s, Einar Hammersten and Torbjörn Caspersson, both among modern cytometry's founding fathers, determined that UV microspectrophotometry of cells could quantify both proteins, which absorbed most strongly near 280 nm, and nucleic acids, which absorbed most strongly at 260 nm, in their native states. Using RNAse, it was possible to separately determine the amounts of DNA and RNA present in eukaryotic cells and bacteria, providing strong evidence that DNA might be the genetic material. The microspectrophotometer, however, was complex and difficult to use, especially with subjects as small as bacteria.

A simpler but equally effective method for DNA and RNA analysis was developed by Jean Brachet, who used methyl green and pyronin, dyes originally described by Ehrlich as blood cell stains, establishing their relative specificity for stoichiometric staining of DNA and RNA using nucleases. Brachet's work antedated the recognition of DNA as the genetic material and the concept of the cell cycle; as flow cytometry gained adherents, interest in the interaction of DNA and RNA during the cycle led Darzynkiewicz *et al.* (1976) to develop a stoichiometric stain for DNA and RNA using a venerable dye, acridine orange, under carefully controlled staining conditions not compatible with viability. The fluorescent Hoechst dyes, which, like methyl green, have a strong affinity for adenine and thymine in DNA, had been used by that time to quantify DNA by flow and image cytometry; I (Shapiro, 1981) added pyronin fluorescence for RNA quantification. A Hoechst 33342/pyronin combination can be applied to living cells and washed out after analysis, leaving the cells viable, and there are now other dyes available for both DNA and RNA staining.

Most cells, both microbial and eukaryotic, contain several times as much RNA as DNA, most of the RNA is ribosomal (rRNA), and a considerable fraction of rRNA contains double-stranded regions. The acridine orange DNA/RNA staining procedure denatures the RNA, allowing dye molecules to crowd closely enough together along each strand to shift their fluorescence emission from green to red, whereas the dye molecules intercalated into the DNA helix are blocked from interacting and retain green fluorescence. In the absence of DNA-binding stains, Pyronin Y will form a fluorescent complex with DNA; in their presence, its fluorescence comes predominantly from a complex with double-stranded rRNA.

Non-proliferating eukaryotic cells, e.g. peripheral blood lymphocytes, contain relatively few ribosomes, and hence little RNA. After treatment with mitogenic lectins or suitably presented immune stimuli, they manufacture RNA and ribosomes for hours before initiating DNA synthesis. Bacterial dormancy is similar to quiescence in eukaryotic cells in that dormant bacteria decrease many metabolic activities, and content of ribosomes and therefore of rRNA decreases. Although there have been a few reports of the use of pyronin and equivalent dyes to measure RNA in bacteria, most people doing microbial cytometry would seem to prefer using sequence probes for rRNA rather than dye to estimate ribosome number.

The sequence probes are, admittedly, invaluable for identifying bacteria; mutations in the DNA coding for rRNA are relatively inconsequential since changes in rRNA sequence usually do not significantly alter ribosome function. Over the eons, mutations have occurred at different points in the branching bacterial evolutionary tree, with the result that some short rRNA sequences can be found in all eubacteria and others appear only in a particular strain of a single species. There are usually multiple copies of rDNA sequences in bacterial genomes, meaning a dozen or so. By contrast, even a dormant bacterium will contain 5000–10,000 ribosomes, and an exponentially growing one will contain 30,000 or more. Binding a couple of rRNA probes to each ribosome will give you signals from at least 10,000 molecules of label, even if there is only one molecule of label per probe and no amplification is used. That's not bad, for a sequence probe.

If you're trying to identify one or a small number of bacteria present at low concentrations, e.g. a few hundred *M. tuberculosis* on a sputum slide, or antibiotic-resistant *Staphylococcus aureus* (MRSA) in pus from a skin infection, rRNA probes can do the job; 'staining,' which in this context obviously involves hybridization, takes no more than a couple of hours, and the results, which can be obtained by fluorescence microscopy, are at least as specific as those obtained by DNA amplification. The Cepheid GeneXpert real-time PCR system can detect 100 CFU of TB bacteria in 1 ml of sputum; staining with rRNA probes could, in theory, reveal one organism in a smear containing 10 µl of sputum, providing the same sensitivity, and probably using less reagent. Those of us who are more comfortable as 'microbe hunters' than as 'genome hunters' would use a DNA dye on the slide, just to be sure that any aggregation of rRNA probes we see is associated with a quantity of DNA comparable to what we would expect to find in a bacterium. I might use two DNA dyes, about which more can be found below.

If you either know what organism you're looking at or do not particularly care, but do want to know whether it is dormant or not, you do not need rRNA probes at all; staining with pyronin, or another RNA dye, combined with a DNA dye will give you a reasonably accurate estimate of ribosome number, as well as telling you that you're looking at bacteria. Staining takes only minutes, the reagents are cheap and stable, and you'll probably get at least a million molecules of DNA dye and hundreds of thousands of RNA dye into a single organism, providing fluorescence signals that are easy to see, or to detect.

Nucleic acid dyes: base composition helps identify organisms

A large number of fluorescent dyes can be used to stain DNA and/or RNA. However, relatively few of them are specific for DNA, and only a minority of these are sensitive to base composition (A+T/G+C ratio). Hoechst 33258 and 33342, as noted above, are A–T specific. Chromomycin A_3, mithramycin, and 7-aminoactinomycin D are G–C specific. Hoechst/chromomycin combinations have been used with dual excitation-beam flow cytometers to discriminate the majority of human chromosomes based on differences in DNA base composition, and to demonstrate differences in base composition among bacterial species; a base-specific DNA dye and one with little or no base preference can also be combined, providing a smaller but still useful degree of discrimination, as shown in Fig. 1.1.

Base composition of bacterial species ranges from near 20% A+T to over 80% A+T; human DNA contains approximately 60% A+T. I noted above that, if I were staining a slide of human sputum with a fluorescent probe for *M. tuberculosis* rRNA, I might want to add a DNA dye to the stain to verify that accumulation of probe occurred in the same place as an amount of DNA in the size range of the bacterial genome. I could further improve specificity by using a base composition sensitive dye pair such as the Hoechst 33342/PicoGreen combination shown in Fig. 1.1. Although there are other bacterial species, some of which might be found in human sputum, with base compositions similar to that of *M. tuberculosis*, the base-sensitive stain would eliminate the possibility that the DNA was either from the human host or from a microorganism differing significantly in A+T content from the tubercle bacillus. The distinction could not be made reliably by a human observer, but could be using relatively simple and inexpensive fluorescence imaging hardware. Probably not a mobile phone camera, but close. Using the base-sensitive dyes in an acid-fast stain could further increase specificity.

Base-sensitive DNA staining should also be useful in malaria diagnosis; the deadliest of the five species of malaria parasites that attack humans, *Plasmodium falciparum*, has an A+T content of over 80%; the A+T contents of the other species are close to that of human DNA (~60%). The

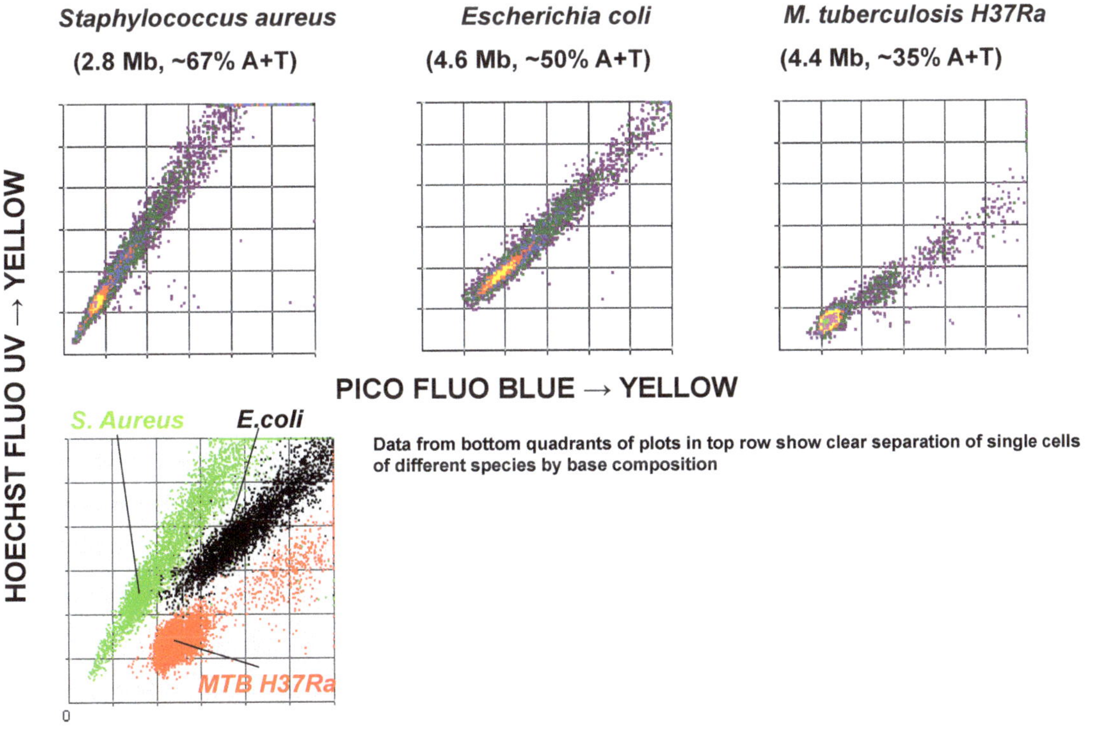

Figure 1.1 Separation of bacterial species based on base composition-sensitive DNA staining. Bacteria were stained with Hoechst 33342 (DNA-specific, AT base preference, UV → yellow fluorescence) and PicoGreen (DNA-specific, no base preference, blue → yellow fluorescence) and measured in a flow cytometer. Since many hundreds of thousands of molecules of each dye are present in a single organism, equivalent measurements could be made (after acid-alcohol washing removes dyes from nonmycobacterial species) by our imaging system, allowing more selective identification of small numbers of MTB on unprocessed sputum slides. This stain could be combined with neutral red, or with Nile red, which detects lipid inclusions in dormant MTB in sputum specimens, to further increase MTB detection specificity.

malarial parasite genomes are all between roughly 23 and 26 Mbp in size, and most parasite-infected cells contain only one or two copies of the genome. Human white cells, normally the only objects in blood with significant DNA content, contain two copies of the human genome, each over 3 Gbp in size. Malaria can readily be diagnosed by a human observer using a fluorescence microscope based on the presence of small amounts of DNA in infected red cells; discrimination of *P. falciparum* from other species would require only the same type of fluorescence imager that could be used for TB diagnosis.

The same type of imaging system could be used to improve precision of assays for microbial growth. There is a long tradition of doing these by turbidity measurement, but that requires bacterial concentrations of millions of organisms/ml, even when a nephelometer is used for readout. More recently, bulk growth assays have been done in microwell plates, using low-background fluorescent DNA dyes such as PicoGreen and SYBR Green; these allow detection of concentrations of a few thousand organisms/ml when the apparatus used measures fluorescence from whole wells. If, instead, a low-magnification fluorescence imager is used to examine whole wells, it becomes possible to detect even a single organism in a well, corresponding to no more than a few tens/ml. This provides significant advantages when it is necessary to monitor slow-growing organisms such as *M. tuberculosis*, which has a generation time of about 18 hours under the best of circumstances and is usually present at levels of only a few hundred CFU/ml in positive sputum specimens.

Protein staining

Total protein staining has not been widely used in flow cytometry, microbial and otherwise, but the decision not to use it is justified because, in general, intensities of signals from total protein stains such as fluorescein (applied as FITC) and sulforhodamine 101 correlate quite well with intensities of SSC signals. Not using the dye does not result in a significant savings of money, but does free up a fluorescence measurement channel for use with a more specific reagent. SSC measurements are not easy to make in simple imaging systems; when it is desirable to have the information they might contain, total protein staining is likely to provide it, and the fluorescence signal, coming as it does from very large numbers of dye molecules, is relatively easy to measure.

Gram staining and membrane integrity ('viability')

The so-called outer membrane of Gram-negative organisms is actually the lipopolysaccharide-containing cell wall, and restricts entry of hydrophobic compounds. In general, dyes that bear alkyl substituents on nitrogen atoms in heterocyclic rings are more hydrophobic when the alkyl chains are longer. Hexidium (HI) is an analogue of ethidium bearing a hexyl side chain rather than an ethyl side chain; it is rapidly taken up by intact Gram-positive organisms but normally excluded from intact Gram-negative ones, and stains both DNA and RNA in the organisms. SYTO 9, which also stains DNA and RNA, is relatively water soluble, and enters both Gram-positive and Gram-negative organisms. When excited at 488 nm, organisms containing both dyes (theoretically Gram-positive) show red fluorescence because SYTO 9 transfers its excitation energy to HI, whereas organisms containing only SYTO 9 (theoretically Gram-negative) fluoresce green. Note that some organisms descended from Gram-negative wild types do not have 'competent' outer membranes, and will let in HI and other hydrophobic dyes; some strains of *E. coli* used in molecular biology fall into this category. The dye mixture has been shown to produce results consistent with Gram staining in several dozen bacterial species, but you'll need to try it if you want to work with one that is not on the short list.

Propidium (PI) contains the same phenanthridinium ring as ethidium and HI, but its side chain is a propyl group with a quaternary ammonium at the end opposite the ring. This gives PI a double positive charge instead of the single charge present in ethidium and HI, making it enter eukaryotic and bacterial cells with difficulty under normal circumstances. Cells that take up enough PI for their nuclei or DNA to stain brightly with it have traditionally been considered to be membrane-damaged enough to be nonviable. During the past decade or so, it has become clear that many bacterial species do not share this opinion. In particular, viable cells in biofilms and cells subjected to the stress of sublethal doses of antimicrobials (which may induce them to form biofilms) have been shown to take up PI and other Quaternary ammonium-bearing dyes while preserving viability. Some cells that do this have been shown to maintain membrane potential, eliminating the possibility that PI is getting into the cell through a large hole in the membrane.

The maintenance of a membrane potential has been shown to correlate at least as well as any other cytometric parameter with reproductive viability in several dozen bacterial species, although the correlation would be expected to be poor if, for example, cells' membrane potentials were measured immediately after they received a lethal dose of radiation. Membrane potential is also lost in dormant cells and spores, but returns very soon after the cells are resuscitated.

The retention of fluorescent products of fluorogenic substrates by organisms containing the relevant enzyme(s) is an indicator that the membrane is relatively intact; cells exhibiting this behaviour are usually termed 'viable,' but see the above paragraph. All of the measurements I have just described could be made in imaging systems as easily as in flow cytometers.

Microscopy versus cytometry: the eyes do not have it

In the early 1960s, even in the richest countries, a great deal of clinical microbial cytometry, including Gram staining of urine and other specimens for bacterial infections, Ziehl–Neelsen acid-fast staining of sputum smears for tuberculosis (TB) and occasional blood smear examinations for malaria parasites, was done by medical students (I was one at the time) as part of their training. We also did most of the blood cell counts; automation came slowly to the city hospitals in which much clinical teaching took place. All of these tests, involving only transmitted light microscopy, were not done any differently in sub-Saharan Africa or any of the other resource-limited areas of the world in which HIV/AIDS, TB, and malaria now take their greatest toll. By 2000, every medical institution with an adequate budget and infrastructure used flow cytometers for laboratory haematology and immunophenotyping, including determination of CD4 antigen-bearing T-helper cells critical for HIV/AIDS management. By contrast, even today, the clinical use of automated cytometry for detection of microbial pathogens is largely limited to those labs equipped with flow cytometric urine analyzers.

There was essentially no immunophenotyping in 1960; the Wright's and Giemsa's stains we used to count different types of white blood cells did not allow us to reliably distinguish lymphocytes, which we had not discovered came in many flavours, from monocytes. Some of the forward-looking haematologists and pathologists were using automated cell counters and thinking about the possibility of automating certain tasks then requiring microscopy, notably the Papanicolaou smear, used in screening for cervical cancer, and the differential white blood cell count. Nobody wanted to try telling cells in tissue apart with an automated microscope; in Pap smears and blood smears, cells were better separated. Microbiology received little attention, although the defence establishment, better financed than most microbiology labs, pursued automated microscopy as an improved means of detecting biowarfare agents. The clinical microbiologists could generally wait a few hours for fast-growing organisms in samples to provide them with more than enough material for diagnoses.

Twentieth-century advances in microscopy increased its attractiveness. By the 1940s, fluorescence microscopy, still an expensive technique, had been demonstrated to offer advantages over Ziehl–Neelsen staining for identification of *M. tuberculosis*, and the newly developed, highly specific fluorescent antibody method had been shown capable of detecting pneumococci in tissue. Then and now, however, the requirement for relatively high-magnification, high-resolution images limited the speed at which an observer could examine a sample and, ultimately, the volume of sample that could be examined. Microscopists do not do well with slides that may contain only 10 or fewer pathogens.

There were also obstacles and disadvantages to automating microscopy. Although video cameras had been used with microscopes by the 1950s, no practical means of storing the data required for automated analysis of a single microscope field existed. In fact, the combined worldwide population of mainframe computers, which cost millions of dollars each, at the time automated blood cell counters first appeared had less processing power and storage capacity than does a single modern smartphone with 'only' 16 GB of memory.

Had storage been available, it would still have been necessary to have high-precision, three-axis motion control hardware and software available to present and focus the hundreds of high-power fields needed for typical applications. That equipment remains expensive; without it, none of the adapted mobile phones now touted for tasks such as water quality analysis and diagnostic microscopy in resource-poor countries will be usable.

The major liability of microscopy, however, was and is the subjectivity and inconsistency of the human observer in the context of the samples, reagents, and apparatus involved.

Both the Papanicolaou stain and the blood stains, although they give a human observer some indication of the protein and nucleic acid content of different regions of cells, are neither highly specific nor stoichiometric, and the human visual system itself limits the accuracy and precision with which an observer can accurately quantify cellular constituents even when specific and stoichiometric stains are used. Identifying cells in blood or bone marrow or Pap smears requires that the observer consider the size and shape and colour of cells and their nuclei, as well as the staining characteristics of the cytoplasm. Humans are wired in ways that make them able to recognize the multilobed 'squiggly' nuclei of polymorphonuclear granulocytes easily, even if no two of them look exactly alike; this is relatively difficult for computers and was more so in 1960.

As I have suggested in the previous sections, the reagents we now use in flow cytometry, including many simple, stable dyes, can potentially provide far more specific identification of blood cells, bacteria, and parasites than can any of the dyes used in the classical stains, especially when quantified with higher accuracy and precision than can be attained by even the best human observer. Humans cannot reliably detect differences of less than 10% in staining intensity using either transmitted light or fluorescence microscopy. Even if they could perceive small differences in fluorescence, they'd have to deal with effects of the bleaching that can occur when stained cells spend more than a few seconds in a moderately intense excitation beam.

A generation of differential white blood cell counters using automated microscopy reached the market in the 1970s; they had been replaced by flow cytometric apparatus, most of which did not use or need to use fluorescent reagents, by the mid-1980s. Haematologists initially suspicious of the new technology accepted it because the fluorescence flow cytometers and sorters in research labs had clearly transformed our understanding of blood cell biology.

Microbiologists are probably at least as conservative and suspicious as are haematologists; they also typically operate on lower budgets. A PubMed search on 'flow cytometry' in early May 2012 returned over 128,000 citations; even accounting for material published outside PubMed's search area, it is unlikely that there have been even 10,000 reports describing cytometry of bacteria, fungi, parasites, or viruses. I would like to believe that finances, rather than lack of trust, have been keeping microbiologists tied to microscopy, and that they would opt to replace century-old microscopy with cytometry rather than automate it if the replacement were affordable. The options have changed considerably in the decades since haematology was automated.

Cytometric options: thinking outside and inside the box

The development of progressively smaller, more powerful, and less expensive computers from the 1950s on has played a central role in making multiparameter cytometers first, possible, and, eventually, practical. We measure increasing numbers of parameters with increasing numbers of reagents and increasingly complex instruments because the biological questions we are trying to answer demand it. A cytometer is now a box into which one puts cells and out of which one gets numbers (and, if the cytometer has sorting capacity, cells). Economic constraints press us to minimize the cost and complexity of our apparatus and procedures, and advances in chemistry, electro-optics, and optics determine how simply and cheaply the box can be made.

Flow cytometers have been the dominant breed for almost 40 years. Although the technology has advanced considerably, the orthogonal 'body plan' of most modern flow cytometers is little changed from that of some instruments first sold in the late 1960s. Lasers still predominate as light sources and photomultipliers (PMTs) as detectors. Computers, which had become inseparable from flow cytometers in the 1980s, have undergone a far more dramatic transformation than the rest of the instrument, but have kept up with the data processing demands as the numbers of light sources and detectors per instrument have increased. During the past 10 years, however, it has become apparent that many of the jobs now done by flow cytometers can be done using simpler and cheaper equipment.

Modern pocket-sized versions of the Gucker aerosol counter, using diode laser light sources and PIN diodes as scatter detectors, with microprocessors for data processing, are now sold by several manufacturers for less than US$2,000. They are able to detect and size particles as small as 0.2 μm in diameter, but, like the original instrument, are single-parameter systems which cannot distinguish particles within a given size range by any other characteristics. Multiparameter flow cytometers measuring one scatter and two fluorescence signals are now appearing on the market at prices between $10,000 and $15,000; they can do many of the simpler jobs now done by instruments costing at least four times as much. There are also fluorescence imaging systems, using LEDs for fluorescence excitation, CCD or CMOS cameras for detection, and microprocessors for data analysis, available for $10,000–$15,000; they can do the same jobs, and at least some can measure additional colours and use wavelengths not available in the bare-bones flow systems.

If you need to sort cells, or measure six or more parameters, you probably need a flow cytometer. You cannot use it to look at tissue, or cells adherent to a substrate or slide, or biofilms, but a flow cytometer that sorts costs at least $100,000, and, if you have the money, you can probably find another $15,000 for an imaging system to do cytometry on slides, culture chambers, etc.

Many problems in microbial cytometry involve looking for very few target organisms in a large volume of sample. A flow cytometer cannot process more than a couple of microliters of prepared sample per second; looking for coliform bacteria in a litre of water, or for one organism in 10 ml of a patient's blood, will require some way of concentrating the targets. Centrifugation and immunoadherent techniques require instruments and/or reagents. Filtration is typically simplest and cheapest. In addition to the Ultramicroscope, Richard Zsigmondy developed filters with very small pore sizes, capable of retaining microorganisms; these were used to concentrate organisms for microscopy in Germany before and during the Second World War. The technology was brought to the USA after the War, and commercialized by a small chemical company that later became known as Millipore, now itself a subsidiary of the German conglomerate EMD.

A well known 1985 paper (Mansour *et al.*, 1985) from Becton-Dickinson described detection of fewer than 100 *E. coli* seeded into 1 ml of blood using an ethidium stain and flow cytometry; the formed elements of blood were lysed using a combination of detergents and enzymes. A less well known earlier paper from the same organization (Sage and Neece, 1984) did the same job using the same lysis technique, but detected equally small numbers of organisms by Gram staining the 2 mm spot into which they had been collected on a filter and observing them with a 40× microscope objective. Fluorescence microscopy and filtration have been combined by many people for many microbiological applications; one can buy a laser scanning instrument costing at least as much as a medium-size flow cytometer to do the job, but an inexpensive LED-illuminated fluorescence imager should be able to handle it.

There are now thousands of small flow cytometers in the field in resource-poor countries, built for the single purpose of counting CD4+ T cells in people infected with HIV. Some of them could probably be adopted for some tasks in diagnostic and public health microbiology, e.g. malaria diagnosis and water quality assessment. None of them could readily be adapted to use presently known reagents for DNA base composition detection, which I have argued could be very helpful in both malaria and TB diagnosis. I believe an imaging system that could use optimal cytometric techniques for both of these tasks could be built to sell for under $10,000, and possibly for considerably less. Such a system would be usable for a wide range of applications in both affluent and resource-poor countries.

The first two editions of my flow cytometry book had plans and instructions for building a four-parameter laser source 'Cytomutt' flow cytometer. Dozens of groups built such instruments, and some have managed to keep them running for decades with no more than a few phone calls and emails to me over the period. Building a maintainable, sustainable, 'green' fluorescence imaging system is much simpler and cheaper. My colleagues and I are trying to figure out how little needs to be put in the box. Stay tuned.

References

Clark, G., and Kasten, F.H. (1983). History of Staining, 3rd edn (Williams and Wilkins, Baltimore, USA).

Darzynkiewicz, Z., Traganos, F., Sharpless, T., and Melamed, M.R. (1976). Lymphocyte stimulation: a rapid multiparameter analysis. Proc. Natl. Acad. Sci. U.S.A. *73*, 2881–2884.

Davey, H.M., and Kell, D.B. (1996). Flow cytometry and cell sorting of heterogeneous microbial populations: the importance of single-cell analyses. Microbiol. Rev. *60*(4), 641–696.

De Kruif, P. (1926). Microbe Hunters. Reprinted with an Introduction by F. Gonzalez-Crussi, 1996 (Harcourt, San Diego, CA, USA).

Gray, M.L., Hoffman, R.A., and Hansen, W.P. (1983). A new method for cell volume measurement based on volume exclusion of a fluorescent dye. Cytometry 3, 428–434.

Gucker, F.I. Jr., O'Konski, C.T., Pickard, H.B., and Pitts, J.N. Jr. (1947). A photoelectric counter for colloidal particles. J. Am. Chem. Soc. *69*, 2422–2431.

Harris, H. (1999). The Birth of the Cell (Yale University Press, New Haven, CT, USA).

Hooke, R. (1665). Micrographia: or Some Physical Descriptions of Minute Bodies Made by Magnifying Glasses with Observations and Inquiries Thereupon (J. Martyn and J. Allestry, Printers to the Royal Society, London, UK).

Lababidi, S.L., Pelts, M., Moitra, M., Leff, L.G., and Model, M.A. (2011). Measurement of bacterial volume by transmission-through-dye imaging. J. Microbiol. Methods *87*, 375–377.

Laflamme, C., Verreault, D., Lavigne, S., Trudel, L., Ho, J., and Duchaine, C. (2005). Autofluorescence as a viability marker for detection of bacterial spores. Front. Biosci. *10*, 1647–1653.

Mansour, J.D., Robson, J.A., Arndt, C.W., and Schulte, T.H. (1985). Detection of *Escherichia coli* in blood using flow cytometry. Cytometry *6*, 186–190.

Müller, S., and Nebe-von-Caron G. (2010). Functional single-cell analyses: flow cytometry and cell sorting of microbial populations and communities. FEMS Microbiol. Rev. *34*, 554–587.

Nebe-von-Caron, G. (2009). Standardization in microbial cytometry. Cytometry A *75*(2), 86–89.

Sage, B.H. Jr., and Neece, V.R. (1984). Rapid visual detection of microorganisms in blood culture. J. Clin. Microbiol. *20*, 5–8.

Shapiro, H.M. (1981). Flow cytometric estimation of DNA and RNA content in intact cells stained with Hoechst 33342 and pyronin Y. Cytometry *2*, 143–150.

Shapiro, H.M. (2003). Practical Flow Cytometry, 4th edn (Wiley-Liss, Hoboken, NJ, USA).

Shapiro, H.M. (2010). The history of flow cytometry and sorting. In: The Microflow Cytometer, Kim, J.S., and Ligler, F.S., eds. (Singapore: Pan Stanford Publishing), pp. 1–24.

'Student' [Gosset, W.S.] (1907). On the error of counting with a haemacytometer. Biometrika *5*, 351–360.

Tzur, A., Moore, J.K., Jorgensen, P., Shapiro, H.M., and Kirschner, M.W. (2011). Optimizing optical flow cytometry for cell volume-based sorting and analysis. PLoS ONE *6*, e16053.

Non-destructive On-chip Imaging Cytometry Assay for Constructive On-chip Cellomics Studies

2

Kenji Yasuda

Abstract

We have developed two series of novel non-destructive imaging cell cytometry assays exploiting microfabrication technology. The on-chip imaging flow cytometry system uses a 1/10,000 of a second ultra high-speed camera with a real-time image analysis unit and a poly(methyl methacrylate (PMMA) based disposable micro-fluidic chip with the following features: (1) a fully automated system for application and collection of samples using 3-D micropipetting; (2) arrangement of the entire fluidic system in a disposable plastic chip for lining-up of biological samples in laminar flow using hydrodynamic focusing; (3) an optical system for direct observation-based cell identification with specific image indexes using phase-contrast/fluorescence microscopy and real-time image processing; (4) a non-destructive wider-size dynamic range sorting procedure using mild electrostatic force in a laminar flow using agarose gel electrodes to prevent loss of electrode or electrolysis bubble formation; (5) a contamination-free re-cultivation micro-reservoir for collected target cells, and (6) removable DNA aptamer labelling. On-chip single cell imaging cytometry has the advantage of being able to measure the 'algebraic' inheritance of epigenetic character of single cells using imaging indexes. Exploiting these advantages, target cells can be successfully separated and their characteristics evaluated by imaging. Results have shown that the on-chip imaging flow cytometry system and the on-chip single-cell imaging cytometry systems are applicable for biological research and clinical diagnostics.

Introduction

In general, cells are minimum units which determine their responses through genetic and epigenetic information such as the history of interactions between them and fluctuations in the environmental conditions which affect them. If we are to understand such topics as variations in cells with the same genetic information, inheritance of non-genetic information between adjacent generations of cells, cellular adaptation processes caused by environmental change, the community effect of cells, we must analyse their epigenetic information. This requires starting from the twin complementary viewpoints of cell regulation as being an 'algebraic' system (with emphasis on temporal aspects; adaptation among generation) and as a 'geometric' system (with emphasis on spatial aspects; spatial pattern-dependent community effect) using identified single cells. We thus commenced a series of studies to analyse epigenetic information among neighbouring generations of cells and in the spatial structures of a cell network in order to expand our understanding of how the fates of living systems are determined. As shown in Fig. 2.1, the strategy behind our on-chip microfabrication method is constructive, involving three steps, which we describe as 'on-chip cellomics'. First, we purify cells from tissue singly in a non-destructive manner using ultra high speed camera-based real time cell sorting, or digestible DNA-aptamer labelling (Yasuda, 2000). We then cultivate and observe them under fully controlled conditions (e.g. cell population, network patterns, or nutrient conditions) using an on-chip single-cell cultivation chip (Hattori *et al.*, 2003; Inoue *et al.*, 2001a,b, 2004; Matsumura

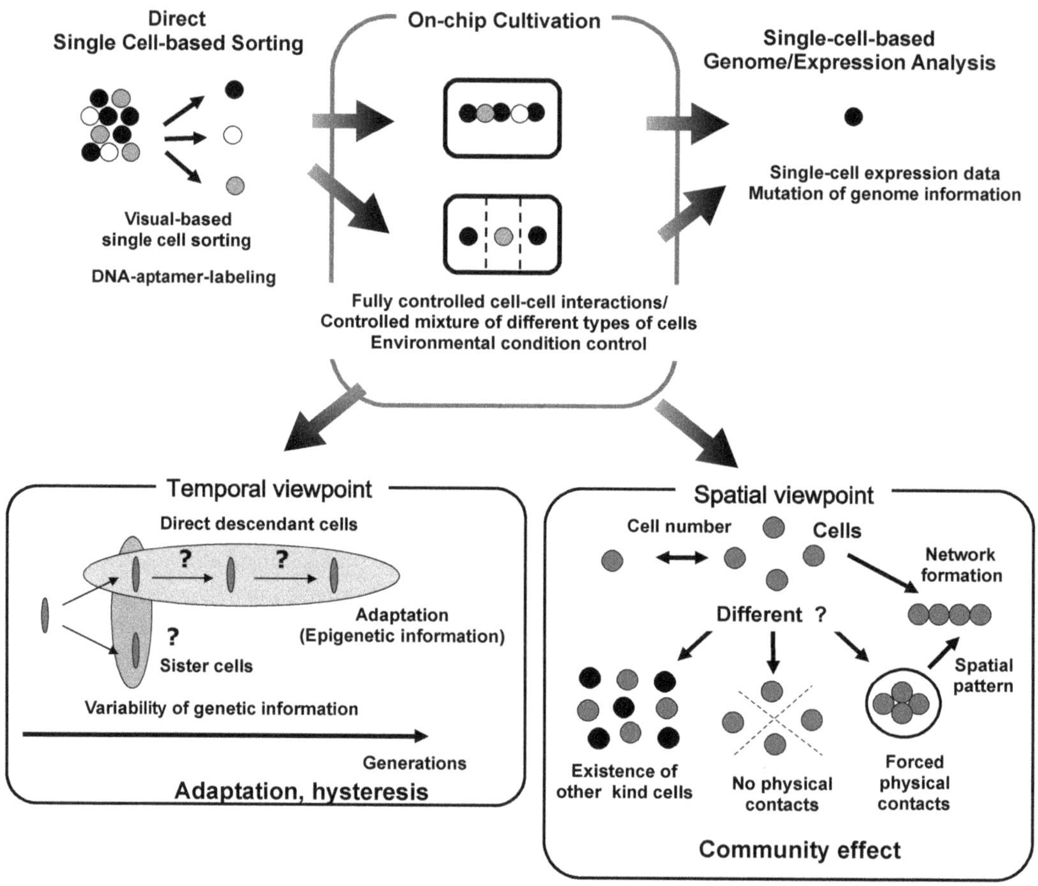

Figure 2.1 On-chip cellomics: constructive on-chip single-cell-based analysis of cellular networks from both temporal and spatial aspects.

et al., 2003a,b; Takahashi *et al.*, 2003; Umehara *et al.*, 2003; Wakamoto *et al.*, 2001, 2003) or an on-chip agarose microchamber system (Hattori *et al.*, 2004; Kojima *et al.*, 2003, 2005; Moriguchi *et al.*, 2002, 2004; Sugio *et al.*, 2004; Suzuki *et al.*, 2004a,b). Finally, we undertake single-cell-based genome/proteome analysis through photothermal denaturation and single-molecule level analysis (Yasuda *et al.*, 2000).

As described above, purification of particular target cells from a crude sample mixture of cells is the first and most important step for further on-chip experiments. However, the operations required for cell purification are a time-consuming and a potentially contaminating process. Efficient and rapid sorting of cells has been accomplished using techniques including: fluorescence-activated cell sorting (FACS) (Ogawa *et al.*, 1994), magnetic activated cell separation (MACS) (Imamura *et al.*, 1995), automated single-cell sorting using dual-beam optical trapping, differential adhesion cell sorting (Hsu *et al.*, 2008), and disposable micro-fabricated fluorescence-activated cell sorting (μFACS) (Azuma *et al.*, 2007; Liu *et al.*, 2011; Yoon *et al.*, 2011). Recent advancements in optofluidic flow cytometers, where optics and microfluidics work together to create novel functionalities on a small chip, holds great promise for the successful development of lab-on-a-chip flow cytometry (Hatayama *et al.*, 1994; Yasuda *et al.*, 2006). Indeed the development of a low-cost, compact, handheld flow cytometer and microfluorescence-activated cell sorter system could have a significant impact in the field of point-of-care diagnostics, thereby improving health care. The traditional flow cytometer and FACS detect and analyse optical signals (angular light scatter or emitted fluorescence) to identify individual cells or biological samples, and are widely used as analytical tools in biomedical research and clinical diagnostics. These tools greatly facilitate the study of both physical properties (e.g. size, shape) and biochemical properties (e.g. cell cycle distribution, DNA contents) of biological samples such as cells. Information about cells of interest is obtained optically in a non-destructive and quantitative manner (Takahashi *et al.*, 2006; Yamashita *et al.*, 2007). However, FACS can damage cells during destructive droplet generation, and the detection method (based on non-direct scattering) has poor cell recognition performance. Other conventional techniques also have disadvantages with regard to their cost, efficiency, response speed, separate resolution, and adaptability. Furthermore, if cells are to be used for cultivation after purification, the damage to cells caused by the sorting process must be minimized. For example, the MACS method allows cells to be separated by incubating with magnetic nanoparticles coated with antibodies against a particular surface antigen. This causes the cells expressing this antigen to attach to the magnetic nanoparticles. Afterwards the cell solution is transferred to a column and placed in a strong magnetic field. In this step, the cells attached to the nanoparticles (expressing the antigen) remain on the column, while other cells (not expressing the antigen) flow through. Although this method is simple and economical, the attached antibodies cannot be removed from the cell surface which sometimes promotes changes in target cells.

In contrast, advantages offered by application of microfluidics include: reduced sample volumes, shorter handling time, lower space requirements and operational costs. In particular, the small size and potentially low cost of a μFACS system allows portability and affordability by individual laboratories and point-of-care clinics. There are three core components of μFACS systems that need to be miniaturized while maintaining performance: (1) the fluidic system for introduction and placement of biological samples; (2) the optical system for illumination of samples and collection of emitted light (both scattered and fluorescent); and (3) the sorting system for deflection of samples of interest, which requires both real-time control and a rapid-response actuation system. The key challenge in developing a high-performance μFACS system is to figure out how to reduce both the volume and cost of these three systems while maintaining adequate performance (e.g. detection sensitivity and sorting throughput). However, even the above μFACS system is difficult to apply for non-invasive cell collection of target cells for applications such as regenerative medicine as they use fluorescent antibodies for labelling and identification of target cells.

To overcome the above problems, we have developed a series of on-chip high speed imaging cell sorting systems (Hattori and Yasuda, 2010; Sawaki *et al.*, 2007; Takahashi *et al.*, 2004). The developed system has the following five core features: (1) a fully automated system for application and collection of samples using 3-D micropipetting; (2) a fluidic system arranged in a disposable plastic chip for lining-up the biological samples in laminar flow using hydrodynamic focusing; (3) an optical system for direct observation-based cell identification with specific image indexes using phase-contrast/fluorescence microscopy, and real-time image processing; (4) a non-destructive wider-size dynamic range sorting procedure using mild electrostatic force in a laminar flow exploiting agarose gel electrode to prevent loss of electrode or any electrolysis bubble formation; (5) a contamination-free re-cultivation microreservoir for collected target cells. The procedure of mild electrostatic force separation is achieved by applying an electrostatic force only to the cells which have to be removed to waste, i.e. the target cells simply flow down the laminar flow and do not receive any stimulation from electrostatic force. In other words, for accomplishing the cell sorter system, we needed to combine different technologies into a system to enable acquisition of particular target cells.

In this chapter we give an overview of our on-chip imaging cytometry assays. First, the automated imaging flow cytometry system using a plastic chip, in which a disposable contamination-free re-cultivation function, included for simple and safe operation in biological research and clinical diagnostics is described. Then our on-chip single-cell imaging cytometer and its applications are outlined in detail to outline the potential of imaging index for the next generation of microbial cell imaging flow cytometers.

A fully automated on-chip imaging flow cytometry system

Design of on-chip imaging flow cytometry system

We have developed a fully automated on-chip imaging flow cytometry system consisting of (Hayashi *et al.*, 2011a); (1) a cell sorting chip attached on to a cell-sorter chip holder; (2) a holder for sample tubes and collection tubes; (3) an automatic 3-D pipetting device; (4) a holder of pipetting chips; (5) an ejection port for used pipetting chips; (6) an air pressure unit connected to a syringe pump; (7) an automatic alignment set of phase-contrast/fluorescent optical microscopy modules; (8) an image-processing unit consisting of a high-speed camera and an image-processing and system control computer; (9) a switching DC power supply (Fig. 2.2a and b).

Design of a disposable contamination-free plastic re-cultivation chip

We have fabricated two types of cell-sorter chips; a two-way-outlet type cell-sorter chip (Fig. 2.3a–c) and a three-way-outlet type cell-sorter chip (Fig. 2.3d–f). Both cell sorter chips have the following parts: inlet pathways for samples and buffers, hydrodynamic focusing pathways, a cell sorting area where sample and buffer flows join together but keep their boundaries, and gel channels (a pair of agarose microelectrodes in a storage space) attaching to the cell sorting area, and outlet pathways of target samples and wastes. The section dimension of the cell-sorting area is 50 μm in width by 25 μm in height. At the agarose microelectrode storage space, these two wide spaces are filled by a 1% agarose solution containing 0.5 M NaCl as electrolyte. The sol–state agarose is injected into the gel inlet on the chip just before each experiment starts. Small acrylic tubes are attached to the cell-sorter chip as sample and collection inlet, and large acrylic tubes are also attached at an upstream inlet area and at a downstream outlet area for buffer reservoir (Fig. 2.3g–h).

In the two-way on-chip cell sorter chip, raw sample is introduced from sample reservoir into one of the two microfluidic pathways fabricated in the chip, and shifts the cells to the other buffer pathway when the voltage is applied to gel electrodes. On the other hand, in the three-way on-chip cell sorter, raw sample is introduced into the centre pathway, and shifted either to the left or right side laminar buffer flow depending on their charges. It should be noted that the thickness of the bottom thin polymethylmethacrylate (PMMA) film is 100 μm (which is within a depth of focus of ×100 obj. lens) and is attached to the microfluidic chip by heat bonding.

Automated sequential process of imaging flow cytometry

Before the automatic operation of cell sorting commences, a cell-sorter chip, sample tubes, a sheath buffer tube, collection tubes, and pipetting chips are set on a cell-sorter system by hand (Fig. 2.4a). Then the system cover is closed and the air inside filtered and cleaned by HEPA filter-based

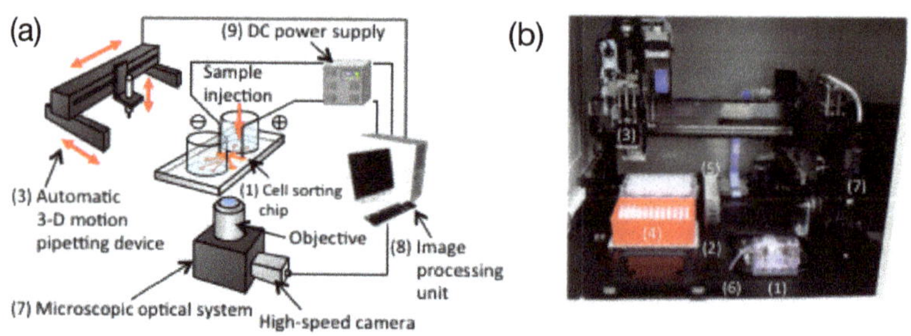

Figure 2.2 A fully automated on-chip imaging flow cytometry system with a disposable contamination-free plastic re-cultivation chip. (a) Schematic diagram of the system; (b) layout of components of the system.

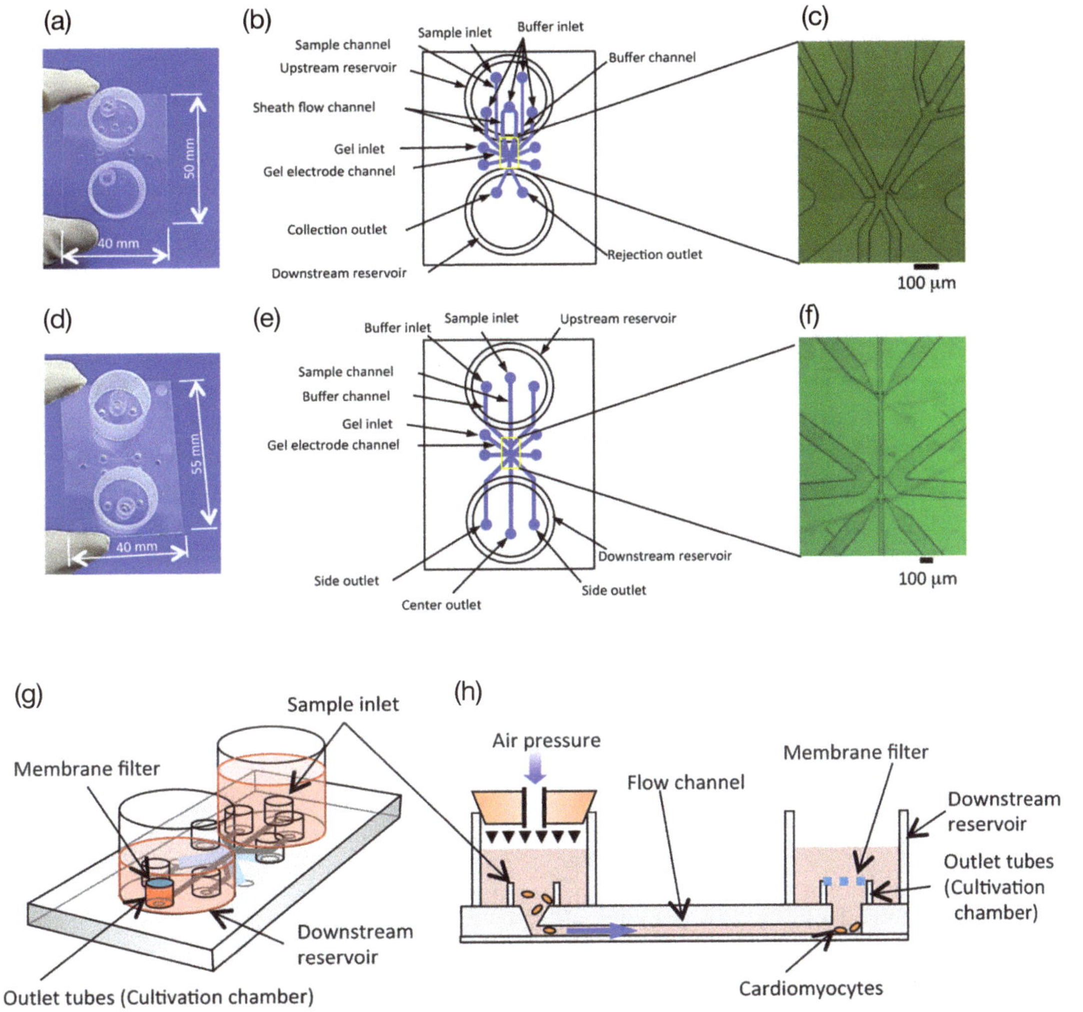

Figure 2.3 Two- (a–c) or three- (d–f) way-outlet type cell-sorter chip. Full view (a and d), schematic diagram (b and e), and close-up (c and f) images of the chips. (g) Full view and (h) side view of a re-cultivation chip.

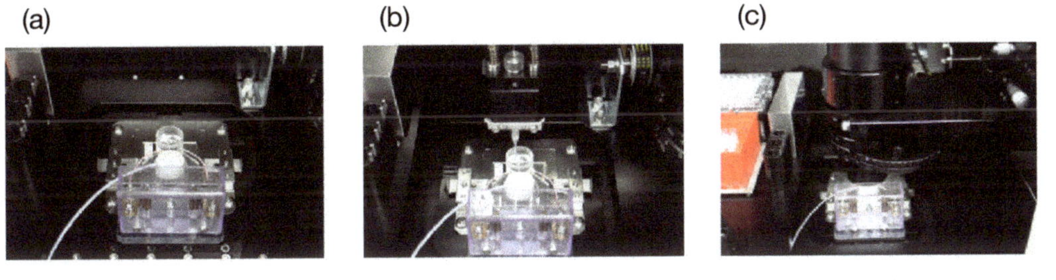

Figure 2.4 Automatic set-up process of the cytometry system. (a) Setting of a PMMA chip on a chip stage; the initial stage of the setup process; (b) loading of sample and sheath buffer on a sample inlet and a reservoir pipe by automatic pipetting device; and then (c) moving of a condenser lens to a measuring position over the chip; the end stage of the process.

air circulation. The initialization process is carried out by the system control unit which resets and memorizes all the positions and conditions. Then an automatic 3-D pipetting device applies the sample into the sample inlet and sheath buffer into the upstream/downstream buffer reservoirs (Fig. 2.4b). An automatic alignment system of the optical microscopy system then adjusts the positions and focus of objective and condenser lens (Fig. 2.4c). Following the above initialization procedure, the pressure is applied to the upstream reservoir to start sample flow and cell sorting process commences. 1/10,000 of a second real time images of sample particles flowing through the cell sorting area are acquired using a high-speed camera, and the image-processing computer recognizes characteristics of each particle according to the parameters of the particle size and shape, and can apply DC voltages into the gel electrodes at a rate of up to 10,000 particle/s real time. All the images of particles and their parameters are stored in the data storage component of the computer.

Real-time recognition of the image of flowing particles

The image processing unit recognizes the image of each flowing particles at a rate of up to 10,000 frames/s and 1/200 of a second analysis intervals as shown in Fig. 2.5a. A background image of observation area is acquired at the end of initialization process. In the sorting process, in order to decide the boundary of particle image, an original image is subtracted by the background image in each frame, and binarized by the preset threshold. The parameters of particle's image, such as a mean intensity, an area, a length of the major and minor axis, the degree of circularity are calculated about the pixel value within the boundary. When one of, or a set of, the parameters satisfies that of the desired or undesired particle, the image-processing unit switches the DC voltage between the gel electrodes of cell-sorter chip by a switching DC power supply in real time.

Fig. 2.5b shows an example of data analysed using the calculated parameters. The analysed roundness and size of each cell image was plotted, and the original images stored on a hard disk drive and cross-referenced to the plots. This enabled detailed images from the plotting to be called up and used for checking detailed shapes of samples. Fig. 2.5c shows examples of acquired micrographs in which multinucleated cells and M-phase cells were identified, illustrating one of the advantages of imaging flow cell sorting compared to conventional diffraction-based cell identification analysis.

Development of a disposable polymethyl methacrylate (PMMA) imaging cell sorter chip

Because of advantages such as transparency for visible light, low auto-fluorescence, flammability after use, use of glue-free heat bonding, flexible change of shape by heating if necessary, and hydrophilicity in microfluidic pathway, we selected poly(methyl methacrylate (PMMA) as the base material of the disposable and economical cell sorter chip. As shown in Fig. 2.6a–c, PMMA has a high transmittance of about 90% at the range of visible light from 400 to 600 nm; the transmittance is approximately the same as a glass slide or a polycarbonate culture dish.

To mould a PMMA-based chip, we first examined a hot embossing method. As shown in Fig. 2.6d, the moulded micro channels showed rounded edges and did not reflect microscopic range edges of microfluidic pathways. This was because PMMA was not soft enough to permeate into the edge of a square-shaped microchannel at the temperature used in the hot-emboss process. Next, we examined an injection moulding method using the same design of micro channels as for the hot emboss. The edge of the microchannel was made as a sharp square shape, however melting of PMMA resin did not permeate to the thin wall structure at the junction of a flow channel and a gel channel (Fig. 2.6e). Next, we redesigned the injection mould to thicken the wall structure by the addition of narrow channels of 15 μm width and 45 μm length between the flow channels and the gel channels. Then we succeeded in fine moulding of a PMMA-based cell sorter chip with the new design; the PMMA resin permeated well into the walls around all the micro channels (Fig. 2.6f). The sectional dimension of flow channel at the cell-sorting area was 50 μm in width by 25 μm in height. The moulded PMMA plate was laminated with a thin PMMA sheet (100 μm thick) using hot lamination.

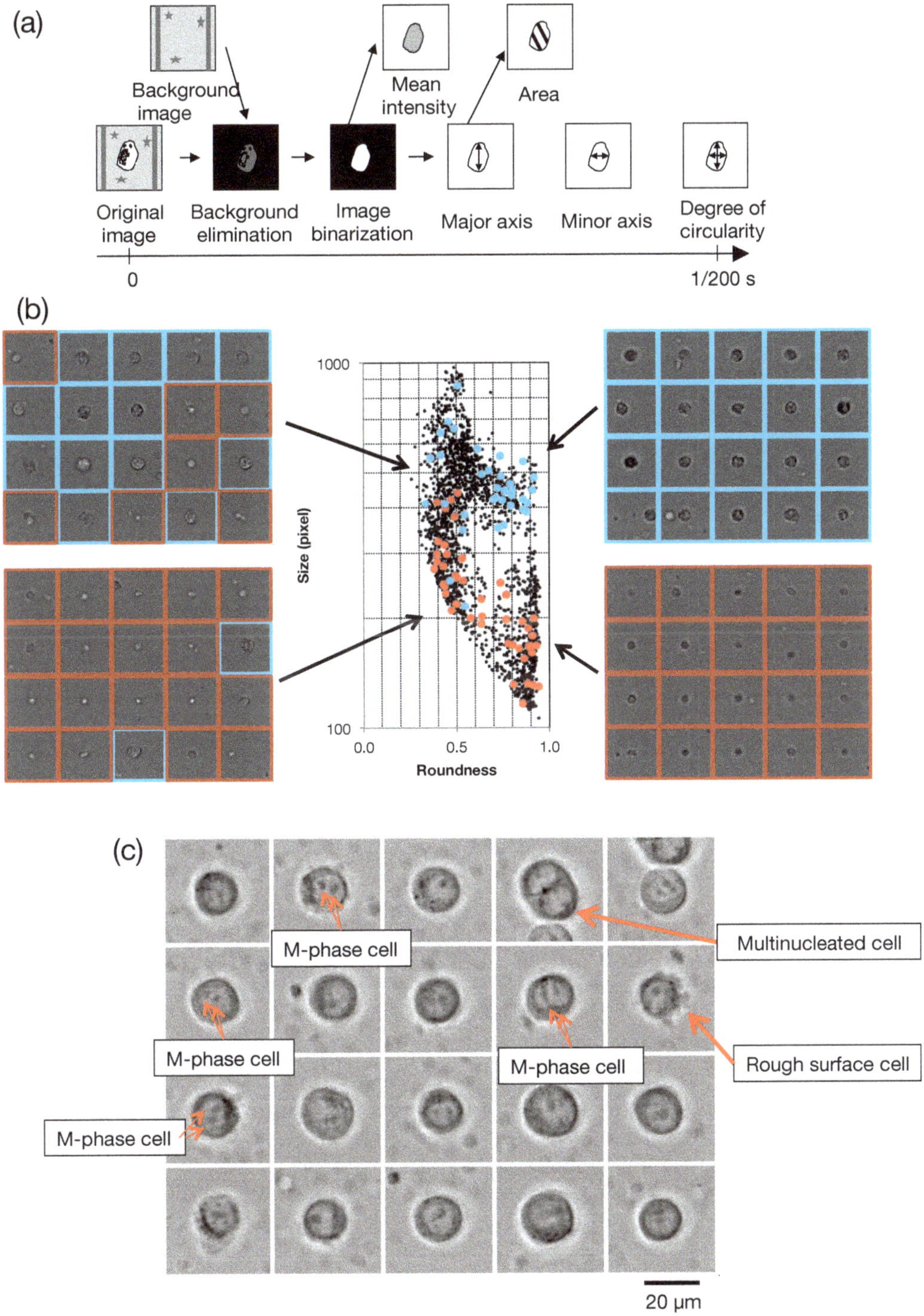

Figure 2.5 Analysis procedure cycle of imaging on-chip imaging cytometry assay and two-dimensional mapping of acquired cell images. (a) Analysis procedure cycle of microscopic images in real time by image processing unit in the case of 1/200 s intervals. (b) Analysed roundness and size (cross-sectional area) of each cell image is plotted on map at centre. Original images are stored and cross-referenced to plots. (c) Magnified micrographs of acquired cell images. Even at a rate of 10,000 images/s, the resolution was sufficient to identify M-phase cells, multinucleated cells, and rough-surface cells.

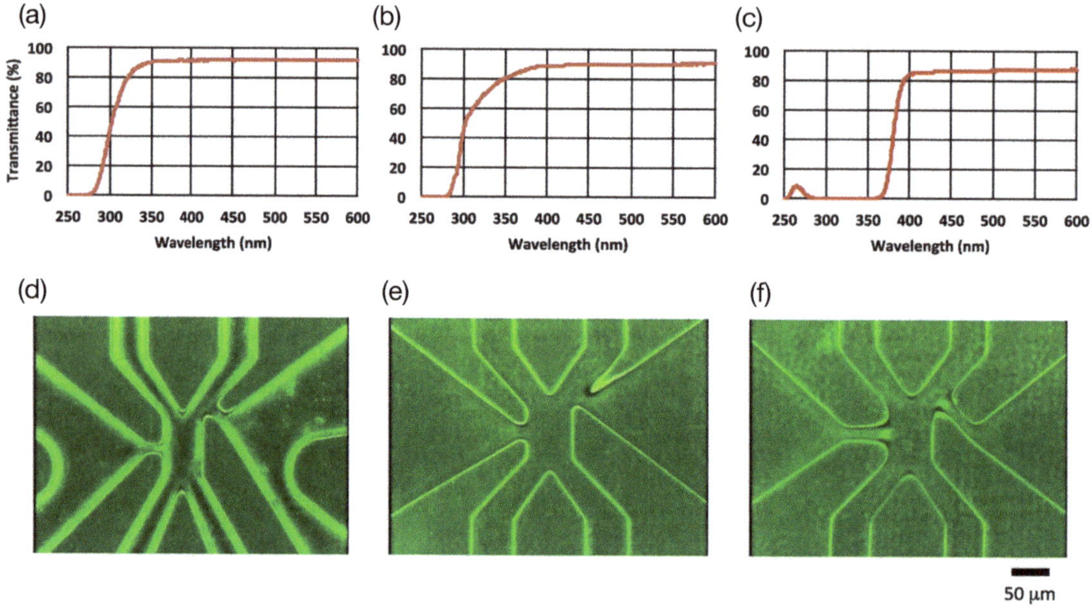

Figure 2.6 Development of a PMMA-based flammable cell sorter chip. (a–c) Transmittance of a glass slide (a), a polystyrene-based culture dish (b), and a PMMA-based cell-sorter chip (c), the thickness of entire sample plate was 1 mm. (d–f) Moulding of PMMA-based cell sorter chips by hot-emboss method (d) and injection moulding method (e and f). The mould design of (f) inserted narrow channel structures between a flow channel and gel channels to the design of (d) and (e).

Spontaneous filling of agarose sol into the gel electrode channel of PMMA-based cell-sorter chip

A PMMA-based chip was preheated on a hot plate at 42°C for 5 minutes. Melting sol containing 1.5% agarose and 0.5 M NaCl was loaded on the gel inlets of the preheated chip. Unlike the PDMS-based microfluidics chip described in a previous report (Hattori and Yasuda, 2010; Sawaki *et al.*, 2007), the melting gel spontaneously permeated into the gel channels by capillary action without additional pressure. The injecting sol moved around the junction of flow channels and gel channels, permeated the narrow channel structure, and stopped at the tip of the narrow channel. The agarose sol turned to a gel by cooling at room temperature keeping the structure attached to flow channels (Fig. 2.7). The spontaneous permeation of melting gel was possible because of the higher wettability of PMMA compared with PDMS.

Laminar flow of sample solution sheathed by hydrodynamic focusing of buffer streams in the microchannel of PMMA-based cell-sorter chip

To examine the laminar flow generation in the micro channel of the PMMA-based cell-sorter chip, we loaded a solution containing dark ink at the sample inlet and a clear buffer into the buffer inlets (Fig. 2.8). The stream of sample solution was narrowed from 50 μm to 5 μm by the hydrodynamic focusing mechanism described in previous reports (Hattori *et al.*, 2011; Hattori and Yasuda, 2010; Sawaki *et al.*, 2007; Takahashi *et al.*, 2004). Laminar flow of sample and sheath solution was observed without turbulence or mixing at the flow rate of the cell-sorting area from 0.1 to 200 mm/s (data not shown). The flow rate of the cell-sorting area was about ten times higher than the sample inlet channel, and the linear density of sample particles of the cell-sorting area was ten times lower than the sample inlet channel, because the ratio

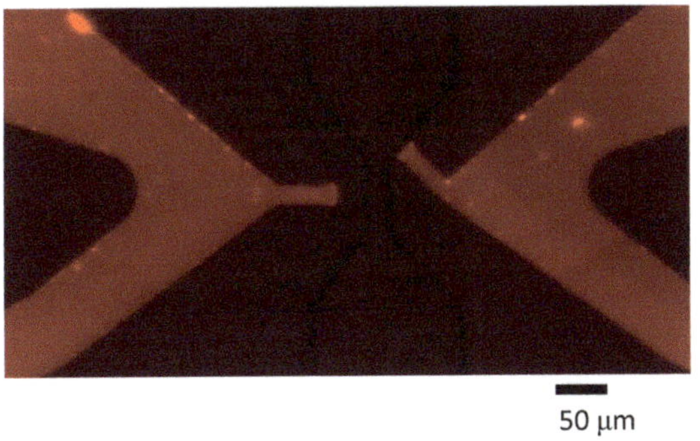

Figure 2.7 Fluorescent image of the cell-sorting area of an injection moulded PMMA-based chip which was filled with the agarose gel and mixed with Rhodamine B.

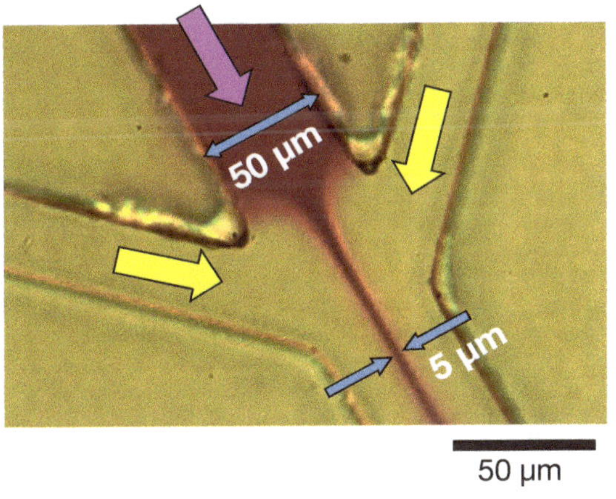

Figure 2.8 Laminar flow generated by a three-forked sheath structure. Sheath buffer containing red ink was applied as sample solution. Flow speed was 1 mm/s.

between the width of the laminar flow from the sample channel and sheath channels were 1:10 as described previously (Sawaki *et al.*, 2007), i.e. the width of the central streamline was less than the size of a cell. Thus the PMMA-based cell-sorting chip had microchannels of satisfactory shape for on-chip cell sorting that was exact as that obtained with a PDMS-based chip.

In addition to particle focusing using a sheath flow, a method based on dielectrophoretic (DEP) force (Hayashi *et al.*, 2011b; Hayashi and Yasuda, 2010) was also examined. The DEP force, F_{DEP}, acting on a particle in a non-uniform electric field is given by the following equations (Hughes, 2002):

$$F_{DEP} = (\pi D^3 \varepsilon_m \mathrm{Re}[f_{CM}] \nabla E^2)/4 \tag{2.1}$$

$$f_{CM} = (\varepsilon^*_p - \varepsilon^*_m)/(\varepsilon^*_p + 2\varepsilon^*_m) \tag{2.2}$$

where D is the particle diameter, ε_m is the permittivity of the medium, f_{CM} is the Clausius–Mossotti (CM) factor, E is the applied electric field, and ε^*_m and ε^*_p are the complex permittivities of the medium and the particle, respectively. The complex permittivity is defined as:

$$\varepsilon^* = \varepsilon - (i\sigma)/\omega \tag{2.3}$$

where ε is the electric permittivity, σ is the electric conductivity, and ω is the angular frequency of the applied electric field. When the real part of the CM factor is positive, a particle moves towards stronger electric fields, and when the real part is negative, it moves towards weaker ones.

To focus the microparticles, including cells, towards the centre of a microfluidic channel using the DEP force, a V-shaped electrode array was placed at the centre of the micro-channel, as shown in Fig. 2.9 (Hayashi *et al.*, 2011b; Hayashi and Yasuda, 2010). An electrode array was placed on the bottom of the microfluidic channel, and alternate electrodes were connected to the opposite terminals of an alternating current (AC) power supply. The principle of particle focusing has a number of features listed below. Application of a voltage causes a non-uniform electric field to form, the strongest region is at the edge of the electrodes at the bottom of the channel, and the weakest region is at the top, immediately above the electrodes. If the DEP force (grey arrows in Fig. 2.9a) on a particle is positive, the particle is pulled towards one edge of the electrodes. In the total of forces between DEP and fluid friction, the particle moves in the tip end direction of the electrode in horizontal and bottom in vertical. If the DEP force on a particle is negative (Fig. 2.9b) the particle moves in the tip end direction in horizontal and top in vertical in the same manner as with a positive charge. In short, all particles are aligned

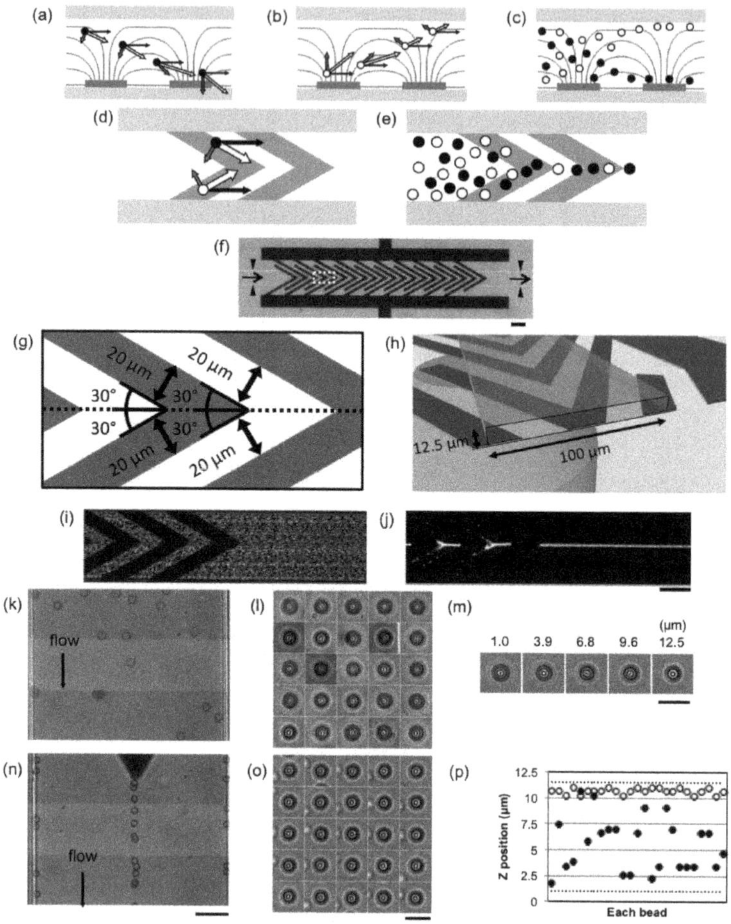

Figure 2.9 Use of DEP force to focus microparticles. (a–e) Schematic images of DEP focusing. Section views (a, b, c) and top views (d, e) of microchannel equipped with V-shaped electrode arrays (grey rectangles). Solid lines in (a), (b), and (c) represent electric flux lines. Direction of flow is from left to right. Circles represent microparticles with positive (black) or negative (white) DEP force applied. Arrows indicate resultant force (white arrows), fluid friction (black arrows), and DEP force (grey arrows). (f–h) Detail of geometry for fabricated V-shaped electrode array. Scale bar: 500 µm. (i, j) Focusing of *Bacillus* spores using V-shaped electrode arrays. Images when AC voltage was (i) not applied and (j) applied are shown. Scale bar: 100 µm. (k–p) Experimental results for PS bead focusing using DEP force generated by a V-shaped electrode array. (k, l) AC voltage was not applied to electrodes; (n, o) AC voltage was applied. (m) Distance of particle from bottom of microchannel was estimated by pattern matching of defocused microscopic images using standard images. Scale bar: 100 µm for (k) and (n) and 20 µm for (l), (m), and (o). (p) Z-positions of PS beads before (filled circles) and after (open circles) flowing through V-shaped electrode array. (Reproduced from Hayashi and Yasuda 2010; Hayashi *et al.* 2011b.)

in the tip-end direction of the V-shaped electrode in the horizontal direction (Fig. 2.9e) and also focused towards the bottom (positive DEP) or the top (negative DEP) of the micro-channel in the vertical direction (Fig. 2.9c). The sign of the DEP force depends on the real part of the CM factor, as shown in Equation 2.1, and it generally depends on the particle type and AC frequency. These properties have been well studied, enabling the DEP force to be used for concentrating and separating particles, including cells (Huang *et al.*, 2011; Unni *et al.*, 2012; Urdaneta and Smela, 2007; Wu *et al.*, 2012).

The focusing ability of a method based on the DEP force was tested using *Bacillus* spores (about 1 µm in diameter) as a cell model with the application of 20 V_{pp} and 1 MHz of AC voltage (Hayashi *et al.*, 2011b; Hayashi and Yasuda, 2010). The spores flowed in the microchannel at a velocity of 1 mm/s. They moved across the stream along the lines of the V-shaped electrode array from the periphery to the centre of the microchannel, as shown in Fig. 2.9j. For the vertical direction, the focusing position of the spores was at the bottom of the microchannel because the spores were subjected to positive DEP forces under the experimental conditions. The vertical distribution of particles was measured using polystyrene (PS) beads (1 μm in diameter) under the same electric field conditions as that for the *Bacillus* spores (Hayashi and Yasuda, 2010). The distribution was measured by pattern matching of their defocused microscopic images (Fig. 2.9k–p). The CM factor of the PS beads was negative (i.e. negative DEP); Fig. 2.9p shows that randomly distributed PS beads (filled circles) were concentrated on the ceiling of the microchannel (open circles).

As described above, we tested two focusing methods: sheath flow focusing and DEP force focusing. Both methods have advantages and disadvantages. Sheath flow focusing is suitable for high-flow velocity because a stable laminar flow forms even when the flow velocity is quite fast. DEP focusing is suitable for a small amount of samples without dilution of sample buffer concentration. The flow velocity is easier to control with DEP focusing because DEP focusing does not require any balance control of flow velocities of additional side sheath buffers. However, the suitability of DEP focusing for high flow velocity is, in principle, limited because of the electrolysis of the solvents used for application of AC voltage to the electrodes placed only on the bottom of the microfluidic channel. Taking these points into account, we are currently using sheath flow focusing for our high-speed cell-sorting system until the problems associated with V-shaped DEP focusing are solved.

Particle sorting and separation using two- or three-way-outlet type cell sorter chip

Two different types of PMMA-based cell sorter chip were manufactured with two or three-way outlet channels (Fig. 2.10). Sample particles were aligned on the centreline of an upstream micro channel using the sheath flow mechanism, and flowed through the exact position of a cell-sorting area between gel electrodes. The outlet channel can be changed using electrophoresis by switching of the DC voltage applied on the electrodes while particles are passing between the tips of gel electrodes. In the case of a two-way-outlet type chip, inlet channels had a symmetrical structure (Figs. 2.3a–c and 2.10a and b).

The sample inlet flow was sheathed with two sheath channels. The sheathed channels and their mirror image-like channels joined at the gel electrode connection point and branched at two symmetrical outlet channels. While the stream of sample particles flowed out to the same side outlet when the DC voltage was turned off (Fig. 2.10a), it flowed out to the other outlet on application of the DC voltage between the gel electrodes (Fig. 2.10b). The separation efficiency using a two-way-outlet type cell-sorter chip on application of continuous DC voltage depended on the voltage and a flow rate (Fig. 2.10l and m). At a constant flow rate, the separation frequency increased at the increasing DC voltage and reached a plateau; efficiency was 90% at a voltage of 45 V and a flow rate of 200 mm/s (Fig. 2.10l). On the other hand, at a constant voltage, the separation frequency decreased with increasing flow rate and no particles were separated; efficiency was below 50% at a voltage of 45 V and a flow rate of 500 mm/s (Fig. 2.9m). Moreover, the two-way-outlet type chip separated small molecules and charged particles, e.g. a sample mixture of polystyrene particles and fluorescein was exclusively separated (Fig. 2.10j and k).

In the case of the three-forked type chip, two sets of the sheathed triplet channels were symmetrically placed as inlet or outlet channels on the chip (Figs. 2.3d–f and 2.10d–g). While the stream of sample particles flowed out to the centre outlet when the DC voltage was turned off (Fig. 2.8d and n). On application of DC voltage between the gel electrodes it flowed out into one of the other outlets according to particle charge and the direction of the electric field (Fig. 2.10e–f, o and p). A mixed suspension containing negative, positive, or non-charged particles was therefore

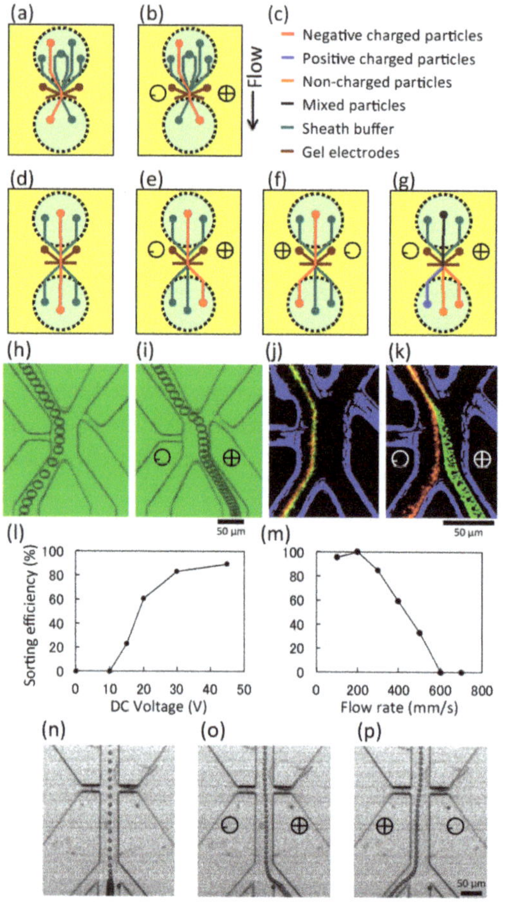

Figure 2.10 Particle separation using two- or three-way outlet type cell-sorter chip. (a–g) Switching mechanism of the two types of cell-sorter chips using electrophoresis by gel electrodes. (a and b) Two-way type or (d–g) three-way type chip. (c) Colour codes of the charged state of sample particles and buffers. Sample particles flow through along a red line without DC voltage as shown in (a) and (d). Application of DC voltage switches the outlet channels according to the sign of particle's charge and the polarity of DC voltage (b, e, and f). Mixed particles of negative, positive, and non-charged particles are sorted exclusively to three outlets by DC voltage (g). (h and i) Tracks of a 20 μm polystyrene particle on a two-way outlet type chip applied without (h) and with (i) DC voltage between gel electrodes. (j and k) Separation of 2 μm polystyrene particles (green track) and fluorescein (red track) applied without (j) and with (k) DC voltage between gel electrodes. The dependence of sorting efficiency on the DC voltage between gel electrodes (l) and the flow rate (m). (n, o, and p) Tracks of a 2 μm polystyrene particle on a three-way outlet type chip applied without (n) and with DC voltage in opposite directions (o and p) between gel electrodes. The direction of electric field is indicated by plus and minus icons on each figure. Flow direction in all figures was from top to bottom.

separated into three outlet channels according to the charged state of each particle (Fig. 2.10g). This is a particular advantage of the three-way type chip for the purification of desired cells from a crude sample containing various kinds of undesired particles such as cells, debris, and inanimate particles.

Real-time size-discrimination of two closely sized particles by microscopic image and improvement of depth of field using zoom lens optics for deeper microfluidic pathways

To examine the image-recognition ability of the cytometry system, we applied a mixed suspension containing two types of closely sized particles (Fig. 2.11). The images of particles were clear but difficult to discriminate for human eyes without a scale or size-comparison of subjects (Fig. 2.11a and b). However, the image recognition unit of

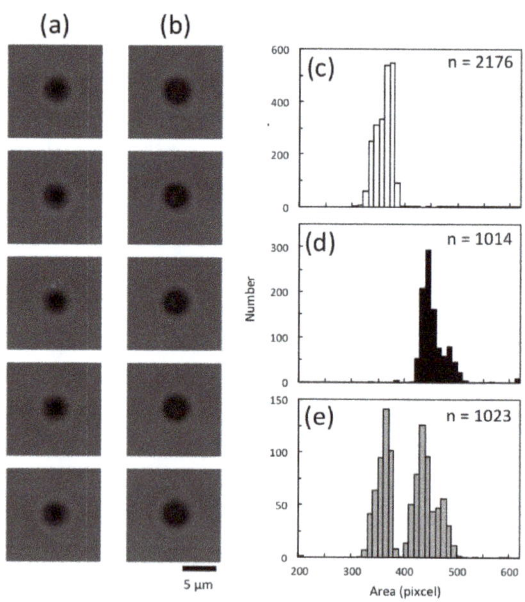

Figure 2.11 Real-time size-discrimination of two closely sized particles by microscopic imaging. (a and b) Bright-field microscopic images of polystyrene particles of 2.5 μm (a) or 3.0 μm (b) diameter. (c–e) Flow rate was 10 mm/s. Histograms of particle area measured by the imaging cytometry system. Sample suspension containing only 2.5 μm particles (c), only 3.0 μm particles (d), or mixed suspension with 2.5 or 3.0 μm particles (e).

the cytometry system precisely discriminated polystyrene particles with 2.5 µm or 3.0 µm in diameter with an error less than 2% at the rate of 200 frames/s in real time at a flow rate of 1 mm/sec (Fig. 2.11c–e). In the case of the sample mixture with 5 µm and 10 µm polystyrene particles, the discrimination error was less than 1% at a flow rate of 200 mm/s. For the mixture containing two kind of particles with given diameters, the discrimination error increased with increasing flow rate as described previously (Hattori *et al.*, 2011; Hattori and Yasuda, 2010). These results suggests that imaging cytometry has many advantages for the precise and high-throughput separation of desired cells from undesired cells of similar size but which cannot be discriminated by the human eye.

We have also examined the resolution of the depth of field when the height of a polystyrene bead was changed using an inverted microscope

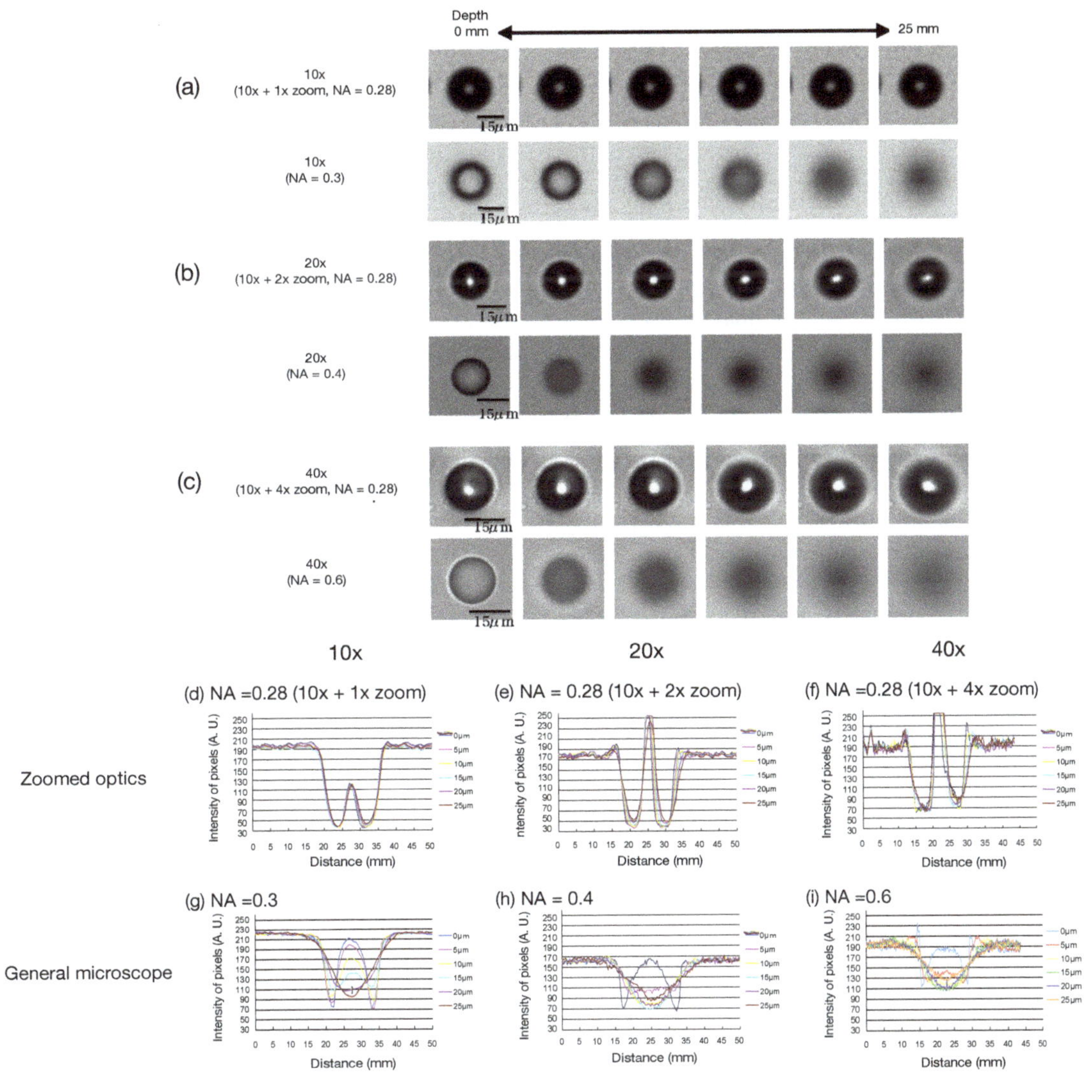

Figure 2.12 Extended depth of field optics for precise image analysis in microfluidic imaging flow cytometry. (a–c) Micrographs of 15 µm polystyrene bead acquired by zoomed optics (upper) and general microscope (lower) every 5 µm of a depth at total magnifications of 10× (a), 20× (b), or 40× (c), respectively. (d–i) Line profiles of zoomed optics (d–f) and regular microscope (g–i) along the line passed through the centre of each image in micrographs (a–c).

(Olympus IX70) with three objective lenses (Olympus UPlanFl 10×, numerical aperture (NA) = 0.3; LCPlanFl 20×, NA = 0.4; LCPlanFl 40×, NA = 0.6). Fig. 2.12a–c shows micrographs of a 15 μm bead acquired by the zoomed optics (10× objective lens and zoom optics) or the inverted microscope every 5 μm depth at total magnifications of 10×, 20×, or 40× (upper micrographs), and also by the objective lenses of 10×, 20×, and 40× objective lenses (lower micrographs), respectively. The total magnifications of optics were adjusted by changing the zoom ratio in the zoomed optics, and exchanging the objective lenses in the general microscope. There is almost no variation of hues of the bead at any depth of bead from 0 to 25 μm in the zoomed optics. In contrast, defocusing was caused by descent of the bead in the regular microscope images. In particular, the amount of focus change was distinguished at magnifications of 20× or 40×. Fig. 2.12d–i shows intensity profiles on the line that passed through the centre of the bead images. The shapes of the intensity profiles of the zoomed optics (Fig. 2.12d–f) were almost similar at any magnification. On the other hand, the amount of change of plots (resolution) at each bead height became larger (Fig. 2.12g–i). As a result of Fig. 2.12, the depth of field was estimated to be 25 μm or greater at any magnification in the zoomed optics. Meanwhile, the depth of field was estimated to be 15 μm at 10×, and 5 μm at 20× and 40× in the regular optical microscope. The results indicated that the improved depth of field was obtained at any magnification from 10× to 40×, and an extended depth of field optics for image-based microfluidic pathways (i.e. wider depth resolution in the deep microfluidic pathway) can be constructed using a low NA objective lens and a zoom lens.

Sealing of semipermeable membrane on the chip reservoir for a contamination-free re-cultivation chip

Transfer of sorted cells from the cell-sorter chip into another culture vessel involves the risks of contamination, damage, or loss of cells. To overcome the risks, we developed an on-chip culture system using a PMMA-based cell-sorter chip. Although the sorted cells in the cell-sorter chip can be cultured by the transfer of whole chip into an incubator, there is a risk of contamination as the medium in outlet tubes are exposed to outside air. Moreover, since the two or three outlet tubes on a cell-sorter chip are linked by culture medium, the waste cells or particles can contaminate the collection outlet by convection or self-propulsion during cultivation. To prevent contamination mediated by culture medium, we tried to separate the outlet tubes and a downstream reservoir using a membrane filter. The membrane filter must allow permeability of nutrients and gasses and impermeability of cells or bacteria. A tight seal was made between the edge of an outlet tube and a membrane by thermal bonding. Since the softening point of PMMA is about 100°C, a membrane filter should have resistance at the temperature for thermal bonding. For tight sealing, the membrane needs no ruck on its surface. The permeability of nutrients and gasses requires a degree of wettability of the membrane for culture medium. The clarity for visible light allowed observation of cells in outlet tubes without breaking the seal. Seven commercial membrane filters were compared based on the criteria described above, and four candidates were selected (Table 2.1).

Next, the permeability of *E. coli* through the membrane filter bonded on the edge of

Table 2.1 Comparison of seven types of commercial membrane filters as a separation filter for on-chip cultivation

	Omnipore	Express	Nylon	Durapore	Isopore	Fluoropore	PVC
Thermal resistance	×	○	○	○	○	×	×
Ruck	○	○	○	○	×	×	×
Wettability	○	○	○	×	×	×	×
Clarity	○	×	×	×	×	×	×
Result	○	×	×	×	×	×	×

outlet tubes was estimated using dummy tubes. Although the Omnipore filter with all pore size and Isopore filter with pore sizes of 0.1 or 0.2 μm was impermeable to *E. coli*, Isopore with pore sizes greater than 0.4 μm and the other filter types had the desired permeability (Fig. 2.13).

Then, we further investigated the properties of a cell-sorter chip whose outlet tubes were sealed with an Omnipore filter (Fig. 2.14). Cardiomyocyte (Rat) showed adhesion to the bottom of membrane-sealed outlets and beating over two days. Contamination of *E. coli* from the downstream reservoir into outlet tubes during cultivation was observed using Omnipore filters with the pore sizes of 0.2 μm or 0.45 μm, but not when using a 0.1 μm pore. Therefore the PMMA-based cell-sorter chip with separation filter on the edge of outlet tubes is useful for on-chip cultivation without contamination.

Application of imaging cell sorting: purification of cardiomyocytes using index of cell surface roughness

For imaging cell sorting, the particular index of cell images is essential for their separation. For example, for separation of cardiomyocytes from the mixture, we have examined and found that most of the smooth round shaped cells should be cardiomyocytes (Kaneko *et al.*, 2011). Hence, we have defined the index of roughness of cell surface *R*, which compares the real perimeter of cell surface and the estimated perimeter of cell surface by approximation of sphere model, to quantitatively distinguish the surface roughness of cardiomyocytes and other cells as

$$R = \frac{l}{\sqrt{4\pi S}} \tag{2.4}$$

where l is the perimeter of cell surface, S is the cross-sectional area of cell (both of l and S were acquired from pixel-based digitized image analysis of cells). For example, a sphere with radius r should have $R[=2\pi r/(4\pi \times \pi r^2)^{1/2}]$ equal to 1, and when the roughness increased, R should be increased in proportion to the increase of the real perimeter. Using the above index R, we can quantify the smoothness/roughness of cell surfaces. For example, typical smooth cells have an index of roughness R less than 1.1 (Fig. 2.15A), and rough cells typically have an R value greater than 1.1 (Fig. 2.15B).

There are two major points for accomplishing non-labelled separation of cardiomyocytes from the mixture of enzyme-treated heart extracts. One is the development of the proper quantified index of surface roughness of cells for judging the target cells and others as described above. The other is the use of the correct enzymatic treatment for appearance of surface shape differences to enable various cells to be distinguished. Regarding cardiomyocyte purification, collagenase-digestion was effective in revealing differences between smooth/rough surface shapes depending on their

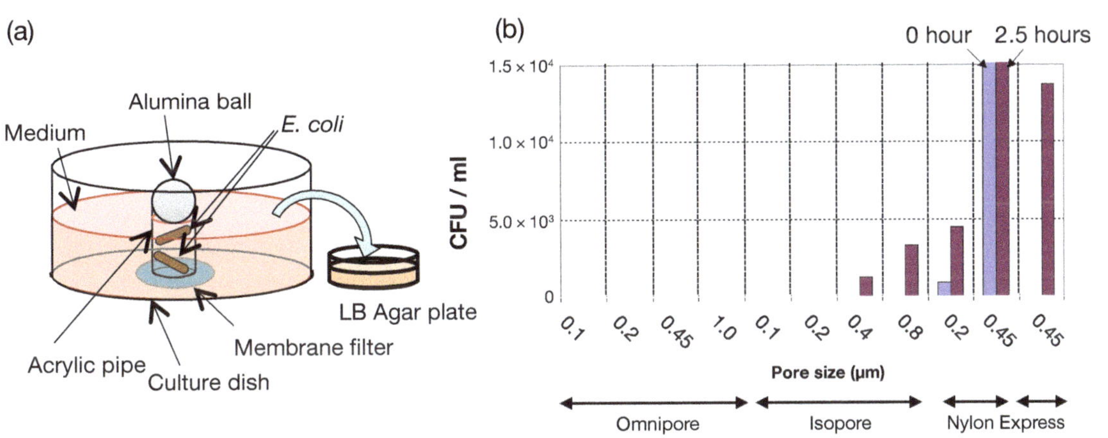

Figure 2.13 Permeability of *E. coli* through membrane filters bonded on a dummy outlet tube by thermal bonding. (a) Experimental setup; (b) Number of colonies recovered on LB agar from the outside medium following injection of an *E. coli* suspension into a dummy outlet tube.

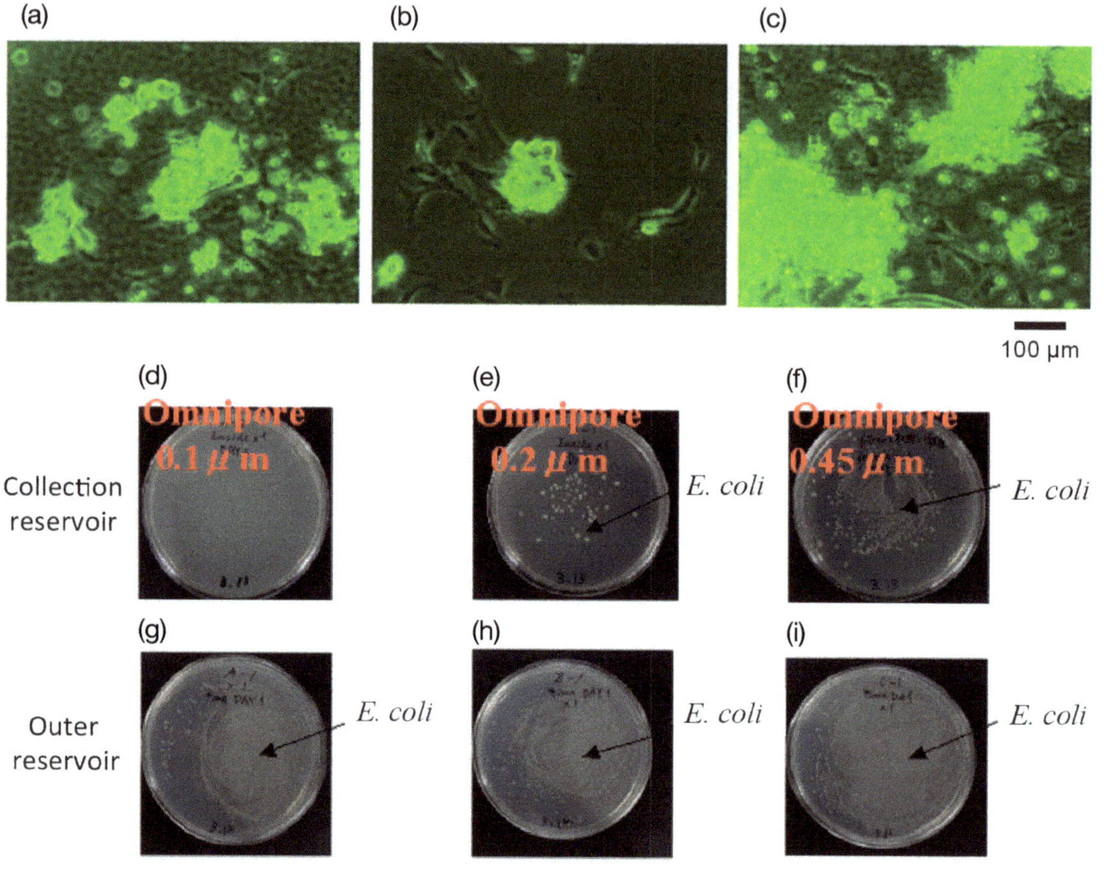

Figure 2.14 Estimation of the viability of cells and the degree of contamination from downstream reservoir using a PMMA-based cell-sorter chip with separation filters. (a–c) Cardiomyocyte cultured in the bottom of an outlet tube after cultivation for two days. LB agar plates of the medium in outlet tubes (d–f) and downstream reservoir (g–i) after cultivation for two days. The cultivation chip was sealed at outlet tubes by Omnipore filters with the pore size of 0.1 µm (a, d, and g), 0.2 µm (b, e, and h), or 0.45 µm (c, f and i).

difference in cell types. Hence, in this experiment, a sample mixture of cells were acquired from collagenase-digested heart isolated from 13-day-old mouse embryos (ICR). It should be noted for this procedure that the collagenase-digestion procedure and subsequent incubation before cell sorting is essential for revealing the differences between of smooth/rough surface of cells. Otherwise, if a conventional trypsination procedure is used for cell preparation, cells do not show apparent smooth/rough surface shape differences. An incubation step after the collagenase-treatment for at least 20 minutes was also essential for revealing shape differences. Without this step, shape differences were not sufficiently distinct on the surface of cells.

Fig. 2.15 shows the typical target smooth surface cells and other rough surface cells acquired from the cell sorting system. Fig. 2.15A shows the micrographs and the field programmable gate array (FPGA) analysed indexes of the collected smooth cells. The R index of all the smooth cells was less than 1.08, indicating that visualized micrograph-based identification of smooth cells is a suitable match to the index $R < 1.1$. Fig. 2.15B also shows the typical rough surface cells acquired by the cell sorting system and as in Fig. 2.15A, each cell indicated the index R was greater than 1.1.

Using the above imaging index, we have succeeded in separating cardiomyocyte cells without labelling by on-chip imaging cell sorting procedures by using an index of their surface roughness.

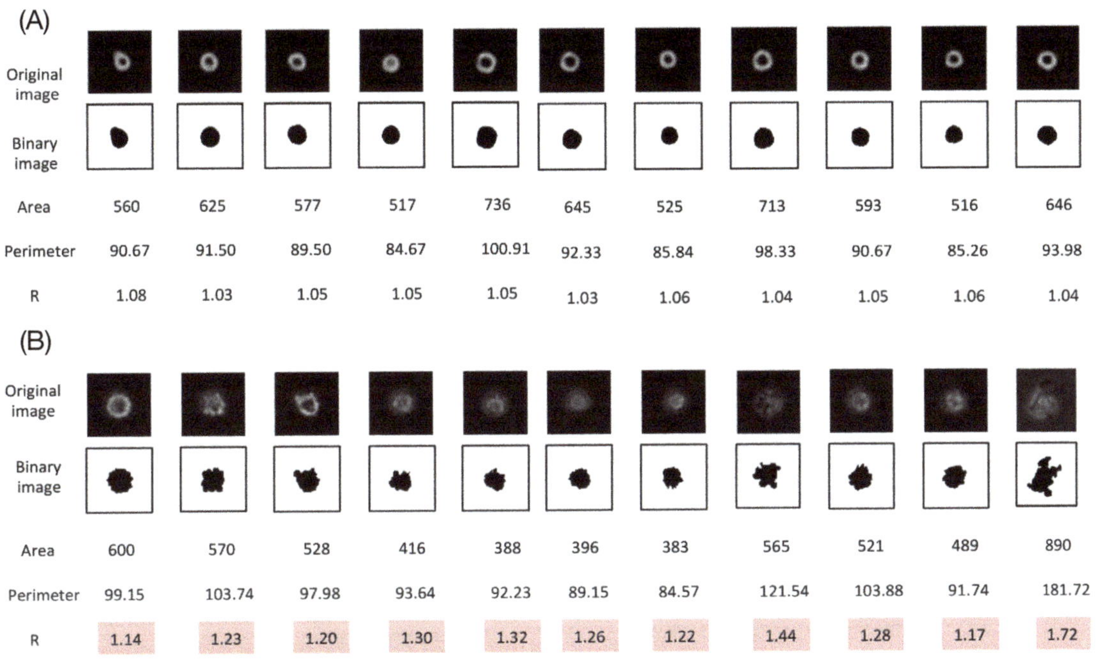

Figure 2.15 Selection of cardiomyocytes with on-chip imaging flow cytometry system. (A) Phase-contrast image and FPGA calculated index of target smooth cells (cardiomyocytes). (B) Phase-contrast image and FPGA calculated index of non-cardiomyocytes.

Data indicated that the more than 99% of healthy re-cultivated cells were cardiomyosine-positive cells. The results suggested that the non-labelling simple phase-contrast image-based cell sorting has potential to precisely separate cardiomyocytes from a mixture of heart cells on the basis of an index of cell surface roughness with less than 1% error rate (Nomura *et al.*, 2011).

As described above, as an example of the effectiveness of the shape-based imaging cell sorting, we showed that the index of roughness of cell surface *R* is one of the most powerful non-labelled imaging index with which to identify target cells such as cardiomyocytes. Although this method has not yet been applied to practical microbial examples, in principle, similar approaches using real time acquired cell shape indexes can be used for identification and separation of targets.

Non-invasive and reversal labelling of live cells using fluorescent DNA aptamers

Separation of target phenotype cells whilst maintaining their native states is important for post genome cell based studies and other practical applications of those cells using on-chip single-cell-based screening assays. Although we have introduced on-chip imaging flow cytometry technologies, characterization of target molecules by antibiotic screening and fluorescence-activated cell sorting (FACS) are still popular because of simplicity and throughput. One of the most serious problems of the FACS and other antibody-based labelling and separation techniques is that strong, often nearly irreversible conjugation, between antibody and antigen molecules on cell surface may affect the function of the targeted cells and their interaction efficacy with other cells (Trickett and Kwan, 2003).

As an alternative to antibodies, nucleotide based probes capable of binding target cell antigens, or aptamer have been developed. Aptamers are single-strand DNAs (ssDNA), RNAs or modified nucleic acids. Systematic Evolution of Ligands by Exponential Enrichment (SELEX) procedures for molecular evolution and aptamer engineering have been reported previously (Ellington and Szostak, 1990; Tuerk and Gold, 1990). The development of ssDNA aptamers for specific binding to live cells (Cell-SELEX) expanded their potential for use in cell separation and labelling (Herr *et al.*, 2006; Ohuchi *et al.*, 2006; Shangguan *et al.*, 2006). In this subsection, we describe a new method for reversible labelling of the cells using fluorescently tagged aptamers (Anzai et al., 2007; Terazono *et al.*, 2010) (Fig. 2.16A).

Our technique uses high affinity ssDNA aptamers specific to surface-exposed target molecules on cells. The target cells are first purified by attaching aptamer-conjugated magnetic microbeads to them, then magnetically separating these labelled target cells from the unlabelled cells, and finally detaching the microbeads from them by nuclease digestion.

We tested our cell separation technique by using cultured human leukaemic lymphoblast (CCRF-CEM) cells (Fig. 2.16B). They were first washed three times with media buffer. Then 5×10^4 cells were dispersed in 25 µl of media buffer, to which 5×10^5 aptamer-magnetic beads in 25 µl of media buffer were added. After the suspension was incubated for 30 minutes at room temperature with gentle mixing, the bead-bound cells were retained using a magnet and rinsed with fresh media buffer to remove unbound and low-affinity cells.

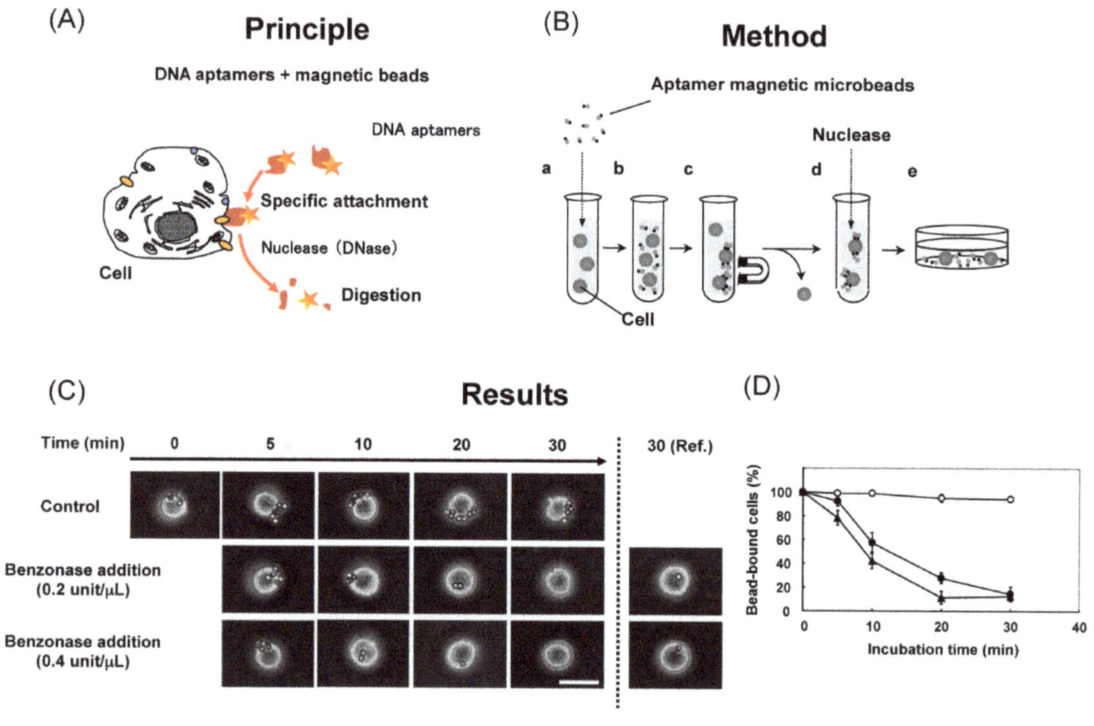

Figure 2.16 (A) Principle of DNA aptamer cell sorting. DNA aptamer probes are attached to biomarkers on cell surface; target cells are then collected using DNA aptamers; and finally attached DNA aptamers are digested and removed by DNase after target cells are collected. (B) Protocols for separating cells using aptamer-conjugated magnetic microbeads and for detaching beads using nuclease. In (a, b) cells are attached to aptamer-magnetic beads in test tube, (c) unbound cells are extracted using a magnet, (d) bead-bound cells are incubated with nuclease, and (e) bead-bound cells are counted. (C) Phase-contrast optical microscope images of bead detachment during incubation with Benzonase nuclease. Upper row: control (absence of Benzonase). Middle row: 0.2 unit/µl Benzonase. Lower row: 0.4 unit/µl Benzonase. Right reference microphotographs show non-specifically attached magnetic beads that could not be removed even after incubation for 30 minutes in 0.2 or 0.4 unit/µl Benzonase. Bar = 20 µm. (D) Growth rate (normalized by that on day 0) of separated cells after nuclease digestion. Filled circles and filled triangles: 0.2 or 0.4 unit/µl Benzonase-treated cells, respectively. Open circles: untreated cells. Error bars indicate S.D. for three measurements. (Anzai *et al.*, 2007).

The separated target cells were incubated at 37°C with gentle mixing in media buffer containing DNase (Benzonase® nuclease), and the number of bead-bound cells was counted periodically using an inverted phase-contrast optical microscope (Fig. 2.16C, micrographs). Prior to Benzonase addition, the beads had aggregated and adhered to the cells, forming a cluster. After the addition of 0.2 or 0.4 unit/µl Benzonase nuclease, the attached beads were progressively cleaved over time; at 30 minutes, most of the beads had detached from the cells. About 20% of the cells, however, still had one or a few beads attached at 30 minutes (Fig. 2.16C). In the absence of Benzonase, the attached bead clusters remained bound to the cells without any change even after incubation for 30 minutes. The time course of the detachment is shown in Fig. 2.16C and plotted in Fig. 2.16D, where 100% represents the number of cells with beads attached at 0 minutes. During incubation in the presence of 0.2 or 0.4 unit/µl Benzonase, the percentage of bead-bound cells decreased with time and reached an equilibrium level within about 30 minutes for both concentrations. In the absence of Benzonase, the percentage of bead-bound cells did not change.

Another non-invasive labelling technique uses ssDNA aptamer-conjugated Qdots (Terazono *et al.*, 2010) (Fig. 2.17). Binding of the fluorescent aptamers to the cells was evaluated by measuring the fluorescence intensity and was confirmed using flow cytometry. The aptamers were removed in ~10 minutes by DNase digestion. Incubation of cells with aptamers or with nucleases resulted in no apparent damage to the cells and did not affect their growth rates.

These two aptamer-based techniques for cell identification and separation are promising alternatives to conventional antibody-based techniques, particularly for non-invasive reversible cell labelling and cell separations when it is important to maintain native cell activity. The digestion of the probe aptamers that attach the magnetic beads to the cells could help solve the problem of attached antibodies affecting cell efficacy (Trickett and Kwan, 2003). The results obtained indicate the potential of multiple-step separation using a series of stepwise different aptamer-magnetic bead separations for precise refinement of target cells when there are multiple target molecules on the cell surface. Our three-step separation method should thus contribute to the rigorous purification of subpopulations and more detailed analysis of phenotypic differences in uniform cell types.

As described above, our labelling procedure produces no noticeable damage to the labelled cells. This method is advantageous over traditional antibody labelling because it allows non-invasive cell separation and reversible cell labelling. It holds great promise for a variety of applications in the advanced imaging flow cytometry technology for microbial cells.

Applications of on-chip imaging cytometry technologies

Phenotypic and behavioural variations from cell to cell have been observed to exist even in a genetically identical population (Gally *et al.*, 1993; Ko *et al.*, 1990; Msadek, 1999; Schwan *et al.*, 1992; Spudich and Koshland, 1976). The resulting heterogeneity in a clonal population may well be important not only for survival (Msadek, 1999), but also for cooperation in a population that must obviously exist and work within multicellular organisms (Shapiro, 1995, 1998; Shapiro and Hsu, 1989). The mechanisms of producing phenotypic variations are explored both theoretically (Arkin *et al.*, 1998; Kepler and Elston, 2001; Kierzek *et al.*, 2001; Levin *et al.*, 1998; McAdams and Arkin, 1997; Thattai and van Oudenaarden, 2001) and experimentally (Blake *et al.*, 2003; Elowitz *et al.*, 2002; Ozbudak *et al.*, 2002) as an intracellular noise-driven process (Rao *et al.*, 2002). McAdams and Arkin (1997) proposed that stochasticity in the process of gene expression could account for the substantially large differences in levels of protein products, which eventually affects the switching mechanisms in individual cells in a group that select between alternative phenotypes. The existence of noise in gene expression processes was shown experimentally by van Oudenaarden and colleagues (Ozbudak *et al.*, 2002). They showed that the resulting expression noise had a strong positive correlation with translational efficiency. As another example of the experiment, Elowitz and colleagues (2002)

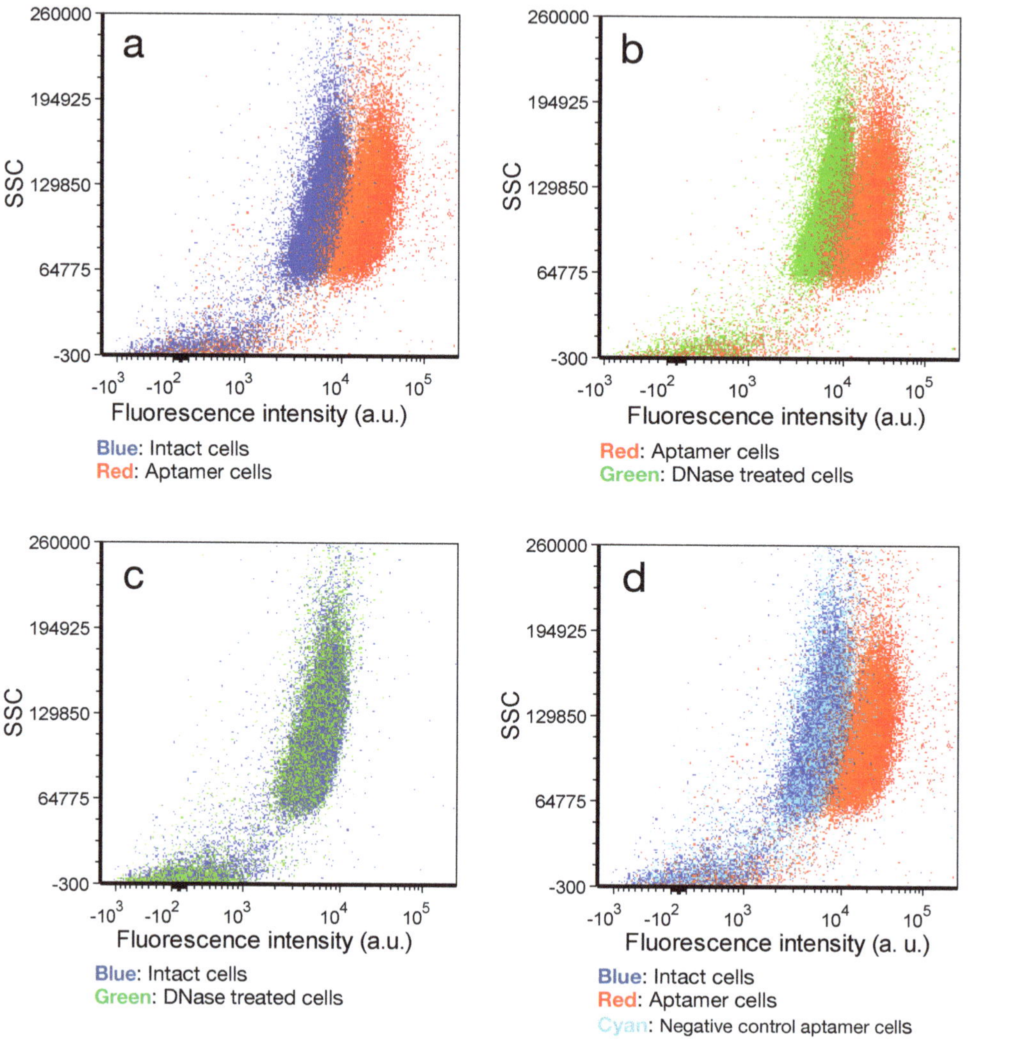

Figure 2.17 Fluorescence intensity of the cells labelled with Qdot-aptamer and treated with nuclease. Fluorescence changes following Qdot-aptamer labelling of intact cells (blue: intact cells, red: aptamer labelled cells) (a). Fluorescence changes following DNase treatment of aptamer cells (red: aptamer labelled cells, green: DNase treated cells) (b). Comparison of unlabelled cells (blue) and DNase treated cells (green) (c). Comparison between unlabelled cells (blue), cells incubated with an irrelevant aptamer (cyan) and cells labelled with cell-specific aptamer (red) (d). Vertical axes show Side Scatter (SSC) values, and horizontal axes show fluorescence intensity measured at 525 nm.

examined the contributions to overall variation from gene expression process and from other cellular components separately, showing that noise in the gene expression process did not uniquely determine the total variability.

These studies are based on the temporal observation of a cell group. The group based observation approach, however, cannot show how an individual cell produces different phenotypes and behaviours in the course of proliferation and whether phenotypes and behaviours specific to an individual cell can be inherited. Conventional techniques such as flow cytometry and direct observation with a microscope do not provide control over cell–cell interactions or selection of cell type. Flow cytometry enables us to obtain the distributions of parameters like concentration, size, shape, DNA content etc. at the single-cell level in a group although it cannot yet be related to epigenetic profiles (Akerlund *et al.*, 1995). The problem for this method is that it cannot continuously track the dynamics of a specific cell because the sample drawn from the culture is discarded after the measurement. Neither can it keep cells under isolated conditions, nor can it identify a particular cell, especially after cell division has occurred. Thus, conventional cytometry can give us information about the average properties of cells as a summation of individual characteristics of different cells in a group, that is, how the group changes including the distribution information of individuals, but it cannot give us tracking information about how the same single cell changes. Direct measurement, with a microscope, of cells in solid media such as cultivation plates (Donachie and Begg, 1996; Elowitz and Leibler, 2000; Gardner *et al.*, 2000; Panda *et al.*, 1999; Shapiro and Hsu, 1989) can identify individual cells, and thus, can track specific cells continuously. However, it is impossible to keep cells isolated especially after cell divisions have occurred and it is impossible to control the interactions between particular cells because the positions of the cells are fixed at the beginning of cultivation. Thus these conventional methods are not a satisfactory means of gaining an understanding of single-cell level interactions of particular cells.

As new techniques are needed to clarify the interactions between genetically identical cells, we have developed an on-chip single-cell-based microculture method exploiting recent microfabrication techniques and conventional *in vivo* techniques for single-cell-based imaging cytometry for 'algebraic' observation of single cells (Fig. 2.18). The advantage of the on-chip technique is that a diverse population development out of one cell can be followed directly. To manipulate cells in microchambers, we use non-contact forces, such as optical tweezers and acoustic radiation force, which have been used to handle cells, organelles, and biomolecules on microscope specimens (Ashkin and Dziedzic, 1989; Ashkin *et al.*, 1986, 1987, 1990; Svoboda and Block, 1994; Wright *et al.*, 1990; Yasuda, 2000).

Especially for microbial aspects, the single-cell-level handling, sorting and monitoring technologies using on-chip technologies are important for understanding the process of adaptation of cells, i.e. whether the adaptation of groups is caused by the selection of phenotypically slightly different cells or the adaptation of individual cells. As shown in Fig. 2.18, if we apply the on-chip single-cell-based cultivation for single cell cultivation, we can distinguish the origin of the cell groups adaptation caused by adaptation of individual cells or selection.

On-chip imaging cytometry can also give us an opportunity to commence the epigenetic research not only based on the expression of molecules, but also based on the imaging indexes such as

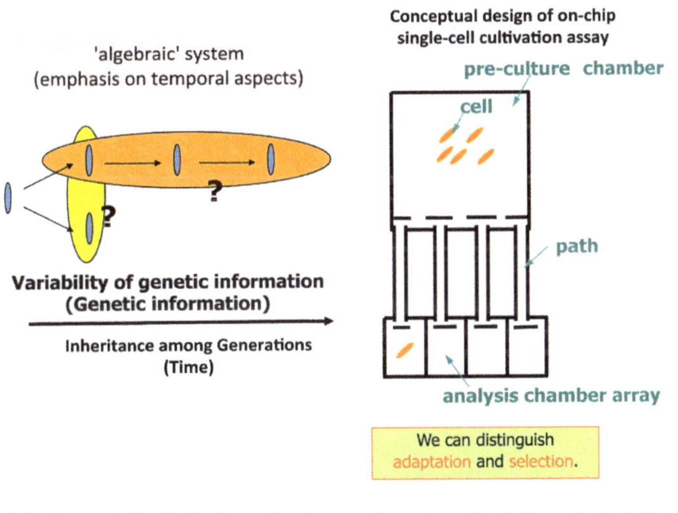

Figure 2.18 'Algebraic' viewpoint of single cell observation, in which the comparison of sister cells and direct descendant cells are accomplished using on-chip cultivation assay.

shapes of cells and their manner of interactions as a part of the non-genetic inheritance information.

In this section, we briefly describe some examples of our on-chip single-cell based microculture methods to explain the microbial possibilities of imaging cytometry.

On-chip imaging cytometry assay for 'algebraic' observation of single cells: on-chip single-cell cultivation system for isolated *Escherichia coli* cells

To understand variations in cells with the same genetic information and observe adaptation processes in cells, we need to directly compare sister or direct-descendant cells. Fig. 2.19 is a schematic drawing of the entire system used for on-chip single-cell-based analysis. It consists of a microchamber array plate (chip), a cover chamber attached to the medium circulation unit, a ×100 phase-contrast/fluorescence microscope, and optical tweezers.

The microchamber array is the microfabricated structure on the glass slide made of thick photoresistant SU-8-5 (Microlithography Chemical Corp., MA) (Fig. 2.20a and b). The height of the microchamber array is 5 μm, in which the cells are enclosed. The microchamber array is sealed with a semi-permeable membrane to prevent cells escaping. The semi-permeable membrane is coated with avidin and the glass slide with biotin to ensure the seal is tight (Fig. 2.20c). With these coatings on both the membrane and slide, it is possible to observe cells in the microchamber without them escaping. The microchamber (compartment) is composed of two main parts. The first is the observation area, which has four subcompartments at the centre of the microchamber. Each subcompartment has a volume of 20× 20× 5 μm. The second part includes the discarding areas at both sides of the four-chamber subcompartments. The first four direct-descendant cells derived from a isolated single cell are placed in one of the four subcompartments individually to keep them isolated. The excess descendant cells are transferred to the two large discarding areas along the white arrow with the optical tweezers through the narrow winding path as shown in Fig. 2.7. As shown in the micrographs, only one cell is enclosed in each of the four subcompartments of

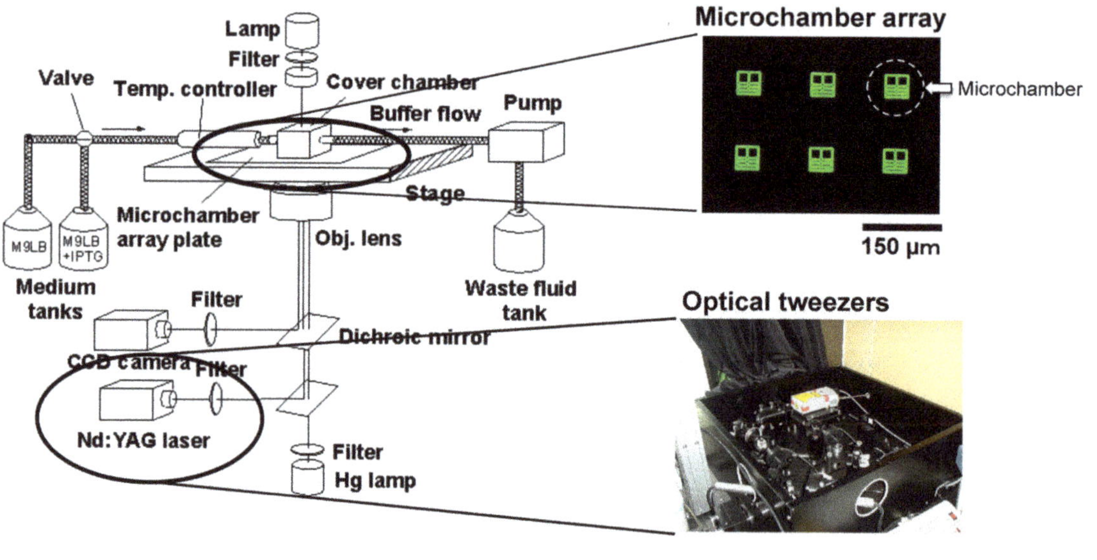

Figure 2.19 Single-cell cultivation in microchambers to determine variability in genetic information. The system consists of three parts; a microchamber array chip with medium supply, optical microscopy with recording equipment, and an optical tweezers. An on-chip single-cell cultivation system enables excess cells to be transferred from the analysis chamber to the waste chamber through a narrow channel that allows a particular cell to be selected from cells in the microfabricated cultivation chamber with non-contact force, optical tweezers.

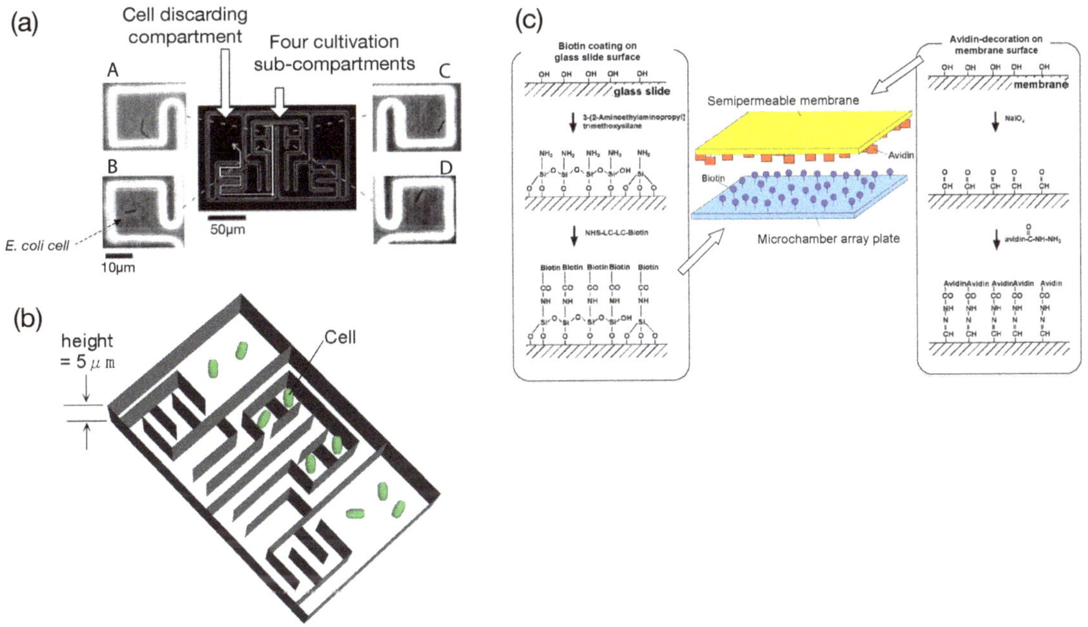

Figure 2.20 Microchamber design. (a) Optical micrograph of micro-chamber (compartment) with four subcompartments. Each single-cell cultivation compartment has four observation subcompartments (A, B, C, and D) at the centre of the microstructure. At both sides of four cultivation compartment, two cell discarding compartments are arranged and connected to the observation subcompartments with narrow pathways. Enlarged views of the four cells in the four subcompartments of the observation area are shown. (b) Schematic three dimensional image of the microchamber. (c) Sealing protocol for semipermeable membrane lid on chip. The semi-permeable membrane is coated with avidin, and the glass slide with biotin. After the introduction of single cells into the micro-chambers the semi-permeable membrane lid seals the top of the micro-chambers with avidin–biotin bonding used to prevent escape of isolated cells from micro-chambers.

the observation area under isolated conditions. Four specific cells in the four subcompartments are simultaneously observed without any disruption by the other cells and without leaving the field of view of the microscope.

Optical tweezers were introduced to enable non-contact handling of the cell specimens. An Nd:YAG laser (wavelength, 1064 nm; T20-8S, Spectra Physics, SpectraPhysics, CA) was guided to the ×100 phase contrast objective lens (Uplan-Apo, Olympus, Tokyo, Japan) in a phase contrast microscope (IX-70, Olympus, Tokyo, Japan) as the light source for the optical tweezers, which are widely used for handling of micron-sized particles and biomaterials (Ashkin and Dziedzic, 1989; Ashkin *et al.*, 1986, 1987, 1990; Svoboda and Block, 1994; Wright *et al.*, 1990). In our protocol, the tweezers were used to transport particular cells within the microchamber. The medium circulation unit utilized a glass box with a volume of 1 ml that had two branches. It was mounted on the microchamber array chip and fresh medium buffer was always circulated in the glass box through the two branches at a rate of 1 ml/min using a peristaltic pump. The bottom of the glass box was open and the condition of the medium around the cells could be constantly maintained by buffer exchanges through the semipermeable membrane. The whole microcultivation part was placed in a thermo control cage (IX-IBM, Olympus, Tokyo, Japan) to maintain the temperature at 37°C throughout observation. The observation images were taken with a CCD camera (CS230, Olympus, Tokyo, Japan) and recorded on digital video cassette. These were analysed on a personal computer (PCV-R73K, Sony, Tokyo, Japan).

Adaptation process for sensor proteins in cells caused by environmental changes – an example of the imaging index for identification of phenotypes

To investigate non-genetic variability in the division cycle and growth of single cells, we measured growth and division times for pairs of *E. coli* daughter cells under isolated conditions using the on-chip single-cell cultivation system described above (Inoue *et al.*, 2001a,b). In the above experiment, the non-labelled imaging index was used to acquire the time course growth and division information continuously and non-invasively. This on-chip single-cell cultivation/analysis system was modified to simultaneously measure the sensor-protein dynamics and motility of identical single cells for several generations (Inoue *et al.*, 2004). This technique revealed the potential of combining the microfabrication technique (single-cell cultivation technique) with molecular biology (single-molecule observation).

E. coli cells are able to respond to changes in environmental chemo-effector concentrations by reversing their flagellar motors (Levit *et al.*, 1998; Manson *et al.*, 1998). Attractants (such as aspartate and serine) promote anti-clockwise rotation of the flagella, resulting in a smooth swimming action, whereas repellents (such as phenol and Ni) promote clockwise rotation, resulting in tumbling. These responses are mediated by membrane-bound, methyl-accepting chemoreceptor proteins (MCPs). Immunoelectron microscopy revealed that MCP–CheW–CheA complexes (CheW and CheA are the Che proteins, which alter the tumbling frequency, and alter the receptors) are clustered *in vivo*, predominantly at the cell poles (Maddock and Shapiro, 1993), and only weaker lateral clusters were observed (Lybarger and Maddock, 1999; Skidmore *et al.*, 2000). Polar-localization changes are expected according to environmental conditions, whereas no evidence concerning the dynamics of localization changes has yet been reported. Conventional cell group cultivation-based experiments do not allow the process of MCP clustering and the effect its change has on consecutive generations in individual cells. This is because their measurements were based on statistics of snap shots of cell groups at a time, and hence, it is difficult to estimate their individual-level time-course change (adaptation), which is essential in estimating the changes occurring during alternation of generations. To understand epigenetic processes such as adaptation and selection, both the protein-dynamics and the cell dynamics of particular single cells should be observed continuously and simultaneously for several generations.

We used an assay of intracellular proteins tagged with green fluorescent protein (GFP) to measure the localization-dynamics of expressed aspartate sensitive sensor protein, Tar. We also modified the shape of the microchambers into a wheel shape to measure the time course for motility (Fig. 2.21A). In the experiment, we first placed a single bacterium in the microchamber and isolated it in the wheel region so that it could swim along the track seal with the semi-permeable membrane lid on the microchamber. Then, the movement of the bacterium around the circle structure was continuously monitored by measuring the tumbling frequency and protein-localization dynamics. When the cell divided into two daughter cells, one of these was picked up with the optical tweezers, transported to the axle area, and confined in this region to stop growth. The bacterium was chemically stimulated by adding aspartate (nutrition for *E. coli*) into the medium.

After the first change of medium, it took more than three generations to recover the original pattern of GFP–Tar localization (Fig. 2.21B(d, e) and Fig. 2.21C(d, e). However, the frequency of tumbling remained higher than the former generations. This may indicate that Tar-localization requires more time to form than to diffuse. Such asymmetric reversibility in protein localization may contribute to cell phenomena being inherited resulting from environmental changes. It also suggests the possibility that the change in Tar localization may be inherited by descendant cells and this can affect their motility and therefore their phenotype.

The results indicate the microbial possibilities of imaging indexes to identify the phenotypical differences of cells, which is potentially applicable for the imaging flow cytometry to separate the different GFP-protein localized cells.

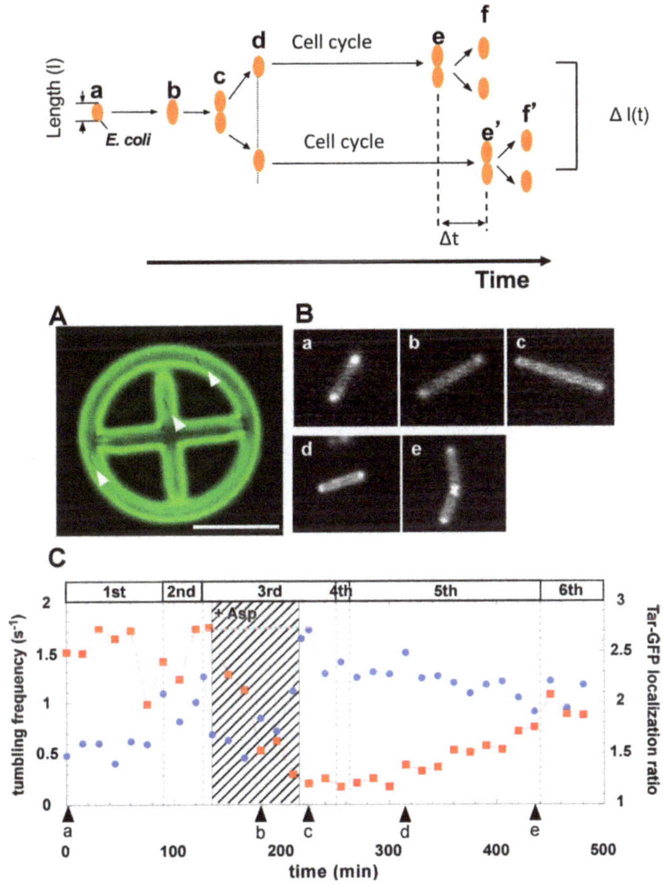

Figure 2.21 Differential analysis of sister cells and direct descendant cells. During on-chip cultivation, both the difference of interdivision time (cell cycle) and lengths at time (t) of sister cells were compared. Simultaneous observation of Tar-GFP localization and motility in identical *E. coli* cell for several generations. When the cultivation started, the tar-localization ratio (red squares) was 2.5 and the tumbling-frequency (blue circles) was 0.5 per second [(a) in A, and arrowhead 'a' in graph in C]. After the second cell division had occurred, a minimal medium containing 1 mM of aspartate was applied to the third generation cell (135 minutes after microcultivation). After the attractant was added, tumbling frequency (filled circles in graph) decreased immediately compared with the previous generation. Localization of the aspartate sensitive sensor protein at two poles in *Escherichia coli* (filled squares) also decreased quickly by half to 45 minutes following the change of medium [B(b) and C(b)]. Finally, after 80 minutes of stimulation with the aspartate, the localized tar had diffused completely. Then, the aspartate was removed from the cultivation medium and the cells were cultivated further to enable the recovery of tar-localization dynamics to be measured [B(c) and C(c)].

Origin of individuality of two daughter cells during division process of *E. coli*

Detecting changes in the swimming behaviour of a particular line of growing cells seems feasible but to date has been challenging to many scientists as it is experimentally difficult to undertake. It is impossible to follow direct descendant cells in a test tube, and even if they were identified they could not be observed continuously as they grow. Clearly identifying transient time-course changes in a particular cell function is difficult, and must be investigated separately as they may provide information about the relationship between growth and motility but not about when the individuality of a particular cell arises.

To evaluate the potential of the imaging index for the measurement of microbial individuality, we have overcome experimental limitations by developing a system that measures growth and motility simultaneously. A single-cell-based dual-recording feature has made it possible to directly compare the cellular growth and motility of a cell with that of its direct descendants (Umehara *et al.*, 2007b).

A typical result of simultaneous measurement of cell length and running speed is summarized in Fig. 2.22. Fig. 2.22a is a record of cell length and running speed for one generation, from the birth of a cell to that of its daughter cells. Throughout its cycle the cell steadily elongated exponentially, while the running speed of the cell gradually decreased. Running speed was calculated on the basis of the recorded positions of the cell. By plotting these positions the swimming path of the cell for an arbitrarily defined time range could be generated (Fig. 2.22b). Each position was determined by an image analysis program driven in real time with 0.1-second resolution. The program recognized positions of the cell based on a successive series of high-resolution images (Fig. 2.22c).

By combining the repeated isolation procedure with this analysis, we can compare the growth and motility of an isolated single cell with the growth and motility of its direct descendants. Fig. 2.23 is an example of simultaneous measurements for seven generations. The cell length increased within each cell cycle (Fig. 2.23a) while running speed decreased throughout each cell cycle (Fig. 2.23b). Within each cell cycle the plots of running speed versus generation time tended to form smooth curves, but the slopes of those curves differed between generations. Tumbling frequency fluctuated between generations and did not show any constant increase or decrease (Fig. 2.23c). Both the decrease in speed and the fluctuation

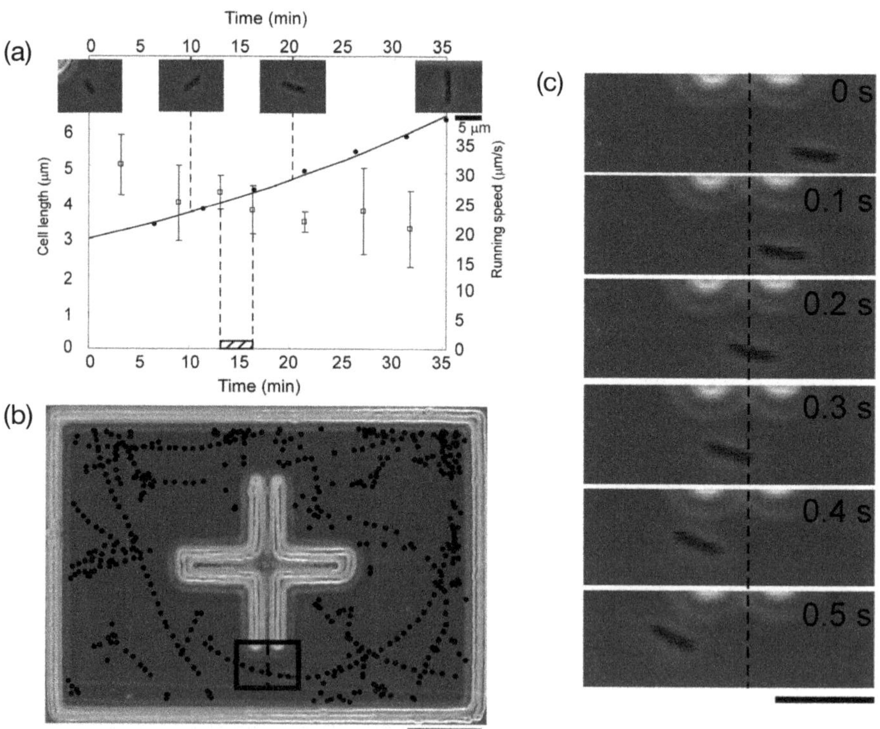

Figure 2.22 Typical example of simultaneous measurement of growth and motility. (a) Time course of single *E. coli* growth and motility for one cell cycle ($t = 0$ to 35.3). Cell length (filled circles with a fitted exponential growth curve) steadily increased as the running speed (open squares with SD) decreased throughout the cell cycle. Photographs show the morphology of the cell at each corresponding time point. (b) Swimming path of the cell. Black dots are recognized positions of the cell by the computer analysis program for the hatched time range shown in (a) ($t = 13.0$ to 16.2). Bar, 20 µm. (c) Photographs of the signified region in (b) at successive 0.1-second frames. The cell was swimming through the region from right to left. Bar, 10 µm.

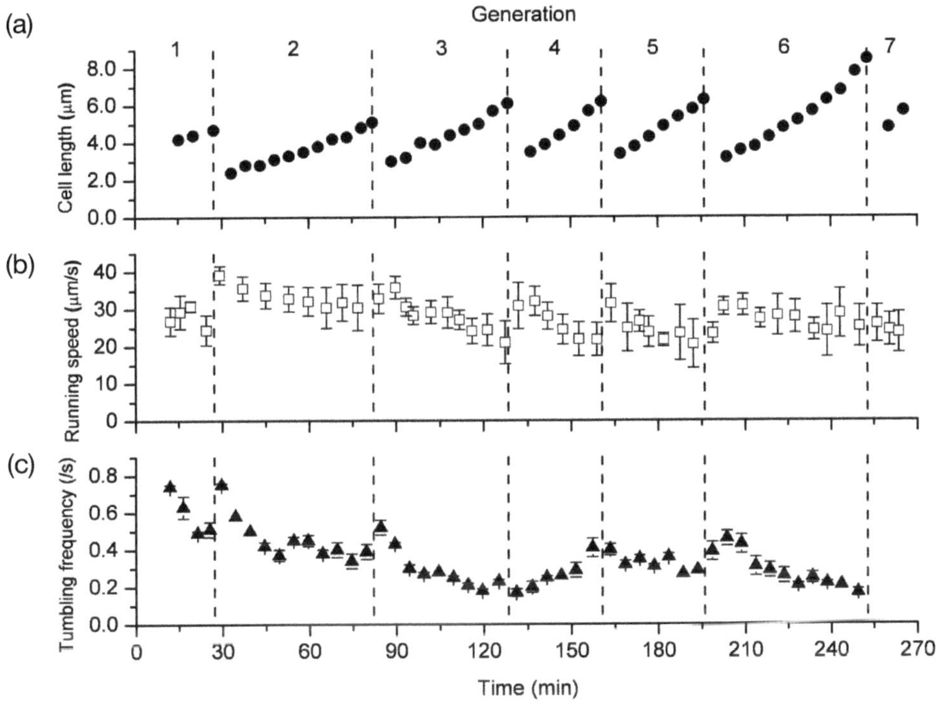

Figure 2.23 Simultaneous measurement of growth and motility for successive generations. Time course of single *E. coli* growth and motility were monitored by the measurement of (a) cell length, (b) running speed (mean ± SD), and (c) tumbling frequency (mean ± S.E.). Vertical dashed lines correspond to the timing of cell divisions.

in tumbling frequency were reproducible among other datasets.

Careful observation revealed that a pre-divisional cell after constriction showed characteristic movement. This featured a longer duration of pausing in motion: typically 10^{-1}–10^{0} s, which is about 10 times longer than that caused by ordinary tumbling (10^{-2}–10^{-1} s). We named this activity *prolonged pausing* in contrast to ordinary tumbling. The duration of pausing remained relatively constant before cell constriction, but it tended to increase after constriction to the level beyond the deviation from the pre-constriction value (generation 5 appeared to be an exception). The increase in pausing duration appeared only after cell constriction, reflecting that only pre-divisional cells exhibit prolonged pausing.

The results of this on-chip measurement were consistent with the results of conventional methods. The measured running speed and tumbling frequency were both acceptable when compared to data already reported for the wild-type strain (25 µm/s and 0.53 s^{-1}) (Alon *et al.*, 1998). This suggests that the micro-chambers confining the cells did not affect swimming behaviour, and that the on-chip measurement was compatible with other conventional methods.

The important fact is that growth and motility could be measured separately yet simultaneously. Conventional methods cannot deal with both of these parameters as growth needs to be observed for a longer term, whereas motility requires close observation for a shorter term. Moreover, conventionally obtained data describes only the average characteristics of a group of cells. The improved on-chip single-cell cultivation system which copes with swimming bacteria has overcome these limitations and has shown a direct relation between growth and motility: cell-cycle dependence of swimming behaviour.

Mother cells and daughter cells are usually defined and distinguished by cell division: a physical process that divides the cell body into two newborn ones. To complete cell division and to become two cells, however, a cell must spend some time preparing at both molecular and cellular levels. Thus there must be a transient state in which two distinct control systems co-exist in a pre-divisional cell. These two systems may independently affect the intracellular mechanism for the whole-cell moving behaviour.

Individuality has also been discussed from the viewpoint of bacterial swimming behaviour by Spudich and Koshland (Spudich and Koshland, 1976). They reported that cellular individuality in bacteria is rather steady throughout their cell cycles. Although the observations in this report may suggest the opposite from our data, it does not contradict this study. This is because these workers regarded cell length as an index of cell cycles, and neglected the fluctuation of synchrony between cell length and cell cycle. Since even direct descendant cells vary in length (Wakamoto *et al.*, 2003), we cannot infer that different cells of the same cell length are at the same stage of their cell cycles. The prolonged pausing we have found here was probably hidden by this large variation, and was revealed by the long-term single-cell monitoring.

In this subsection, we have introduced an another non-labelled and non-invasive imaging index, *prolonged pausing*, as an pre-divisional microbial cell cycle marker, in which not only the shape of cells but also the behaviour such as time course spatial movement and tumbling of cells are applied for their identification. This particular imaging index also has the potential to be included into imaging flow cytometry technology.

Asynchrony in growth and motility responses to environmental changes by individual bacterial cells

Various physiological states of bacteria are related to their survival strategy under environmental stresses. One example of the phenotypic diversity is a non-growing but metabolically active state called quiescence, in which cells channel available metabolic resources towards particular cellular functions, such as the expression of specific genes (Rowe and Summers, 1999).

In this subsection, as an example of the imaging index, the method and the index used to study how the growth and motility of a single cell responds to a change in external nutrient conditions are described. Simultaneous monitoring of growth and motility with an on-chip single-cell cultivation system revealed asynchrony in the growth and motility responses to nutrient starvation. Cell growth stopped quickly after starvation began, whereas cell motility was initially maintained for several hours and then gradually lost. Discussing these results, we considered the energetics within an individual cell, the flagellation state of starved cells, and the possible mechanisms responsible for the observed response asynchrony. (Umehara *et al.*, 2007a).

For this study an *E. coli* strain AW539 was transformed with a plasmid coding a GFP-tagged aspartate receptor. This study demonstrated the feasibility of long-term, trans-generational observation of single-cell growth and motility by sequential use of multiple medium conditions of choice. Nutrient condition changes were applied to an isolated single cell to measure the response of its movement and growth after the environmental changes. In each experiment a motile cell was chosen and then continuously monitored, swimming freely but only within its chamber. Trans-generational observation was achieved using optical tweezers to repeatedly remove sister cells produced by cell division. During the repeated removal the laser power was increased so that unwanted cells were ‘killed’ and permanently fixed to the dead-ends of the chamber. The isolated cell condition was maintained during medium exchange. The series of above mentioned procedures provided data related to the time-course change of cell length and running speed for multiple generations.

The simultaneous observation of cell growth and motility revealed asynchrony between these two functions when a single cell responded to nutrient starvation (Fig. 2.24). The cell cycle of the single cell was first checked in the constant LB (nutrient-rich) medium condition. During each of the first few generations, the growth curve increased monotonically and the speed curve showed a general tendency to decrease. After a few generations, the LB medium was completely exchanged for a 0.9% NaCl aqueous solution (nutrient-free) so that the cell experienced nutrient starvation. Cell growth stopped abruptly (in

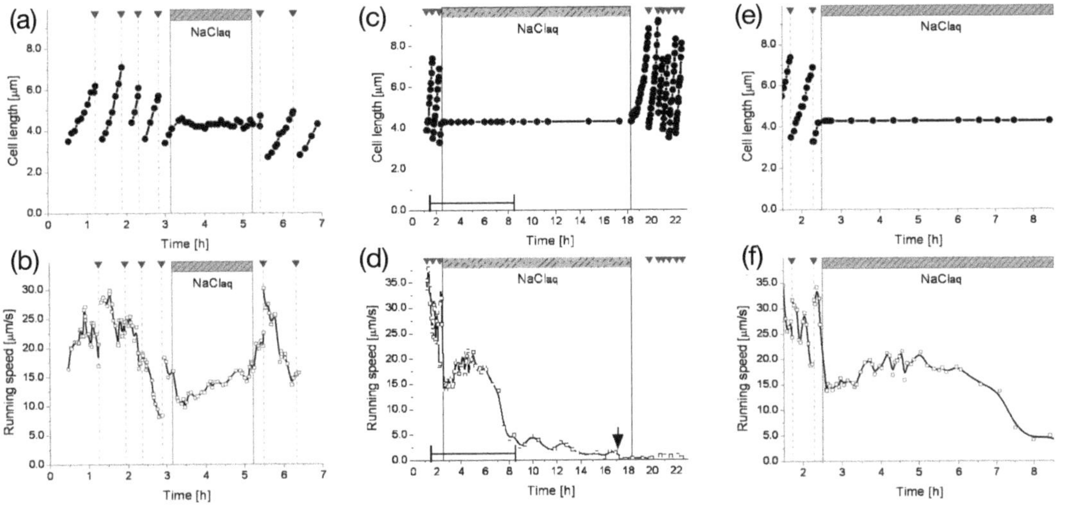

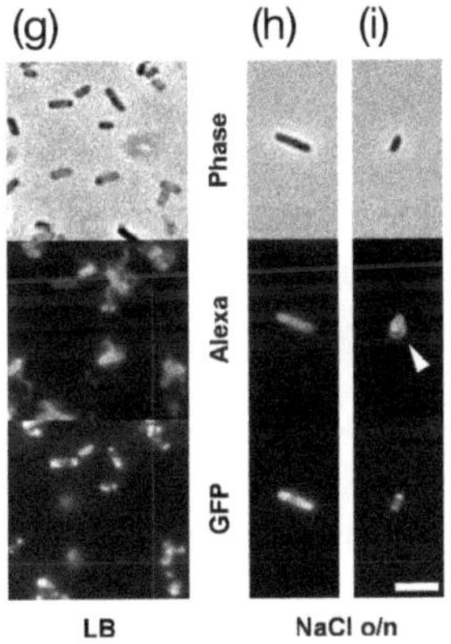

Figure 2.24 The effect of 2-hour starvation on single-cell growth and motility (a and b). The medium was changed from LB to NaCl solution (hashed) and then back to LB again. (a) The growth curve of a single *E. coli* cell and its direct descendants. (b) The speed curve of the same cells. The triangle symbols with vertical dotted lines indicate the timing of cell division. The effect of 16-hour starvation on the single-cell growth and motility (c-f). The medium was changed from LB to NaCl solution (hashed) and then back to LB again. (c) The growth curve of a single cell and its direct descendants. (d) The speed curve of the same cells. The triangle symbols indicate the timing of cell division. The arrow in (d) indicates when the cell completely lost its motility and started to show Brownian motion. (e) A magnified graph of (c) showing the region between 1.5 and 8.5 hours. (f) A magnified graph of (d) showing the region between 1.5 and 8.5 hours. Flagellation state of starved cells (g–i). (g) Control cells cultured in LB medium. Most of the cells had multiple flagella. (h and i) Cells that experienced overnight (20-hour) nutrient starvation. Almost all were deflagellated as that shown in (h), but a very few had a short single flagellum (indicated by the arrowhead) as that shown in (i). Phase: phase-contrast images. Alexa: fluorescent images of Alexa 546 labelled flagella and the entire cell body. GFP: fluorescent images of membrane chemoreceptor protein Tar-GFP for identification purposes. Bar, 5 μm.

10–20 minutes), after which cell length (and its shape and volume) remained the same. On the other hand, the cell remained motile with occasional tumbling during starvation. The running speed initially dropped in the first 5 minutes but then started to increase slightly. After the two-hour starvation period, the saline medium was exchanged for the nutrient-rich LB medium. The cell resumed its growth in about 10 minutes and then proliferated normally. Running speed increased sharply (in 5 minutes) before the first cell division, and then the cell exhibited the same tendency of decrease in speed noted prior to starvation.

Whereas the cell maintained its motility during starvation, it lost motility during an extended starvation period (Fig. 2.24c–f). When nutrient-free medium replaced nutrient-rich medium, the initial response was similar to that seen during short-term starvation: the cell stopped growing but kept moving, and the running speed dropped within 5 minutes but then started to increase slightly. A few hours after the medium was exchanged, running speed reached a local maximum and started decreasing. During this slowing-down process the cell did not swim as smoothly as before, frequently stopping and rotating. Eventually the cell lost motility and started to demonstrate Brownian motion (indicated by the arrow in Fig. 2.24d). Motility was not recovered after re-introduction of LB medium, however growth and proliferation resumed.

A possible explanation for the loss of motility could be starvation-induced deflagellation (depolymerization of flagellar filaments). To visualize the flagellar filaments, we prepared population samples cultured in LB medium, washed them with nutrient-free NaCl solution by centrifugation, and continued to culture them in the NaCl solution overnight. Although the percentage of motile cells in the samples decreased from 91% to 12% when cells had been cultured in the NaCl solution for 20 h, immediately after a 2-h centrifugation step, it was still 81%. This slow decrease suggests that the loss of motility was caused not by the centrifugation step but by nutrient starvation. The swimming behaviour of cells remaining motile after overnight starvation was similar to that observed in the single-cell measurement: reduced swimming speed and a rotating motion of the cell body. Isolated single *E. coli* cells in a nutrient-free condition abruptly stopped growing but remained motile for several hours. As the abrupt stop in growth in response to nutrient deprivation was also observed in our previous study with non-motile, flagella-deficient samples (Umehara *et al.*, 2003), it may represent a starvation response common to *E. coli* strains, with or without motility. The functional elements responsible for cell growth and cell division appear to be preserved under starvation conditions so that growth and division can be readily resumed when nutrients become available.

Because growth-suspended cells remain motile, we believe that cellular activity gets channelled into selected functions. Similar to non-growing but metabolically active quiescent cells (Rowe and Summers, 1999), starved cells may use resources normally allocated to growth and division to maintain the transmembrane proton gradient that drives flagellar rotation. This could be related to a foraging strategy in *E. coli* (Liu *et al.*, 2005) where starving cells take a risk to increase their likelihood of survival by redistributing available energy.

The molecular basis of the maintenance of motility remains to be revealed. Related physiological factors include pH, osmotic pressure, and chemotactic attractants present in the medium. For example, gene regulation is known to be affected by external pH (Maurer *et al.*, 2005). The effects of these factors can be investigated individually by appropriate combinations of medium conditions.

What is behind the observed response asynchrony? Or, what is behind the apparent irreversibility in cell motility? Mutations are usually unlikely to alter the genetic information of direct descendant cells, since the per-genome mutation rate in *E. coli* is only 0.0025 per genome replication (Drake, 1999), but the mutation rate in extremely severe environments may be considerably higher due to a stress response (Saint-Ruf and Matic, 2006). The apparent irreversibility might also be due to epigenetic mechanisms that either work within an individual cell, or, depend on the presence of other cells. Intercellular communication such as quorum sensing (Park *et al.*, 2003) or contact-dependent regulation (Aoki *et al.*, 2005) could possibly be the mechanism.

This subsection has demonstrated an example of the dual measurement of two imaging indexes, cell length, for identification and characterization of individual microbial cells, which is only appropriate for imaging cytometry technology.

Quantitative evaluation of cell-to-cell communication effects in cell group class using on-chip individual-cell-based imaging cytometry

Cells possess the ability to adapt their phenotypes through cell-to-cell communication, which underlies the cooperative behaviour among interacting cells within a group (Bassler, 2002; Basu *et al.*, 2005; Gurdon, 1988; Miller and Bassler, 2001; Shapiro, 1995, 1998; Shapiro and Hsu, 1989; Shapiro and Dworkin, 1997; Sturme *et al.*, 2002; You *et al.*, 2004). Studies on cell-to-cell communication are therefore of great importance in revealing how cooperative behaviour and multicellularity emerge within the class of cell group. To understand the role of a particular mode of cell-to-cell communication from temporal and spatial aspects, the effects from all the other possible modes must be carefully excluded. Hence the sole effect from the specific type of cell-to-cell communication on cellular phenotype becomes measurable and understandable. Cell-to-cell communication can be divided into at least two essentially different modes: communication through diffusible signals and through cell-to-cell direct contact. Effects from these two modes on cellular phenotype should be separately examined by observing the responses of cells to the imposed mode of communication. Conventional methods of cell observation, however, are unable to distinguish between these two modes; cells interact with each other extensively in complicated ways in normal cell cultures (Aoki *et al.*, 2005; Fuqua *et al.*, 1994; You *et al.*, 2004). Therefore, a new method to separately control the mode of cell-to-cell communication is desired in order to take a logical approach towards understanding of the exact role of cell-to-cell communication on cellular phenotype.

For the purpose of developing a new cell observation method in which the mode of cell-to-cell communication is controlled, we came up with the idea to apply a kind of microfluidic system called an 'on-chip single cell cultivation system' as a method of on-chip imaging cytometry. In this system, cells are enclosed in 'micro-chambers' and continuously observed for many generations under isolated and strictly controlled environmental conditions. Direct cell-to-cell contact can be avoided by the use of optical tweezers. Moreover, the medium conditions surrounding isolated cells can be changed at arbitrary timings; responses to the changes of medium conditions can be investigated at the individual cell level. Therefore, by extracting medium from a cell culture at high cell density and imposing it on isolated cells, the effect of cell-to-cell communication only through diffusible signals is measurable (Wakamoto and Yasuda, 2006)

Thus we have examined the sole effect of the cell-to-cell communication through diffusible signals under the conditions where the cell-to-cell direct contact is strictly avoided, using *E. coli* (EJ2848) as a model organism. We then showed how individual isolated bacterial cells changed their cellular states in response to diffusible signals and that the coordinated entry of the population into stationary phase were explicable from the characteristics of single cell responses to diffusible communication signals.

All cell cultivations in the preparation were done at 37°C with M9 minimal medium (Qbiogene) supplemented with 0.2%(w/w) glucose and 1/2 MEM amino acids (Invitrogen). For acquiring stationary phase medium, first, a 5-μl volume of glycerol stock of *E. coli* EJ2848 was inoculated into 1 ml of M9 medium, and cultured overnight at 37°C with shaking. A volume of 100 μl of the fully grown culture was diluted in 40 ml of M9 medium and cultured with shaking for 20 hours. The culture under these conditions reached stationary phase approximately 10 hours after dilution. Then, the stationary phase culture was centrifuged at 2500 × g for 10 minutes. The supernatant was filtered through a 0.2-μm syringe filter to obtain the stationary phase medium.

The medium containing potential cell communication signals was extracted from the stationary phase cultures of *E. coli* cultivated in M9 minimum medium (Qbiogene) supplemented with 0.2%(w/w) glucose and an amino acids mixture (1/2 MEM amino acids, Invitrogen). The stationary phase medium was prepared from cultures that entered stationary phase approximately 10 hours before the extraction. Cells in the mid-exponential phase of batch cultures were loaded into the microchambers shown in Fig. 2.7. The microchamber was made of a thick negative photo-resist with high aspect ratio, SU-8-5 (MicrochemCorp), whose height was adjusted to 5 μm in the fabrication process. Cells in any of the four rooms of microchambers in the observation area were continuously observed by maintaining them under isolated conditions using optical tweezers. Cells removed from the observation area were transferred and enclosed in the discarding areas at the both sides of the micro-chamber. The micro-chamber was covered by a cellulose semipermeable membrane (MW 50000, SpectraPor®) to avoid cells escaping. The membrane was coated with streptavidin while the glass slide was coated with biotin, making a strong bond between the membrane lid and the glass slides. The media were pumped constantly at the rate of 1 ml/min. The medium in the microchamber could be exchanged rapidly by diffusion with that flowing above the membrane lid because the height of microchamber (h) was as low as 5 μm; the exchange speed of glucose (diffusion constant: $D = 6.7 \times 10^{-10}\,m^2/s$), for instance, was calculated as 19 seconds ($h^2/2D = 19\,m/s$).

With this system, exponentially growing isolated *E. coli* cells were examined for their response to the stationary phase environment, which should contain abundant diffusible communication signals. First, a cell was cultured in a microchamber in the constant flow of M9 medium using the on-chip single cell cultivation system. Then, the flowing media was rapidly exchanged from M9 medium to the stationary phase medium 240 minutes after the observation started, by flushing the stationary phase medium for 30 seconds at the rate of 30 ml/min. Flow rate was maintained at 1 ml/min thereafter. Fig. 2.25a shows one example of the response of a single cell to the stationary phase medium. The cell stopped its growth and division instantly in response to the stationary phase medium. Moreover, it exhibited no growth and division in the stationary phase medium. The flowing medium was exchanged from the stationary phase medium to M9 medium 360 minutes after the stationary phase medium incubation started; the cell resumed its growth without delay, assuring the preservation of its viability in the stationary phase medium. We confirmed that the rapid recovery of growth to the original level after the incubation in the stationary phase medium for almost all of the observed cells (14 cells out of the 15 observed cells).

There still remains the possibility that the growth stop in Fig. 2.25a was caused by a deficiency of essential nutrients in the medium. Thus we next supplemented the stationary phase medium with glucose and amino acids with the same amount that was added to prepare fresh M9 medium, and imposed this supplemented medium on exponentially growing cells (Fig. 2.25b). The cell also stopped its growth and division in response to the supplemented medium even though it contained sufficient nutrients; the growth stop was independent of nutrients. Therefore, it is possible that the growth stop was caused by diffusible communication signals in the stationary phase medium.

To characterize the speed of the response to the change of the flowing media, the transitions of growth rate every ten minutes around the time of the medium exchange from M9 medium to the stationary phase medium were examined (Fig. 2.25c). The growth rate, v, from time t minutes

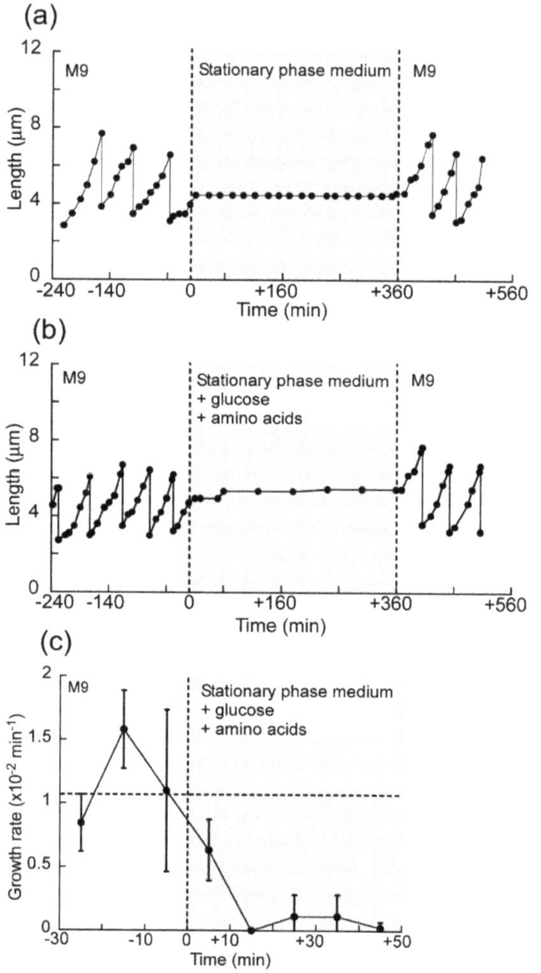

Figure 2.25 Response of exponentially growing *E. coli*. cells to inclusion of stationary phase medium (a). Response to stationary phase medium without glucose and amino acids supplements. Medium was exchanged from M9 medium to the stationary phase medium in 240 minutes from outset of the observation. Medium was also exchanged from the stationary phase medium to M9 medium in 360 minutes since the previous medium exchange, (b). Response to stationary phase medium with supplemented glucose and amino acids, (c). Growth rate transition around the medium exchange from M9 medium to the nutrient-supplemented stationary phase medium. The transition of the growth rate average of every 10 minutes of the four isolated single cells around the medium exchange was plotted. The error bars indicate the standard deviations of the growth rates of the four cells. The horizontal dotted line denotes 1.07×10^{-2} per minute, which is the averaged growth rate over one hour before the change from-M9-to-stationary medium.

to $(t+10)$ minutes was defined as $v=(1/10) \ln [L(t+10)/L(t)]$ (per minute), where $L(t)$ is cell length at the time of t, assuming that the length growth of *E. coli* occurs exponentially as, $L(t)=L(t_0)\ e^{v(t-t_0)}\ (t>t_0)$. Fig. 2.25c shows the transition of the growth rate average of the four isolated cells in the four chambers of the observation area around the medium exchange. Growth rate between the 10 minutes or the 20 minutes following the medium exchange decreased to zero, indicating that all the cells ceased growth within 10 min in response to the signals in the stationary phase medium. This reveals the ability of *E. coli* to adapt its growth rate quickly to diffusible communication signals.

To investigate the growth suppression mechanism through diffusible signals, we next diluted nutrient-supplemented stationary phase medium with M9 medium at various concentrations and imposed it on exponentially growing cells in micro-chambers. Fig. 2.26a–d shows the responses of single isolated cells to 90% (Fig. 2.26a), 80% (Fig. 2.26b), 70% (Fig. 2.26c) or 60% (Fig. 2.26d) added stationary phase media. The results show that growth was suppressed instantly in response to the diluted media as was also found

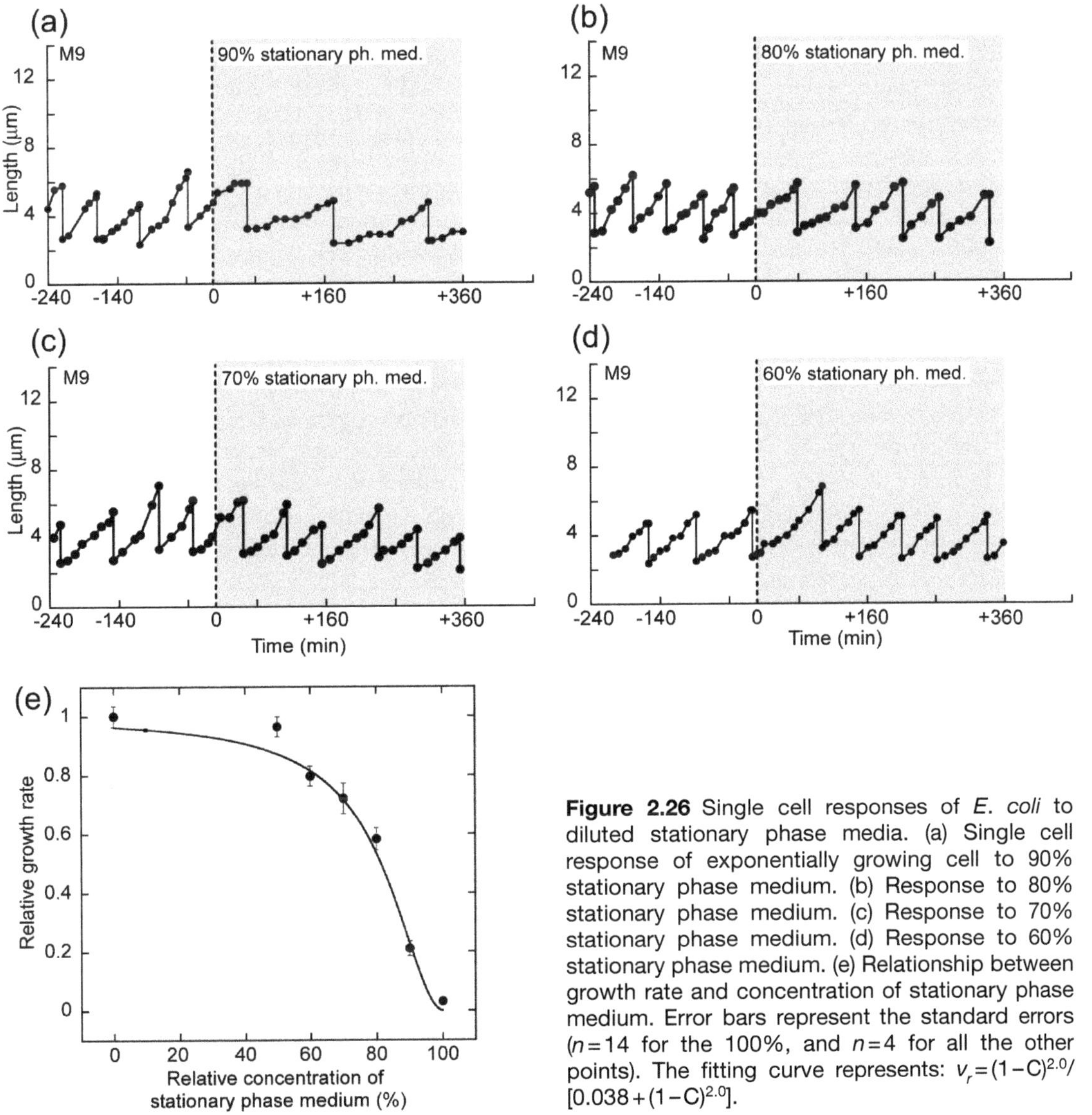

Figure 2.26 Single cell responses of *E. coli* to diluted stationary phase media. (a) Single cell response of exponentially growing cell to 90% stationary phase medium. (b) Response to 80% stationary phase medium. (c) Response to 70% stationary phase medium. (d) Response to 60% stationary phase medium. (e) Relationship between growth rate and concentration of stationary phase medium. Error bars represent the standard errors ($n=14$ for the 100%, and $n=4$ for all the other points). The fitting curve represents: $v_r=(1-C)^{2.0}/[0.038+(1-C)^{2.0}]$.

in the responses to the 100% stationary phase medium. This extent of suppression decreased with concentration of the stationary phase medium. Moreover, cells in the diluted media maintained suppressed-level growth rates without gradual decreases. These results suggest that cells quickly adjusted their growth rate according to the concentrations of the communication signals present around them.

The relationship between growth rate and concentration of stationary phase medium is shown in Fig. 2.26e. The graph reveals that the growth suppression started from the 60% concentration of the medium and increased dramatically. This non-linear relationship was fitted with a Hill equation: $v_r = (1 - C)^{2.0}/[0.038 + (1 - C)^{2.0}]$, where v_r is a relative growth rate to that in M9 medium (the 0% stationary phase medium concentration) and C is a concentration of stationary phase medium.

The next question is whether the characteristics of a population are explicable from the previously noted behaviour of single cells. To explore this possibility we measured the responses of batch culture cell populations to the stationary phase media. For this measurement, we inoculated *E. coli* cells into a 20 ml of M9 medium. Cells were cultured at 37°C with shaking for 5 hours to reach mid-exponential phase (optical density at 600 nm was ca. 0.1). Cultures were then centrifuged at 2500 × g for 10 minutes, the supernatant being discarded thereafter. Twenty millilitres of nutrient-supplemented media at various stationary phase medium concentrations was added, and the cells in the precipitate were diffused by pipetting. Cells were again cultured at 37°C with shaking and growth in the various initial concentrations of the stationary phase medium were measured.

Fig. 2.27a shows the growth of the cell populations measured by optical density at 600 nm. Relative growth after the medium exchange was increased in Fig. 2.27b, re-plotting with the initial cell concentrations normalized. Growth of the population in the 100% stationary phase medium stopped instantly and stayed constant thereafter as was expected from the single cell observation in Fig. 2.26. In the diluted supernatants, the populations exhibited suppressed-level growth rates in response to the medium exchanges, which was also expected from the single cell observation in Fig. 2.26. However, an intriguing and unexpected characteristic was shown in the population growth: they entered stationary phases at different cell concentrations without reaching maximum cell concentration of the population whose medium was exchanged with the 0% stationary phase medium (M9 medium).

As another characteristic of the growth of the populations, the time to enter the stationary phase was compared between populations (Fig. 2.27c). The result shows that the population whose medium was exchanged with the 80% stationary phase medium exhibited a significant delay entering stationary phase.

To examine the effect on growth of populations from cell-to-cell direct contact, we altered cell concentrations of batch cultures in mid-exponential phase, exchanging the media from 20 ml M9 medium to 10 ml, 20 ml or 40 ml of the 70% stationary phase media or M9 medium. This allowed a comparison of the growth of populations at differing initial cell concentrations whereby cell-to-cell contact frequencies differ. Fig. 2.27d shows growth of these cell populations. Relative growth after medium exchanges were increased in Fig. 2.27e with the initial cell concentration of each population normalized. Growth rates over the first 1.5 hours were measured in Fig. 2.27f, showing that growth rates in the identical concentration of the stationary phase medium were similar, despite differences in initial cell concentrations, that is, under the conditions where the cell-to-cell direct contact frequencies were different. This means that cell-to-cell direct contact does not affect the growth rate; only communication through diffusible signals appears to affect growth rates.

Results of the single cell observations are summarized as follows: (1) a cell determines its growth rate according to the concentration of the diffusible signals in the immediate environment present (Figs. 2.25a and b, and 2.26a–d), (2) the relationship between growth rate and the concentration of the diffusible signals can be described by the Hill equation (Fig. 2.26e).

As shown in above, we have demonstrated the potential of the imaging cytometry for quantitative measuring of the sole effect of microbial cell-to-cell communication through diffusible

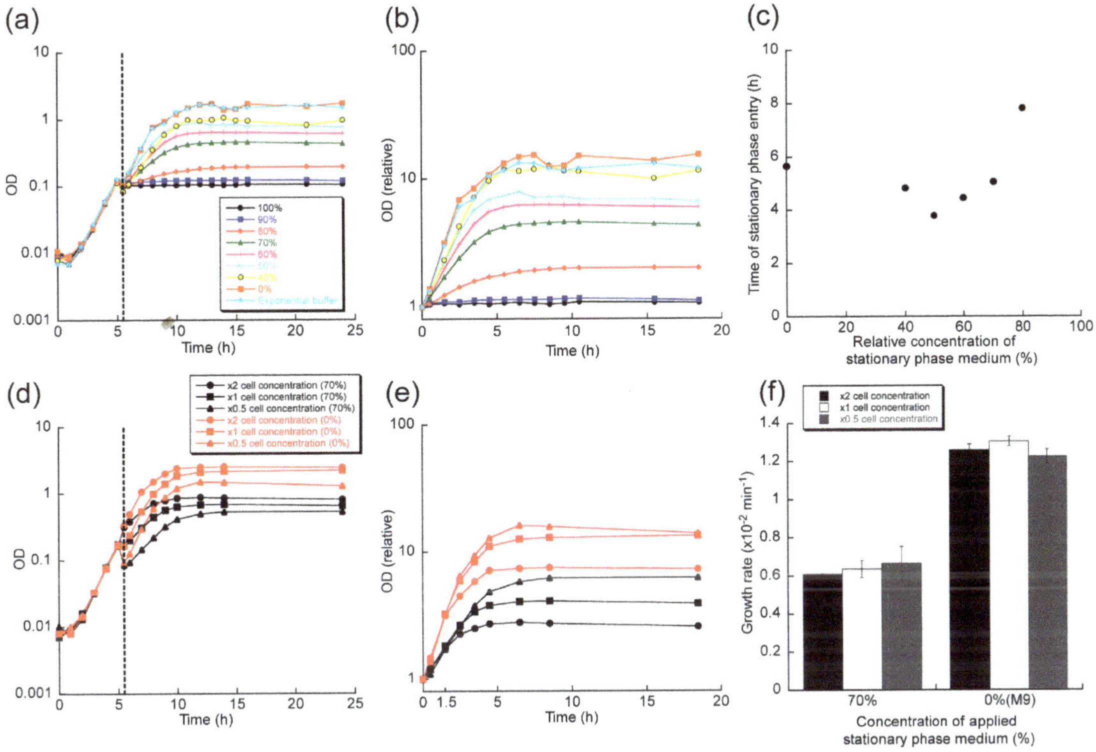

Figure 2.27 Response of cell populations of *E. coli*: (a) Responses to stationary phase medium at various concentrations. Stationary phase media at various concentrations were imposed on the cell populations in mid-exponential phase; growth in the imposed media were thereafter observed. Growth was measured by optical density at 600 nm. (b) Growth of populations after medium exchanges. Growth in the imposed media was increased with normalized initial cell concentrations. (c) Time of stationary phase entry. The time of entering stationary phase (see text for the definition) was plotted against the concentration of the imposed stationary phase medium. (d) Effect of cell-to-cell contact. Cell populations in the mid-exponential phase were imposed with the 70% or the 0% stationary phase media at ×2, ×1, or ×0.5 cell concentration relative to that prior to the medium exchanges. (e) Growth at different initial cell concentrations. The growth of the populations in (d) after the medium exchanges was enhanced with the initial cell concentrations normalized. (f) Growth rate at different initial cell concentrations. The growth rates of the populations in (d) for the first 1.5 hours were compared. Error bars represent standard deviations ($n = 3$).

signals under conditions where cell-to-cell direct contact is strictly avoided. Owing to the ability of this method to separate the two types of cell-to-cell communication, growth suppression observed for the single cell is clearly attributable to communication through diffusible signals in the stationary phase medium. In conventional methods of measuring the effect of cell-to-cell communication such as observing cells on gel plates (Shapiro and Hsu, 1989) and measuring the growth rates at high cell density in liquid media (Aoki *et al.*, 2005), the two modes of cell-to-cell interactions, through diffusible signals and through cell-to-cell contact, could not be strictly separated. Therefore, the exact contribution of each type of communication to the determination of individual cell phenotypes and to co-ordination of cellular behaviour were not measurable.

Centrifuged double Y-shape microfluidic platform for simple continuous cell environment exchange

In above sub-sections, we demonstrated the importance of rapid exchange of environmental cultivation media at single cell level. However, for the expansion of imaging cytometry into imaging flow cytometry, we need to examine the effectiveness of continuous medium exchange technique for single cell-level flow cytometry.

Recently, microfluidic platforms (Bhagat *et al.*, 2008; Yasuda, 1995a) for the continuous separation of biomolecules and particles have been proposed. Some of these use passive working principles (Yasuda, 1995b; Yasuda and Endo, 1995; Yasuda and Ishiwata, 1995), others apply dielectrophoresis (Ihara *et al.*, 1995; Yasuda, 1995c), acoustic standing waves (Yasuda, 2000; Yasuda *et al.*, 1995, 1997; Yasuda and Kamakura, 1997), segmented flow (Bell *et al.*, 1995; Ishizuka *et al.*, 1994; Yasuda *et al.*, 1994a), or magnetophoresis (Yasuda and Morita, 1994; Yasuda *et al.*, 1994b). Applying centrifugal force is one of the most powerful candidates for the passive steady-non pulsatile-flow in microfluidic systems (Schurch *et al.*, 1994) for continuous fully automated processing such as cell arrangement (Salonga *et al.*, 1994), DNA concentration (Breder *et al.*, 1994), biomolecular separation (Suzuki *et al.*, 1994), and also a compact disc (CD) microfluidic chip-based enzyme-linked immunosorbent assay (ELISA) (Hikawa *et al.*, 1994), whole blood glucose analysis (Majima *et al.*, 1994), and hybridization assay for phenylketonuria (PKU) screening (Yasuda, 1994). In particular, extraction of plasma from whole blood exploiting microfluidic pathways with centrifugal pumping (Yasuda, 1995a) and cell trapping exploiting reagent-impregnated agarose-made micro holes with centrifugal liquid pumping (Hatayama *et al.*, 1994) were examined for practical applications. The majority of the proposed platforms are at the proof-of-concept stage of the system integration, and none of them have examined the ability of microfluidic platforms for quick exchange of cell environments at single cell level.

In this subsection, we introduce the continuous medium exchange unit for single cells within the simple Y-shape microfluidic chip with passive centrifugal force with charge coupled device (CCD) camera image analysis. The continuous exchange of a medium condition surrounding single cells was performed by the following three processing steps: cell introduction into the Y-shape microfluidic pathway, a medium exchange step that removes the environmental medium buffer by cells' shift from one laminar flow to the other through the boundary of two laminar flows, and cell collection by another reverse-Y-shape microfluidic pathway. On our microfluidic platform the above three steps are implemented in a microfluidic structure contained in a monolithic chip. A rotating platform provides the required passive centrifugal force field for cell motion within the chip. For the constant centrifugal force generation, a rotating stage with a stepping motor and a control unit was mounted in the optical microscope system (Fig. 2.28a).

The microfluidic chip comprises five parts (Fig. 2.28b–d): (1) two inlets (inner one for sample cell solution and the other outer one for the washing buffer); (2) Y-shape microfluidic pathway (flow channels) guiding two buffers into the channel junction; (3) the channel junction where two laminar flows are brought into contact via a laminar–flow interface (boundary) in order to transfer the cells from one buffer to the other; (4) reverse Y-shape pathway to split into two laminar flows once again; and (5) two outlets (inner one for sample cell buffer and the other outer one for the cells with washing buffer). The 2D-layout of the microfluidic structure was fabricated in the polydimethylsiloxane (PDMS) (Hattori and Yasuda, 2010; Kubota *et al.*, 1994) attached to a 1-mm-thick glass slide.

The rotating stage was set at the position of the focal plane of the microscope, and the chip was set on the stage (4 cm from the centre of rotation), providing sufficient centrifugal force for cell transportation inside the microfluidic chip. The role of centrifugal force in this system is to generate the driving force for injecting buffers into the microfluidic flow with the same pressure and velocity and to transport cells from one laminar flow to the other.

In order to maintain the two laminar flows at the junction area, the velocity must be synchronized precisely at this area. In this system, each of the two inlets and two outlets were designed at the same distance from the centre of rotation (Fig. 2.28e). The acceleration in microfluidic chip was determined by $r\omega^2$, where r is the distance of the chip from the centre of rotation (4 cm), ω is angular velocity of rotator. In this case, the accelerations were 1 g for 2.5 Hz rotation, and 36.3 g for 15 Hz.

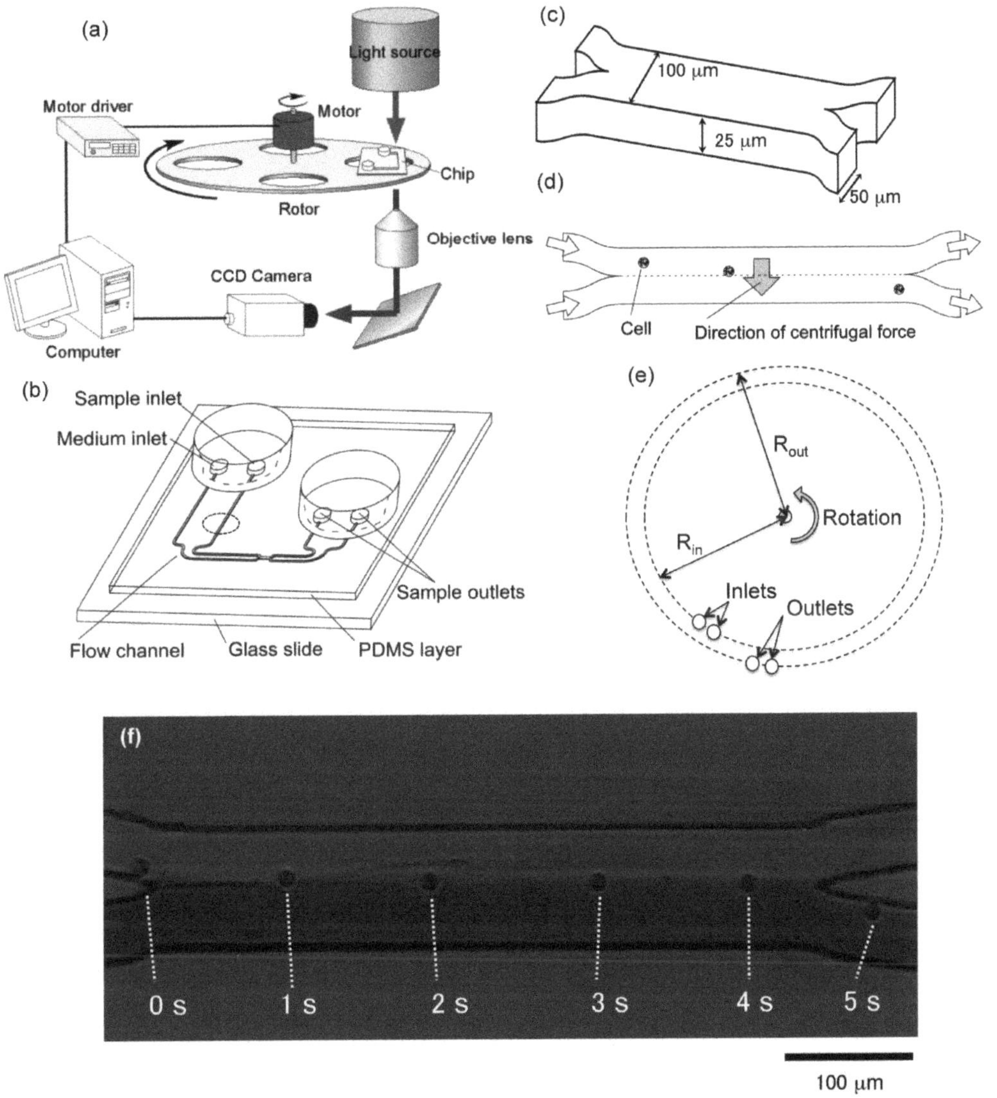

Figure 2.28 Schematic view of the system setup, microfluidic structure of microfluidic chip and experimental result. (a) System set-up; (b) microfluidic chip design; (c) and (d) gathering part of double Y-shape microfluidic pathway. (e) Schematic images explaining spatial arrangements of chip inlets/outlets arrangements and chip rotation for centrifugal force generation. (f) Time course tracking of single HeLa cell transportation from one laminar flow to the other under 1 *g* acceleration with 2.5 Hz rotation.

Fig. 2.28f shows the results of the above two subjects, i.e. cell transportation and medium mixture. As shown in this figure, a HeLa cell with a 100 μm/s flow velocity, shifted successfully from the first laminar flow to the other along the 400 μm length of two laminar flow boundaries at the gathering area. In contrast, the medium buffers maintained their boundaries, and no mixture was observed either visibly or in fluorescence detection of collected buffer (the fluorescent dye Rhodamine B was added into one of two buffers for quantitative evaluation of mixture).

In this subsection, we introduced the experimental results of a fast and simple cell medium exchange using a double Y-shape microfluidic chip and centrifugal force. Based on the simple rotating platform, we demonstrated that cells can shift effectively from one pathway to the other without any mixture of two laminar flows of different buffers even using 1 *g* acceleration. This device can be combined with the imaging flow cytometry for the applications of medium exchange measurement of microbial cells.

Summary

We have developed two types of on-chip imaging cytometry assays; on-chip imaging flow cytometry, and on-chip single-cell-based imaging cytometry. The advantages of on-chip high-speed imaging flow cytometry system are: (1) optical image-based real time cell sorting; (2) all cell sorting procedures are included within an economical and disposable (flammable after use) plastic chip; (3) cell separation electrodes consist of stable and bubble formation-resistant gel-electrodes; (4) contamination-free re-cultivation reservoir; (5) fully automated operation; and (6) DNA aptamer labelling and its non-invasive digestion. The results also indicated that cells and fluorescent dyes can be separated using this system. The required time for cell sorting from the sample with 1×10^7 particles and 20 μl is within 30 minutes. We also have investigated the on-chip single-cell-based imaging cytometry to evaluate the potential of imaging index for characterization of 'algebraic behaviour' of microbial cells from single cell level. The advantages of the on-chip single-cell-based imaging cytometry system are: (1) control of isolation/interaction of single cells; (2) direct observation and recording of behaviour and characteristics of single cells for comparison with their sister cells and direct descendant cells for more than 10 generations; and (3) complete control of environmental nutritional conditions. By exploiting these advantages, the inheritance of behaviour of cells was successfully measured and demonstrated the effectiveness of the imaging index. Therefore the combination of these two kinds of simple imaging cytometry technologies with the above advantages may provide practical and useful methodologies in biological research and clinical diagnostics. In the near future, the imaging index developed in the latter single-cell-based imaging cytometry could be included into the advanced integrated imaging flow cytometry system.

References

Akerlund, T., Nordstrom, K., and Bernander, R. (1995). Analysis of cell size and DNA content in exponentially growing and stationary-phase batch cultures of *Escherichia coli*. J. Bacteriol. *177*, 6791–6797.

Alon, U., Camarena, L., Surette, M.G., Aguera y Arcas, B., Liu, Y., Leibler, S., and Stock, J.B. (1998). Response regulator output in bacterial chemotaxis. EMBO J. *17*, 4238–4248.

Anzai, Y., Terazono, H., and Yasuda, K. (2007). Simple non-invasive cell separation method using magnetic aptamer-conjugated microbeads and nuclease digestion. J. Biol. Phys. Chem. *7*, 83–86.

Aoki, S.K., Pamma, R., Hernday, A.D., Bickham, J.E., Braaten, B.A., and Low, D.A. (2005). Contact-dependent inhibition of growth in *Escherichia coli*. Science *309*, 1245–1248.

Arkin, A., Ross, J., and McAdams, H.H. (1998). Stochastic kinetic analysis of developmental pathway bifurcation in phage lambda-infected *Escherichia coli* cells. Genetics *149*, 1633–1648.

Ashkin, A., and Dziedzic, J.M. (1989). Internal cell manipulation using infrared laser traps. Proc. Natl. Acad. Sci. U.S.A. *86*, 7914–7918.

Ashkin, A., Dziedzic, J.M., Bjorkholm, J.E., and Chu, S. (1986). Observation of a single-beam gradient force optical trap for dielectric particles. Opt. Lett. *11*, 288.

Ashkin, A., Dziedzic, J.M., and Yamane, T. (1987). Optical trapping and manipulation of single cells using infrared laser beams. Nature *330*, 769–771.

Ashkin, A., Schutze, K., Dziedzic, J.M., Euteneuer, U., and Schliwa, M. (1990). Force generation of organelle transport measured *in vivo* by an infrared laser trap. Nature *348*, 346–348.

Azuma, C., Yasuda, K., Tanabe, Y., Taniguro, H., Kanaya, F., Nakayama, A., Chen, Y.M., Gong, J.P., and Osada, Y. (2007). Biodegradation of high-toughness double network hydrogels as potential materials for artificial cartilage. J. Biomed. Mater. Res. A *81*, 373–380.

Bassler, B.L. (2002). Small talk. Cell-to-cell communication in bacteria. Cell *109*, 421–424.

Basu, S., Gerchman, Y., Collins, C.H., Arnold, F.H., and Weiss, R. (2005). A synthetic multicellular system for programmed pattern formation. Nature *434*, 1130–1134.

Bell, G.I., Yasuda, K., Kong, H., Law, S.F., Raynor, K., and Reisine, T. (1995). Molecular biology of somatostatin receptors. Ciba Found. Symp. *190*, 65–79; discussion 80–68.

Bhagat, A.A., Kuntaegowdanahalli, S.S., and Papautsky, I. (2008). Continuous particle separation in spiral microchannels using Dean flows and differential migration. Lab Chip *8*, 1906–1914.

Blake, W.J., KÆrn, M., Cantor, C.R., and Collins, J.J. (2003). Noise in eukaryotic gene expression. Nature *422*, 633–637.

Breder, C.D., Hazuka, C., Ghayur, T., Klug, C., Huginin, M., Yasuda, K., Teng, M., and Saper, C.B. (1994). Regional induction of tumor necrosis factor alpha expression in the mouse brain after systemic lipopolysaccharide administration. Proc. Natl. Acad. Sci. U.S.A. *91*, 11393–11397.

Donachie, W.D., and Begg, K.J. (1996). 'Division potential' in *Escherichia coli*. J. Bacteriol. *178*, 5971–5976.

Drake, J.W. (1999). The distribution of rates of spontaneous mutation over viruses, prokaryotes, and eukaryotes. Ann. N. Y. Acad. Sci. *870*, 100–107.

Ellington, A.D., and Szostak, J.W. (1990). *In vitro* selection of RNA molecules that bind specific ligands. Nature *346*, 818–822.

Elowitz, M.B., and Leibler, S. (2000). A synthetic oscillatory network of transcriptional regulators. Nature *403*, 335–338.

Elowitz, M.B., Levine, A.J., Siggia, E.D., and Swain, P.S. (2002). Stochastic gene expression in a single cell. Science *297*, 1183–1186.

Fuqua, W.C., Winans, S.C., and Greenberg, E.P. (1994). Quorum sensing in bacteria: the LuxR-LuxI family of cell density-responsive transcriptional regulators. J. Bacteriol. *176*, 269–275.

Gally, D.L., Bogan, J.A., Eisenstein, B.I., and Blomfield, I.C. (1993). Environmental regulation of the fim switch controlling type 1 fimbrial phase variation in Escherichia coli K-12: effects of temperature and media. J. Bacteriol. *175*, 6186–6193.

Gardner, T.S., Cantor, C.R., and Collins, J.J. (2000). Construction of a genetic toggle switch in *Escherichia coli*. Nature *403*, 339–342.

Gurdon, J.B. (1988). A community effect in animal development. Nature *336*, 772–774.

Hatayama, T., Yasuda, K., and Nishiyama, E. (1994). Characterization of high-molecular-mass heat shock proteins and 42 degrees C-specific heat shock proteins of murine cells. Biochem. Biophys. Res. Commun. *204*, 357–365.

Hattori, A., and Yasuda, K. (2010). Comprehensive study of microgel electrode for on-chip electrophoretic cell sorting. Jpn. J. Appl. Phys. *49*, 06GM04.

Hattori, A., Umehara, S., Wakamoto, Y., and Yasuda, K. (2003). Measurement of incident angle dependence of swimming bacterium reflection using on-chip single-cell cultivation assay. Jpn. J. Appl. Phys. *42*, L873.

Hattori, A., Moriguchi, H., Ishiwata, S.i., and Yasuda, K. (2004). A 1480/1064 nm dual wavelength photo-thermal etching system for non-contact three-dimensional microstructure generation into agar microculture chip. Sensor Actuat. B: Chem. *100*, 455–462.

Hattori, A., Kaneko, T., and Yasuda, K. (2011). Improvement of particle alignment control and precise image acquisition for on-chip high-speed imaging cell sorter. Jpn. J. Appl. Phys. *50*, 06GL06.

Hayashi, M., and Yasuda, K. (2010). Simple microfluidic continuous concentration of microparticles with different dielectric constants using dielectrophoretic force in a v-shaped electrode array. Jpn. J. Appl. Phys. *49*, 097002.

Hayashi, M., Hattori, A., Kim, H., Terazono, H., Kaneko, T., and Yasuda, K. (2011a). Fully automated on-chip imaging flow cytometry system with disposable contamination-free plastic re-cultivation chip. Int. J. Mol. Sci. *12*, 3618–3634.

Hayashi, M., Kaneko, T., and Yasuda, K. (2011b). Continuous concentration and separation of microparticles using dielectrophoretic force in a v-shaped electrode array. Jpn. J. Appl. Phys. *50*, 06GL03.

Herr, J.K., Smith, J.E., Medley, C.D., Shangguan, D., and Tan, W. (2006). Aptamer-conjugated nanoparticles for selective collection and detection of cancer cells. Anal. Chem. *78*, 2918–2924.

Hikawa, Y., Maeda, M., Tanaka, N., Ishikawa, S., Imamura, M., Inoue, Y., Saito, Y., and Yasuda, K. (1994). [Use of Bullard intubating laryngoscope in emergency room]. Masui *43*, 1761–1765.

Hsu, C.H., Di Carlo, D., Chen, C., Irimia, D., and Toner, M. (2008). Microvortex for focusing, guiding and sorting of particles. Lab Chip *8*, 2128–2134.

Huang, C.T., Weng, C.H., and Jen, C.P. (2011). Three-dimensional cellular focusing utilizing a combination of insulator-based and metallic dielectrophoresis. Biomicrofluidics *5*, 04401.

Hughes, M.P. (2002). Strategies for dielectrophoretic separation in laboratory-on-a-chip systems. Electrophoresis *23*, 2569–2582.

Ihara, Y., Yamada, Y., Fujii, Y., Gonoi, T., Yano, H., Yasuda, K., Inagaki, N., Seino, Y., and Seino, S. (1995). Molecular diversity and functional characterization of voltage-dependent calcium channels (CACN4) expressed in pancreatic beta-cells. Mol. Endocrinol. *9*, 121–130.

Imamura, M., Aoki, H., Eya, K., Murakami, T., and Yasuda, K. (1995). Balloon angioplasty before Wheat's operation in a patient with Turner's syndrome. Cardiovasc. Surg. *3*, 70–72.

Inoue, I., Wakamoto, Y., Moriguchi, H., Okano, K., and Yasuda, K. (2001a). On-chip culture system for observation of isolated individual cells. Lab Chip *1*, 50–55.

Inoue, I., Wakamoto, Y., and Yasuda, K. (2001b). Non-genetic variability of division cycle and growth of isolated individual cells in on-chip culture system. Proc. Jpn Acad. Ser. B Phys. Biol. Sci. *77*, 145–150.

Inoue, I., Shiomi, D., Kawagishi, I., and Yasuda, K. (2004). Simultaneous measurement of sensor-protein dynamics and motility of a single cell by on-chip microcultivation system. J. Nanobiotechnology 2, 4.

Ishizuka, T., Yamamoto, M., Kajita, K., Nagashima, T., Taniguchi, O., Wada, H., Itaya, S., and Yasuda, K. (1994). Phorbol ester and insulin stimulate protein kinase C isoforms in rat adipocytes. Diabetes Res. Clin. Pract. *26*, 91–99.

Kaneko, T., Nomura, F., and Yasuda, K. (2011). On-chip constructive cell-network study (I): contribution of cardiac fibroblasts to cardiomyocyte beating synchronization and community effect. J Nanobiotechnology 9, 21.

Kepler, T.B., and Elston, T.C. (2001). Stochasticity in transcriptional regulation: origins, consequences, and mathematical representations. Biophys. J. *81*, 3116–3136.

Kierzek, A.M., Zaim, J., and Zielenkiewicz, P. (2001). The effect of transcription and translation initiation frequencies on the stochastic fluctuations in prokaryotic gene expression. J. Biol. Chem. *276*, 8165–8172.

Ko, M.S., Nakauchi, H., and Takahashi, N. (1990). The dose dependence of glucocorticoid-inducible gene expression results from changes in the number of transcriptionally active templates. EMBO J. *9*, 2835–2842.

Kojima, K., Moriguchi, H., Hattori, A., Kaneko, T., and Yasuda, K. (2003). Two-dimensional network formation of cardiac myocytes in agar microculture chip with 1480 nm infrared laser photo-thermal etching. Lab Chip 3, 292–296.

Kojima, K., Kaneko, T., and Yasuda, K. (2005). Stability of beating frequency in cardiac myocytes by their community effect measured by agarose microchamber chip. J. Nanobiotechnology 3, 4.

Kubota, A., Yamada, Y., Kagimoto, S., Yasuda, K., Someya, Y., Ihara, Y., Okamoto, Y., Kozasa, T., Seino, S., and Seino, Y. (1994). Multiple effector coupling of somatostatin receptor subtype SSTR1. Biochem. Biophys. Res. Commun. *204*, 176–186.

Levin, M.D., Morton-Firth, C.J., Abouhamad, W.N., Bourret, R.B., and Bray, D. (1998). Origins of individual swimming behavior in bacteria. Biophys. J. *74*, 175–181.

Levit, M.N., Liu, Y., and Stock, J.B. (1998). Stimulus response coupling in bacterial chemotaxis: receptor dimers in signalling arrays. Mol. Microbiol. *30*, 459–466.

Liu, M., Durfee, T., Cabrera, J.E., Zhao, K., Jin, D.J., and Blattner, F.R. (2005). Global transcriptional programs reveal a carbon source foraging strategy by *Escherichia coli*. J. Biol. Chem. *280*, 15921–15927.

Liu, P., Meagher, R.J., Light, Y.K., Yilmaz, S., Chakraborty, R., Arkin, A.P., Hazen, T.C., and Singh, A.K. (2011). Microfluidic fluorescence *in situ* hybridization and flow cytometry (muFlowFISH). Lab Chip *11*, 2673–2679.

Lybarger, S.R., and Maddock, J.R. (1999). Clustering of the chemoreceptor complex in *Escherichia coli* is independent of the methyltransferase CheR and the methylesterase CheB. J. Bacteriol. *181*, 5527–5529.

McAdams, H.H., and Arkin, A. (1997). Stochastic mechanisms in gene expression. Proc. Natl. Acad. Sci. U.S.A. *94*, 814–819.

Maddock, J.R., and Shapiro, L. (1993). Polar location of the chemoreceptor complex in the *Escherichia coli* cell. Science *259*, 1717–1723.

Majima, T., Yasuda, K., Yamamoto, N., Kaneda, K., and Hayashi, K. (1994). Deterioration of mechanical properties of the autograft in controlled stress-shielded augmentation procedures. An experimental study with rabbit patellar tendon. Am. J Sports Med. 22, 821–829.

Manson, M.D., Armitage, J.P., Hoch, J.A., and Macnab, R.M. (1998). Bacterial locomotion and signal transduction. J. Bacteriol. *180*, 1009–1022.

Matsumura, K., Yagi, T., and Yasuda, K. (2003a). Differential analysis of cell cycle stability in Chlamydomonas using on-chip single-cell cultivation system. Jpn. J. Appl. Phys. *42*, L784.

Matsumura, K., Yagi, T., and Yasuda, K. (2003b). Role of timer and sizer in regulation of Chlamydomonas cell cycle. Biochem. Biophys. Res. Commun. *306*, 1042–1049.

Maurer, L.M., Yohannes, E., Bondurant, S.S., Radmacher, M., and Slonczewski, J.L. (2005). pH regulates genes for flagellar motility, catabolism, and oxidative stress in *Escherichia coli* K-12. J. Bacteriol. *187*, 304–319.

Miller, M.B., and Bassler, B.L. (2001). Quorum sensing in bacteria. Annu. Rev. Microbiol. *55*, 165–199.

Moriguchi, H., Wakamoto, Y., Sugio, Y., Takahashi, K., Inoue, I., and Yasuda, K. (2002). An agar-microchamber cell-cultivation system: flexible change of microchamber shapes during cultivation by photo-thermal etching. Lab Chip *2*, 125–132.

Moriguchi, H., Takahashi, K., Sugio, Y., Wakamoto, Y., Inoue, I., Jimbo, Y., and Yasuda, K. (2004). On-chip neural cell cultivation using agarose-microchamber array constructed by a photothermal etching method. Electr. Eng. Jpn *146*, 37–42.

Msadek, T. (1999). When the going gets tough: survival strategies and environmental signaling networks in *Bacillus subtilis*. Trends Microbiol. *7*, 201–207.

Nomura, F., Kaneko, T., Hattori, A., and Yasuda, K. (2011). Label-free shape-based selection of cardiomyocytes with on-chip imaging cell sorting system. J. Bioproces. Biotechniq. *S3*, 003.

Ogawa, T., Sakata, S., Nakamura, S., Takuno, H., Matsui, I., Sarui, H., and Yasuda, K. (1994). Thyroid hormone autoantibodies in patients with Graves' disease: effect of anti-thyroid drug treatment. Clin. Chim. Acta *228*, 113–122.

Ohuchi, S.P., Ohtsu, T., and Nakamura, Y. (2006). Selection of RNA aptamers against recombinant transforming growth factor-beta type III receptor displayed on cell surface. Biochimie *88*, 897–904.

Ozbudak, E.M., Thattai, M., Kurtser, I., Grossman, A.D., and van Oudenaarden, A. (2002). Regulation of noise in the expression of a single gene. Nat. Genet. *31*, 69–73.

Panda, A.K., Khan, R.H., Rao, K.B., and Totey, S.M. (1999). Kinetics of inclusion body production in batch and high cell density fed-batch culture of *Escherichia coli* expressing ovine growth hormone. J. Biotechnol. *75*, 161–172.

Park, S., Wolanin, P.M., Yuzbashyan, E.A., Silberzan, P., Stock, J.B., and Austin, R.H. (2003). Motion to form a quorum. Science *301*, 188.

Rao, C.V., Wolf, D.M., and Arkin, A.P. (2002). Control, exploitation and tolerance of intracellular noise. Nature *420*, 231–237.

Rowe, D.C., and Summers, D.K. (1999). The quiescent-cell expression system for protein synthesis in *Escherichia coli*. Appl. Environ. Microbiol. *65*, 2710–2715.

Saint-Ruf, C., and Matic, I. (2006). Environmental tuning of mutation rates. Environ. Microbiol. *8*, 193–199.

Salonga, J.P., Matsumura, H., Yasuda, K., and Yamabe, Y. (1994). Bond strength of adhesive resin to three nickel-chromium alloys with varying chromium content. J. Prosthet. Dent. *72*, 582–584.

Sawaki, J., Tsutsui, H., Hayashi, N., Yasuda, K., Akira, S., Tanizawa, T., and Nakanishi, K. (2007). Type 1 cytokine/chemokine production by mouse NK cells following activation of their TLR/MyD88-mediated pathways. Int. Immunol. *19*, 311–320.

Schurch, B., Yasuda, K., and Rossier, A.B. (1994). Detrusor bladder neck dyssynergia revisited. J. Urol. *152*, 2066–2070.

Schwan, W.R., Seifert, H.S., and Duncan, J.L. (1992). Growth conditions mediate differential transcription of fim genes involved in phase variation of type 1 pili. J. Bacteriol. *174*, 2367–2375.

Shangguan, D., Li, Y., Tang, Z., Cao, Z.C., Chen, H.W., Mallikaratchy, P., Sefah, K., Yang, C.J., and Tan, W. (2006). Aptamers evolved from live cells as effective molecular probes for cancer study. Proc. Natl. Acad. Sci. U.S.A. *103*, 11838–11843.

Shapiro, J.A. (1995). The significances of bacterial colony patterns. Bioessays *17*, 597–607.

Shapiro, J.A. (1998). Thinking about bacterial populations as multicellular organisms. Annu. Rev. Microbiol. *52*, 81–104.

Shapiro, J.A., and Dworkin, M. (1997). Bacteria as Multicellular Organisms (Oxford University Press, Oxford, UK).

Shapiro, J.A., and Hsu, C. (1989). *Escherichia coli* K-12 cell–cell interactions seen by time-lapse video. J. Bacteriol. *171*, 5963–5974.

Skidmore, J.M., Ellefson, D.D., McNamara, B.P., Couto, M.M., Wolfe, A.J., and Maddock, J.R. (2000). Polar clustering of the chemoreceptor complex in *Escherichia coli* occurs in the absence of complete CheA function. J. Bacteriol. *182*, 967–973.

Spudich, J.L., and Koshland, D.E. (1976). Non-genetic individuality: chance in the single cell. Nature *262*, 467–471.

Sturme, M.H., Kleerebezem, M., Nakayama, J., Akkermans, A.D., Vaugha, E.E., and de Vos, W.M. (2002). Cell to cell communication by autoinducing peptides in Gram-positive bacteria. Antonie Van Leeuwenhoek *81*, 233–243.

Sugio, Y., Kojima, K., Moriguchi, H., Takahashi, K., Kaneko, T., and Yasuda, K. (2004). An agar-based on-chip neural-cell-cultivation system for stepwise control of network pattern generation during cultivation. Sensor Actuat. B: Chem. *99*, 156–162.

Suzuki, E., Yasuda, K., Miyazaki, S., Takeda, N., Inouye, H., Omawari, N., and Miura, K. (1994). 1H-NMR analysis of nerve edema in the streptozotocin-induced diabetic rat. J. Lab. Clin. Med. *124*, 627–637.

Suzuki, I., Sugio, Y., Jimbo, Y., and Yasuda, K. (2004a). Individual-cell-based electrophysiological measurement of a topographically controlled neuronal network pattern using agarose architecture with a multi-electrode array. Jpn. J. Appl. Phys. *43*, L403–L406.

Suzuki, I., Sugio, Y., Moriguchi, H., Jimbo, Y., and Yasuda, K. (2004b). Modification of a neuronal network direction using stepwise photo-thermal etching of an agarose architecture. J. Nanobiotechnology *2*, 7.

Svoboda, K., and Block, S.M. (1994). Biological applications of optical forces. Annu. Rev. Biophys. Biomol. Struct. *23*, 247–285.

Takahashi, H., Yotsuyanagi, H., Yasuda, K., Koibuchi, T., Suzuki, M., Kato, T., Nakamura, T., Iwamoto, A., Nishioka, K., Iino, S., *et al.* (2006). Molecular epidemiology of hepatitis A virus in metropolitan areas in Japan. J. Gastroenterol. *41*, 981–986.

Takahashi, K., Orita, K., Matsumura, K., and Yasuda, K. (2003). On-chip microcultivation chamber for swimming cells using visualized poly(dimethylsiloxane) valves. Jpn. J. Appl. Phys. *42*, L1104.

Takahashi, K., Hattori, A., Suzuki, I., Ichiki, T., and Yasuda, K. (2004). Non-destructive on-chip cell sorting system with real-time microscopic image processing. J. Nanobiotechnology *2*, 5.

Terazono, H., Anzai, Y., Soloviev, M., and Yasuda, K. (2010). Labelling of live cells using fluorescent aptamers: binding reversal with DNA nucleases. J. Nanobiotechnology *8*, 8.

Thattai, M., and van Oudenaarden, A. (2001). Intrinsic noise in gene regulatory networks. Proc. Natl. Acad. Sci. U.S.A. *98*, 8614–8619.

Trickett, A., and Kwan, Y.L. (2003). T cell stimulation and expansion using anti-CD3/CD28 beads. J. Immunol. Methods *275*, 251–255.

Tuerk, C., and Gold, L. (1990). Systematic evolution of ligands by exponential enrichment: RNA ligands to bacteriophage T4 DNA polymerase. Science *249*, 505–510.

Umehara, S., Wakamoto, Y., Inoue, I., and Yasuda, K. (2003). On-chip single-cell microcultivation assay for monitoring environmental effects on isolated cells. Biochem. Biophys. Res. Commun. *305*, 534–540.

Umehara, S., Hattori, A., Inoue, I., and Yasuda, K. (2007a). Asynchrony in the growth and motility responses to environmental changes by individual bacterial cells. Biochem. Biophys. Res. Commun. *356*, 464–469.

Umehara, S., Inoue, I., Wakamoto, Y., and Yasuda, K. (2007b). Origin of individuality of two daughter cells during the division process examined by the simultaneous measurement of growth and swimming property using an on-chip single-cell cultivation system. Biophys. J. *93*, 1061–1067.

Unni, H.N., Hartono, D., Yung, L.Y.L., Ng, M.M.L., Lee, H.P., Khoo, B.C., and Lim, K.M. (2012). Characterization and separation of Cryptosporidium and Giardia cells using on-chip dielectrophoresis. Biomicrofluidics *6*, 12805–1280514.

Urdaneta, M., and Smela, E. (2007). Multiple frequency dielectrophoresis. Electrophoresis *28*, 3145–3155.

Wakamoto, Y., and Yasuda, K. (2006). Quantitative evaluation of cell-to-cell communication effects in cell group class using on-chip individual-cell-based cultivation system. Biochem. Biophys. Res. Commun. *349*, 1130–1138.

Wakamoto, Y., Inoue, I., Moriguchi, H., and Yasuda, K. (2001). Analysis of single-cell differences by use of an on-chip microculture system and optical trapping. Fresenius J. Anal. Chem. *371*, 276–281.

Wakamoto, Y., Umehara, S., Matsumura, K., Inoue, I., and Yasuda, K. (2003). Development of non-destructive, non-contact single-cell based differential cell assay using on-chip microcultivation and optical tweezers. Sensor Actuat. B: Chem. *96*, 693–700.

Wright, W.H., Sonek, G.J., Tadir, Y., and Berns, M.W. (1990). Laser trapping in cell biology. IEEE J. Quantum Elect. *26*, 2148–2157.

Wu, L.Q., Yung, L.Y.L., and Lim, K.M. (2012). Dielectrophoretic capture voltage spectrum for measurement of dielectric properties and separation of cancer cells. Biomicrofluidics *6*, 14113–1411310.

Yamashita, R., Fujiwara, Y., Ikari, K., Hamada, K., Otomo, A., Yasuda, K., Noda, M., and Kaburagi, Y. (2007). Extracellular proteome of human hepatoma cell, HepG2 analyzed using two-dimensional liquid chromatography coupled with tandem mass spectrometry. Mol. Cell. Biochem. *298*, 83–92.

Yasuda, K. (1994). Development and clinical use of ultrasonic probes. Endoscopy *26*, 816–817.

Yasuda, K. (1995a). [Clonazepam, nitrazepam, diazepam]. Nihon Rinsho Jpn. J. Clin. Med. *53*(Su Pt 1), 911–914.

Yasuda, K. (1995b). [Pantothenic acid]. Nihon Rinsho Jpn. J. Clin. Med. *53*(Su Pt 1), 697–700.

Yasuda, K. (1995c). [Voiding dysfunction in brain disease]. Nihon Hinyokika Gakkai Zasshi *86*, 51–54.

Yasuda, K. (2000). Non-destructive, non-contact handling method for biomaterials in micro-chamber by ultrasound. Sensor Actuat. B: Chem. *64*, 128–135.

Yasuda, K., and Endo, N. (1995). [Vitamin B6 (pyridoxine)]. Nihon Rinsho Jpn. J. Clin. Med. *53*(Su Pt 1), 701–703.

Yasuda, K., and Ishiwata, Y. (1995). [Vitamin B2 (riboflavin)]. Nihon Rinsho Jpn. J. Clin. Med. *53*(Su Pt 1), 694–696.

Yasuda, K., and Kamakura, T. (1997). Acoustic radiation force on micrometer-size particles. Appl. Phys. Lett. *71*, 1771–1773.

Yasuda, K., and Morita, H. (1994). [Important points in diagnosis and therapy: Cushing's syndrome]. Nihon Naika Gakkai Zasshi *83*, 2064–2068.

Yasuda, K., Inagaki, N., Yamada, Y., Kubota, A., Seino, S., and Seino, Y. (1994a). Hamster gastric inhibitory polypeptide receptor expressed in pancreatic islets and clonal insulin-secreting cells: its structure and functional properties. Biochem. Biophys. Res. Commun. *205*, 1556–1562.

Yasuda, K., Murashita, T., Takigami, T., Yamauchi, H., Shiiya, N., Sasaki, S., Matsui, Y., and Sakuma, M. (1994b). [Experience of reoperations followed by repairs of abdominal aortic aneurysms]. Rinsho Kyobu Geka *14*, 480–485.

Yasuda, K., Umemura, S.-i., and Takeda, K. (1995). Concentration and fractionation of small particles in liquid by ultrasound. Jpn. J. Appl. Phys. *34*, 2715.

Yasuda, K., Haupt, S.S., Umemura, S.-i., Yagi, T., Nishida, M., and Shibata, Y. (1997). Using acoustic radiation force as a concentration method for erythrocytes. J. Acoust. Soc. Am. *102*, 642–645.

Yasuda, K., Okano, K., and Ishiwata, S. (2000). Focal extraction of surface-bound DNA from a microchip using photo-thermal denaturation. Biotechniques *28*, 1006–1011.

Yasuda, K., Roneker, K.R., Miller, D.D., Welch, R.M., and Lei, X.G. (2006). Supplemental dietary inulin affects the bioavailability of iron in corn and soybean meal to young pigs. J. Nutr. *136*, 3033–3038.

Yoon, Y.E., Chang, S.A., Choi, S.I., Chun, E.J., Cho, Y.S., Youn, T.J., Chung, W.Y., Chae, I.H., Choi, D.J., and Chang, H.J. (2011). The absence of coronary artery calcification does not rule out the presence of significant coronary artery disease in Asian patients with acute chest pain. Int. J. Cardiovasc. Imaging *28*, 389–398.

You, L., Cox, R.S. 3rd, Weiss, R., and Arnold, F.H. (2004). Programmed population control by cell–cell communication and regulated killing. Nature *428*, 868–871.

Application of Flow Cytometry to Environmental Biotechnology

3

Mónica Herrero and Mario Díaz

Abstract

The number of publications applying flow cytometry (FCM) as an analytical technique is increasing in the field of environmental biotechnology. This field is defined 'as the development, use and regulation of biological systems for remediation of contaminated environments (land, air, water) for environmental-friendly processes'. FCM complements existing technologies and enables new information to be obtained from microbial ecosystems. Difficulties with analysis and differentiation of bacterial cells due to their small size and similar morphologies can be addressed with recent technical advances in equipment handling, improved instrument cell-sorting capacities, development of fluorescent dyes and multi-parametric analysis. FCM allows the use of taxonomic and physiological probes to detect and characterize microbial communities or single cells, even target cells expressing specific gene functions. Indeed many of the microorganisms present in natural samples or involved in environmental bioprocesses are not cultivable and often undergo changes in their conditions that rapidly affect their population dynamics. Characterization of microbial physiological states is a key application for environmental bioprocesses optimization. FCM has been used to ensure effectiveness of drinking water quality, wastewater treatments, to check the hazardous effects of anthropogenic toxic substances released to the environment or to assess bacterial viability in soils, facilitating rapid statistically relevant data acquisition, within almost real-time conditions.

Introduction

For decades, biological reactors used in environmental engineered bioprocesses have been considered as 'black boxes', relaying for their functionality only on empirical knowledge. Current rapid and molecular techniques allow for a deeper knowledge of these biological systems, such that technicians and engineers can utilize data for new applications, to design novel bioreactors or to optimize operating conditions. The use of these interesting techniques including FCM, is enormously valuable to diagnose and prevent operating problems in biotreatment plants. The research objectives are not only undertaken to investigate the systematic and taxonomic structure of the microbial consortia taking part in environmental bioprocesses, but also to gain an understanding into how these biological systems respond to changes, how the microbial population dynamics evolved and in what way these changes can be related to bioprocess efficiency.

Microbial consortia are essential to maintain sustainable environmental engineered bioprocesses (e.g. bioremediation, solid waste treatment, water treatment, biodegradation of industrial recalcitrant effluents, energy bioconversion). However, biological treatments are often difficult to control and predict. A complex mixture of compounds is found in contaminated environments such as organics (biodegradable and refractory, either of which may be toxic at certain concentrations), inorganics or heavy metals. It should be highlighted that, although available biotreatments could be suitable for different classes of contaminants when found separately, when present in mixtures, each can inhibit or slow down the removal of others. Moreover, as wastes usually fluctuate widely in terms of their composition and volume, this adds problems in terms of the biological treatment required.

Bioprocess efficiency depends on the ratio of biologically active/sensitive microorganisms to the contaminants present in the waste/effluent for treatment. However, bioreactor performance is only assessed by applying material balances. In order to establish effective large scale bioreactors for treating chemically mixed wastes it is essential that they harbour and maintain stable key microbial consortia with sufficient active biomass to degrade each of the contaminant compounds present. To this aim, advanced bioreactor design is focussed on compartmentalization (Barber and Stuckey, 1999) and on the use of membranes (Akram and Stuckey, 2008) to confine the key microbial consortium within the process while extending their contact with the stream for treatment to be completed. Identification of microbial communities with desired catabolic traits and the ability to colonize contaminated niches provides opportunities for developing specialized inocula that can be exploited by bioaugmentation (van der Gast *et al.*, 2002). Advances in environmental microbiology enable taking advantage of microbial specialization, searching for the efficiency of microbial consortia specifically defined and designed to remediate a contaminated environment, or, to optimize environmental-friendly processes. In this context, one of the advantages of using so-called 'Heirloom' microorganisms (strains showing exceptional abilities, easily cultured, fast growing, thoroughly characterized and accessible worldwide in culture collections) for specific environmental purposes is that they possess available phenotypic and genotypic detailed meta-dossiers which can be used (Singer *et al.*, 2005) for effective biotreatment.

It should be noted that, taking into account that 90–99% of microbes living in natural environments do not respond to the conventional culturing methods (Stenuit *et al.*, 2008), the emergence of culture-independent methods has provided the opportunity to explore and exploit microbial diversity. Culture-independent techniques resemble an environment without resource competition, thereby allowing those microorganisms that grow very slowly to be monitored without interference from fast-growing microorganisms. Methods based on growth-dependent competition are also limiting when dealing with microbial contamination events affecting public health. At this level, FCM is a useful technique for detecting and monitoring water-borne pathogens (Jolivet-Gougeon *et al.*, 2006). Although FCM is restricted to liquid sample analysis, soil and sediment samples can also be processed after suitable pre-treatments including suspension/extraction in the liquid phase, sonication, permeabilization, physical separation of individual cells from others and from the matrix material in which they are adsorbed. Since FCM is a single-cell method, in principle is not ideal for complex macroparticle analysis (e.g. biofilms, activated sludge flocs or anaerobic granules) unless non-lethal sample pre-treatment is undertaken.

FCM may be further improved with the continuous development of taxonomic probes based on hybridization of rRNA and on probes targeting functional gene markers. One drawback of applying only taxonomic DNA-based molecular techniques to study microbial communities is that it is not straightforward to find out the physiological properties of the microorganisms based solely on their phylogenetic relationship. This is especially true for environmental samples in which interesting technological traits are widespread within a broad variety of phylogenetically unrelated organisms (as the case of denitrification ability). In addition, within engineered bioprocesses, microbial communities are often subjected to rapid changes in various parameters in their environment (temperature, pH, salinity, substrate availability, exposure to toxins or to xenobiotic organic molecules). The complex nature of microbial ecosystems involved in environmental bioprocess responds rapidly to such changes, provoking stress conditions, damage or alterations to the population dynamics. The short-term effects of these environmental factors over microbial consortia cannot always be detected by taxonomic methods since the microbial response may be reflected at physiological levels. Additionally, it should be borne in mind the difficulties involved in PCR-based methods when claiming that sequencing results accurately reflect the whole microbial diversity in ecosystems (Wang *et al.*, 2010).

Characterization of microbial physiological states (metabolically active, reproductive growing, membrane integrity, permeabilized (dead) cells) (Nebe-von-Caron *et al.*, 2000) by fluorescence dyes used as physiological probes is a key opportunity to advance environmental bioprocesses efficiency. Multi-parametric analysis (multi-colour specific labelling which increases the information available) of the community structure, can be extended to physiological analysis at a single-cell level (by measuring biological parameters such as nucleic acid content, respiration rates, intracellular enzyme activity or cytoplasmic and outer membrane activity). FCM when combined with other techniques, such as antibody staining, fluorescence *in situ* hybridization (FISH) or the use of autofluorescence protein reporters also allows discrimination of target from non-target cells based on certain phylogenetic, biochemical or structural properties. Moreover, by using fluorescence-activated cell sorting (FACS), cells showing features of interest can be physically separated form the rest of the microbial population. If the stains used do not kill cells, stained microorganisms can be used for cultivation-based downstream analysis. Care should be taken when using 5-cyano-2,3-ditoryl tetrazolium chloride (CTC) because tetrazolium salts have toxic effects on cellular metabolism (Ullrich *et al.*, 1996). Sorted cells can be used for subsequent culture-based (Kalyuzhnaya *et al.* 2008; Wang *et al.* 2009) or culture-independent analyses (Hoefel *et al.*, 2005; Fujii and Hiraishi 2009). It should be highlighted that there is a huge interest in using FACS for facilitating physical isolation of environmental microorganisms of interest, to allow further biotechnological implementations. Although '-omic' techniques such as metagenomics in principle overcome the need for cultivation, when seeking strains for industrial or field biotechnological applications, culturing is a key requirement. FCM applications for environmental samples have included (but not have been limited to) bacterial identification and enumeration, assessment of cell viability, measurement of DNA content, estimation of biovolume and ecological studies on biodiversity. Using FCM, it is now possible to overcome limitations of conventional microscopy or plating techniques. Moreover, this technique facilitates assessment and monitoring of microbial cell function and physiology, exploration of novel catabolic activities for potential exploitation in advanced environmental biotechnology.

A very important feature of FCM is that it enables acquisition of large data sets, with statistically representative cell numbers in nature from environmental samples, which permits extrapolation and correct data interpretation. Scaling is a problem in microbial ecology (Woodcock *et al.*, 2007) on the basis that extrapolation is typically based on very small sample quantities and that microbial systems are always studied at a scale much smaller than is required to characterize them. For these reasons, the disparity between sample size and community size is enormous and far greater than any comparable sampling issues in ecology. Consequently, patterns are perceived through a sparse, often distorted map of the microbial world (Sloan *et al.*, 2007). FCM analysis makes it possible to rapidly analyse thousands of cells, about 10,000–20,000 cells per second, up to 100,000 cells per second in specialized cytometers, with sorting facilities (in high-speed sorters) of up to 40,000 cells per second. In this way, the statistical robustness of FCM is ideal for quantitative measurements, while the multi-parametric versatility allows cytometers to simultaneously detect dozens of fluorescence parameters at one time.

The potential of FCM in microbiology is still far from being fully realized (Wang *et al.*, 2010) and until the mid-1990s it was regarded as being in its infancy. Along with machine costs, data acquisition and the lack of proper dyes were probably the major limitations. Although the first steps in FCM analysis are relatively straightforward, the use of proper controls, data validation and interpretation can be complex. Nevertheless with technical advances in equipment, lab-on a chip devices, development of new dyes and probes, software evolution for data acquisition and analysis, FCM is becoming more frequently used in the field of environmental biotechnology. Results obtained by FCM provide opportunities to advance bioprocess control by having access to rapid automated information, in almost real-time conditions, with statistically reliable data from heterogeneous and complex samples. These analytical and data collection features can facilitate a decision making process. Because each method has various advantages and limitations, FCM in combination with other molecular techniques may be a reasonable way to achieve a better understanding and higher efficiency for engineered environmental bioprocesses (Fig. 3.1).

Application to drinking water quality

There are significantly more microbial cells in drinking water than can be cultured on semi-synthetic growth media. Nonetheless, cultivation of heterotrophic plate counts and faecal indicator bacteria are used worldwide as quality parameters for drinking water treatments and distribution systems. Total bacterial cell concentrations are normally not considered during drinking water treatment as a legislative parameter, mainly because easy and rapid methods for quantification of total bacterial cell concentrations have up until now, not been available. But recently it has been shown that total cell concentration measured by FCM is a rapid, easy, sensitive descriptive parameter of several widely applied drinking water treatment processes (Hammes *et al.*, 2008). Several FCM developments have been focussed on water-borne pathogens of public health concern (Jolivet-Gougeon *et al.*, 2006). It should noted that this technique is useful for the measurement of yeast cells, algae, protozoa and viruses (which despite their small size, can be detected by advanced, specialized cytometers) in water environments.

Enumeration of total bacterial cells in treatment processes for drinking water by FCM provides a powerful substitute for time-consuming microscopic methods. FCM has another advantage over microscopy in that it is possible to enumerate bacteria at concentrations as low as 10^3–10^4 cells/ml. Table 3.1 shows several stains used in combination with FCM for drinking water analysis.

The application of FCM with nucleic acids staining has advanced knowledge regarding the

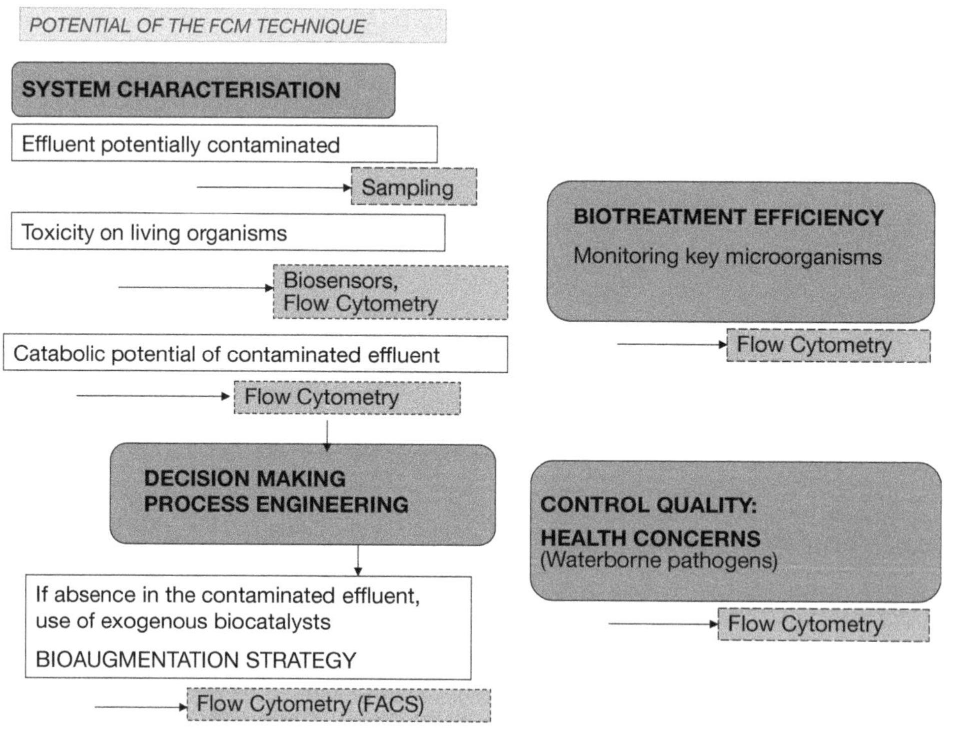

Figure 3.1 Possible application of FCM (combined with other techniques such as FISH, physiological probes or gene targeting, in analysis or sorting modes) to enhance environmental bioprocess efficiency.

Table 3.1 Examples of stains used in drinking water applications

Stain	Target	Reference
Propidium Iodide (PI)	Cytoplasmic membrane integrity	Hoefel *et al.* (2003)
SITO 9	Nucleic acids	Hoefel *et al.* (2003)
Carboxyfluorescein diacetate acetoxymethyl ester (cFDA-AM)	Esterase activity	Hoefel *et al.* (2005)
4′6,-Diamidino-2-phenylindole (DAPI)	Nucleic acids	Freese *et al.* (2006)
DiBAC4(3), bis-(1,3-dibutylbarbituricacid)trimethine oxonol	Membrane potential	Bosshard *et al.* (2009)
2-[N-(7-Nitrobenz-2-oxa-1,3-diazol-4-yl)amino]-2-deoxy-D-glucose (2-NBDG)	Glucose uptake	Bosshard *et al.* (2009)
Ethidium bromide (EB)	Efflux pump activity	Bosshard *et al.* (2009)
SYBR Green I	Nucleic acids	Hammes *et al.* (2008)
SYBR Green II	Nucleic acids	Phe *et al.* (2004)
5-Cyano-2,3-ditolyl tetrazoliumchloride (CTC)	Respiratory activity	Falcioni *et al.* (2006)

growth potential of natural microbial consortia in drinking water distribution systems (Hammes *et al.*, 2005). Hoefel *et al.* (2005) sorted active microbial populations from drinking water samples using 5-carboxyfluorescein diacetate acetoxymethyl ester (cFDA-AM), and the LIVE/DEAD BacLight bacterial viability kit. Furthermore, sorted populations were characterized by denaturing gradient gel electrophoresis (DGGE) and sequencing analysis, with the detection of active ammonia-oxidizing bacteria (AOB) present within the drinking water distribution system. Successful application of total cell counting by FCM in a drinking water treatment pilot plant that processes lake water has been described (Hammes *et al.*, 2008). The total cell number appears to be a good indicator of changes in water quality during various treatments, including rapid sand filtration, granular activated carbon, ozonation and slow sand filtration. Data were compared with adenosine tri-phosphate (ATP) measurements and conventional cultivation-based heterotrophic plate counts performed on the same water samples. The authors demonstrated that the impact of all three major treatment steps in the system could accurately be described on the basis of total cell counting. Ozonation caused chemical destruction of the bacterial cells; granular active carbon filtration facilitated significant regrowth of the microbial community and membrane ultrafiltration physically removed bacterial cells from water. FCM typically detected 1–2 log units more than conventional plate counts, while ATP measurements were prone to interference from extracellular ATP released during the ozonation step (Hammes *et al.*, 2008). More work was carried out on a range of drinking water samples including non-chlorinated household tap water, untreated natural spring water, and commercially available bottled water. Using a combination of cultivation-independent methods including fluorescence staining (for membrane integrity, membrane potential and esterase activity) with FCM and total adenosine tri-phosphate (ATP) measurements, it was concluded that the esterase-positive cell fraction and the high nucleic acid (HNA) bacterial fraction (from SYBR Green I staining) were most representative of the active/viable population in the water samples (Berney *et al.*, 2008). These rapid methods have emerged as an alternative method to assess the general microbial quality of drinking water as well as specific events that can occur during treatment and distribution, with possibilities for application in both research and routine analysis.

The effectiveness of various disinfection treatments was evaluated by: UVA irradiation (Berney *et al.*, 2007), solar disinfection of enteric pathogens (Bosshard *et al.*, 2009), peracetic acid treatment effects on the virulence of *S.* Typhimurium LT2 (Jolivet-Gougeon *et al.*, 2006) and exposure to heat or UVC (Falcioni *et al.*, 2006). However, care should be taken considering the effects of certain treatments on microbial integrity and their staining potential. Several studies have shown that bacterial staining by fluorochromes (DAPI, SYBR Green II, and TOTO-1) is altered by chlorination. The fluorescence intensity of PI-stained (Propidium Iodide) DNA and RNA in solution dramatically decreased with an increase in the chlorine concentration applied (Phe *et al.*, 2007), when tested on *Escherichia coli* in suspension and also as biomolecules in solution (DNA, RNA, BSA, palmitic acid and dextran).

FCM has also been applied to determine the kinetics of oxidation and disinfection processes during ozonation with different dosages in a full-scale reactor treating secondary wastewater effluent (Zimmerman *et al.*, 2011). Total cell counts were analysed by FCM using staining with SYBR Green. Assimilable organic carbon concentration (AOC) was measured by inoculation of wastewater samples with a natural microbial consortium for 4 days. Further determination of cell numbers was performed by FCM (using a correlation to allow the conversion of the growth of the microbial consortium into AOC concentrations), monitoring of total cell counts and *Escherichia coli* inactivation to test ozonation effectiveness. The evaluation of FCM as an accepted test for water quality in Switzerland is outlined in greater detail later in this book.

Application to wastewater treatments

Modern molecular tools have permitted investigations to identify and compare the most abundant bacterial taxa in wastewater treatment plants (WWTPs), and an examination of the stability and reproducibility of this diversity. Some reports have revealed that the general bacterial community is highly variable, never settles down to a stable community, and varies between bioreactors. However, it has been reported that other groups, notably the *Archaea* in anaerobic systems and the AOB in activated sludge, tend to be more reproducible among bioreactors and less variable within a given treatment plant (see Curtis *et al.*, 2003).

In municipal WWTPs, the activated sludge process degrades organic contaminants under aerobic conditions. FCM in combination with fluorescent dyes has been used for counting bacteria in activated sludge (Falcioni *et al.*, 2006). Activated sludge flocs (ranging from small (< 150 μm) to large (> 500 μm) in diameter) are formed by microorganisms, colloids, organic polymers and cations. Since FCM analysis demands a homogeneous cell suspension, detachment of bacteria from flocs is required. It was observed that application of sonication to sludge contributed to the dispersion of aggregates and solubilization of particulate matter with an increase in biodegradability, but it may have damaged microorganisms arising from the loss of cellular membrane integrity. Sonication and homogenization were compared to determine the optimal pre-treatment method for activated sludge samples for bacterial cytometric analysis. Bacterial viability was tested by FCM using a nucleic acid double-staining (NADS) protocol (Barbesti *et al.*, 2000). Each method showed good efficiency in terms of bacterial detachment, based on both viability data and rapidity of analysis. Based on the degree of cell detachment and viability, maximum disaggregation of sludge flocs was obtained using sonication.

Bacterial biomass in activated sludge (usually called active biomass) is generally estimated by theoretical calculations based on substrate mass balances using kinetic and stoichiometric parameters. For quantifying biomass in wastewater treatments, the content in suspended solids is used, and is expressed as Total Suspended Solids (TSS) or Volatile Suspended Solids (VSS). However, VSS does not accurately reflect the microbial biomass since it also includes bacterial debris and organic non-biotic particulate matter present in the sample. Nevertheless, accurate information regarding the real active microbial biomass in the biotreatment is a key parameter to determine process efficiency. Direct fluorescence staining of bacteria in activated sludge enables them to be distinguished from other constituents of biological flocs such as non-biotic organic and anaerobic particles, which constitutes most of the total solid content.

Recently, FCM has been applied for direct biomass quantification not only in activated sludge but also in wastewater (influent and effluent) within a WWTP (Foladori *et al.*, 2010a). Bacterial biomass in raw wastewater influent in a WWTP is rarely quantified and its content is often neglected in mass balances, due to its low dry mass. In treated effluents, generally only quantification of faecal indicators is considered, and then only to evaluate disinfection performance. In this work (Foladori *et al.*, 2010a), FCM was coupled with fluorescence staining (SYBR-Green I plus PI to discriminate cell integrity and cell permeabilization, and the fluorogenic substrate BCECF-AM to identify enzymatically active bacteria arising from non-specific intracellular esterases). Additionally, a procedure was used to convert the number of bacterial cells into an equivalent biomass, expressed as dry weight, taking into account the bacterial biovolume estimated by FCM. This conversion permitted calculation of viable and active bacterial biomass in wastewater, activated sludge and effluent, expressed as VSS or particulate chemical oxygen demand (COD). The FCM method was applied to several samples of raw and pre-settled wastewater, activated sludge and effluents and subsequently compared with conventionally obtained parameters (TSS and COD). Results indicated that viable bacterial biomass represented only a small part of particulate COD in raw wastewater, settled wastewater, activated sludge and effluent. Active bacterial biomass accounted for 30–47% of the viable bacterial biomass within the various stages of the WWTP. By applying the FCM procedure to rapidly assess the bacterial biomass in activated sludge or wastewater samples, (about 30 minutes), analysis of many samples within a working day could be achieved.

Park *et al.* (2005) also reported the use of a combination of FACS, cultivation and molecular genetics approach to develop a new method to access and characterize biodiversity in activated-sludge samples. Although previous reports for the application of FACS to characterize environmental samples involved fluorescent probes targeting DNA, their use could reduce cell viability to values as low as 2% (Resina-Pelfort *et al.*, 2001). Fluorescent dyes have different affinities for compounds present on the external surfaces of microbial cells, such as lipids, proteins, and carbohydrates. This involves staining the cells with various lipid-specific fluorescent dyes as N-(fluorescein-5-thiocarbamoyl)-1,2-dihexadecanoyl-sn-glycero-3-phosphoethanolamine, 'DHPE', triethanolamine salt or 3,3′- dioctadecyloxacarbocyanine perchlorate, 'DiO', $DiOC_{18}$. Thereby it was possible to selectively alter the fluorescence characteristics of some of the species in mixed bacterial cell populations. This approach allowed particular bacterial types/species to be distinguished from the bulk. Thus, bacterial populations of interest could be quantified and/or recovered by FACS. Microbial subpopulations obtained by FACS differed substantially from the original population, as demonstrated by DGGE and 16S rDNA gene sequences. These subpopulations could be subsequently used to inoculate growth media, allowing isolation of microbial species different from those that could be readily cultivated from the original sample on the same growth media. Comparative analysis of 16S rDNA sequences showed that FACS enabled detection of numerous bacterial species, including previously unknown species, not readily detectable in the original sample due to their low relative abundance.

Multi-parametric FCM has been applied for monitoring population dynamics of thermophilic *Bacillus licheniformis* CCMI 1034 in batch or continuous cultures (Reis *et al.*, 2005). Isolation of cultures from thermophilic aerobic wastewater treatment reactors found that thermophilic *Bacillus* spp. were the dominant microorganisms present. Information regarding the population dynamics and the physiological responses to a starvation period followed by a glucose pulse within the system may allow improved process efficiency as high numbers of dead or dormant cells present will obviously have a detrimental effect on the biodegradation process. Using a mixture of two specific fluorescent stains, $DiOC_6$ (3) (3,3′-dihexylocarbocyanine iodide), and PI, FCM analysis revealed cell physiological heterogeneity. Identification of four cell subpopulations could be achieved, based on differential fluorescent staining, corresponding to: healthy cells stained with $DiOC_6$(3), cells or spores with depolarized cytoplasmic membranes, unstained cells, cells with a permeabilized depolarized cytoplasmic membrane stained with PI, and permeabilized cells with a disrupted cytoplasmic membrane or 'ghost cells' double stained with $DiOC_6$(3) and PI. At the individual cell level, and with the aim of enhancing process efficiency, a progressive inherent fluctuation in physiological heterogeneity was noted in response to changing environmental conditions.

FCM has been also used to assess the dynamics of polyphosphate-accumulating bacteria (PAOs) within microbial communities in a WWTP with enhanced biological phosphorus removal (EBPR) using a dual-staining technique (Günter *et al.*, 2009). A novel staining protocol (using 4′,6′-Diamidino-2-Phenylindole (DAPI) and tetracycline labelling) was applied to identify not-cultivated bacteria capable of polyphosphate accumulation within highly diverse communities. This staining technique uses the bright green fluorescence of the antibiotic tetracycline which is generated when it complexes the divalent cations acting as a countercharge in polyphosphate granules. Allowing fast and convenient monitoring, the dynamics of cellular DNA contents and cell sizes as growth indicators were determined in parallel to detect the most active polyphosphate-accumulating sub-communities. Their phylogenetic affiliation was also determined by cell sorting, as a basis for process control optimization. Clone library (16S rRNA genes) and the lengths of fluorescent terminal restriction fragments (T-RFLP) analyses were performed to verify the specificity of the staining technique. This approach appears to overcome the limitations of using specific oligonucleotide FISH probes for typing phylogenetic groups known to be PAOs. These organisms do not always produce polyphosphates, and besides, their low rRNA content can render them difficult for detection by conventional FISH techniques. A major pitfall of FISH is that cells must be fixed, so the method destroys the structural integrity of the cell, preventing further downstream cultivation when combining with sorting. Moreover, unlike the FISH technique, a fluorescence-based approach using FCM with a simple and reliable staining technique can also include PAOs of unknown phylogeny.

Another interesting study developed methodology to reveal the denitrification ability of the provisionally named '*Candidatus* Accumulibacter phosphatis', a putative PAO which has yet to be isolated (Miyauchi *et al.*, 2007). This organism is abundant in acetate-fed laboratory-scale bioreactors under cyclic anaerobic/aerobic or anaerobic/anoxic conditions, and particularly under conditions of efficient phosphorus removal. This microorganism was first concentrated from sludge samples (withdrawn from a laboratory-scale sequential batch reactor (SBR) by FCM sorting, using FISH probes specific for this organism 5′-labelled with 6-carboxyfluorescein. The purity of the target cells was about 97% of total cell count in the sorted, double-stained FISH/DAPI subpopulation. Subsequently, cells were screened for functional genes. PCR amplification of the nitrite reductase genes (*nirK* and *nirS*) from unsorted and sorted cells was performed. Although *nirK* and *nirS* were amplified from unsorted cells, only *nirS* was detected from sorted cells, indicating that '*Ca.* Accumulibacter phosphatis' contains *nirS*. Furthermore, *nirS* fragments were cloned from unsorted and sorted cells and classified by restriction fragment length polymorphism analysis (RFLP). The most dominant clone in the unsorted clone library, which represented 62% of the total number of clones, was not found in the sorted clone library.

In contrast, the dominant clone in the sorted clone library, representing 59% of the total number of clones, comprised only 2% of the total number of clones in the other library, indicating that this clone could be that of '*Ca.* Accumulibacter phosphatis.' The sequence of this *nirS* clone exhibited less than 90% similarity to the sequences of known denitrifying bacteria in the database. The recovery of the *nirS* genes makes it likely that '*Ca.* Accumulibacter phosphatis' behaves as a denitrifying PAO capable of utilizing nitrite instead of oxygen as an electron acceptor for phosphorus uptake.

FCM combined with FISH has also been used for tracking the composition and dynamics of nitrification and denitrification of the microbial community within a biofilm reactor by PCR–DGGE (Xia *et al.*, 2010). A compact suspended carrier biofilm reactor (SCBR) was operated at three different carbon/nitrogen (C/N) ratios focusing on reactor performance and microbial community composition of nitrifying and denitrifying bacteria. The process was capable of greater than 90% COD removal and greater than 83.3% of simultaneous nitrification and denitrification

(SND) efficiency. FISH–FCM analysis showed that the biofilm was composed of 27% AOB, 15% nitrite oxidizing bacteria (NOB) and 21% heterotrophic bacteria. However, nested analysis of variance assay (ANOVA) demonstrated that the C/N ratio did not significantly affect the shift of populations in different groups of nitrifying and denitrifying bacteria detected by FISH–FCM. In this study, as a pretreatment protocol, 0.5 g samples of biofilm biomass were collected from every time period, detached by sonication and filtered to remove large cell aggregates before fixing with paraformaldehyde.

Application to solid waste treatments

FCM has been used effectively for evaluating different sludge reduction technologies (Foladori *et al.*, 2010b). Most sludge reduction technologies (physical, mechanical, thermal, chemical, oxidative treatments) aim at solids solubilization, disintegration of the biological floc structure and disruption of bacterial cells. The ultimate goal is that cells undergo lysis such that intracellular compounds can be released for further degradation in biological reactors. Foladori *et al.* (2010b) investigated how the various sludge reduction technologies affected the integrity and permeabilization of bacterial cells. FCM was used for rapid and statistically accurate quantification of intact, permeabilized or disrupted bacteria in the sludge, by using a double fluorescent DNA-staining (SYBR-Green I and PI) instead of conventional plate counts or microscopic observations. Data indicated that physical/mechanical treatments (ultrasonication and high pressure homogenization) caused moderate effects on cell integrity with significant cell disruption occurring only at high specific energy levels. Conversely, thermal treatment caused significant damage of bacterial membranes even at moderate temperatures. Ozonation significantly affected cell integrity, even at low dosages, causing an increase in permeabilized or disrupted cells. At higher dosages, the compounds solubilized after cell lysis acted as scavengers in the competition between soluble compounds and particulate cells. It could be concluded, by comparison of the applied specific energy for each treatment, that achieving complete disruption of bacterial cells is not always an economic advantage. On the contrary, it was indicated that is probably more advantageous to cause sludge floc disintegration and only partially damage the bacterial cells. This may promote better interaction between bacteria, enzymes and substrates and favour the biodegradation process.

Immunofluorescent labelling and staining with PI with FCM analysis have also been used to enumerate nitrifiers in various sewage plants (Volsch *et al.*, 1990). Two different serotypes of the genus *Nitrosomonas* were isolated from samples and a test for the determination of the activity of AOB was also developed using sodium chlorate as a selective inhibitor of nitrite oxidizers. Activities were calculated from the linear increase of nitrite during the incubation time. Concentrations and activities of ammonia oxidizers were then measured for a period of 6 months in the sewage plant.

Another interesting application should be highlighted. Disposal of wastes to landfill remains the most common waste management procedure globally. Leachates are generated by the excess of rainwater percolating through the waste layers and their toxicity is classically determined by physico-chemical parameters. The effects of toxic compounds (including dissolved organic matter, inorganic macro-components, heavy metals or xenobiotic organic compounds) in landfill leachates on human health are not fully clarified yet. However this issue is of increasing interest. Although applications of FCM to toxicity studies of wastes will not be reviewed here since they generally use mammalian cells or model organisms such as *Drosophila melanogaster*, an application for bacterial toxicity evaluation using FCM has recently been published (Kumar *et al.*, 2011). There is a notable lack of information regarding the effects of engineered nanomaterials on human health and their environmental impact. The use of nanoparticles (NPs) in electronics, tyres, fuel cells, filters and even in personal-care products such as cosmetics and sunscreens (listed among the so-called emerging contaminants) is leading to their inadvertent release in surface and subsurface environment through landfills and other waste disposal methods. Earlier studies by the same group with human epidermal cells have shown that ZnO NPs possessed some potential for causing damage to DNA. On the basis of the important role played by bacteria in maintaining the ecosystem, the accumulation of NPs in ecosystems was determined by evaluating their threat to bacteria as non-specific targets. The study investigated the effect of ZnO and TiO_2 NPs on *Salmonella* Typhimurium as a model bacterium. The uptake of ZnO and TiO_2 bearing NPs in the nanometre range in *S.* Typhimurium was demonstrated by the internalization and uniform distribution of NPs inside the cells using transmission electron microscopy (TEM) analysis. Additionally, FCM was used to assess size and granularity by forward scatter and side scatter intensities, and also demonstrated that both ZnO and TiO_2 NPs were significantly internalized in *S.* Typhimurium cells in a concentration-dependent manner. At the concentrations tested, weak mutagenic potential was exhibited as a cytotoxic effect which could be detected in the test strains (*S.* Typhimurium TA98, TA100, TA1535 TA1537 *and E. coli* (WP2 uvrA), as determined by colony forming units (CFU) count and FCM. Data obtained reinforced the need for re-evaluation of the environmental toxicity of ZnO and TiO_2 NPs which have been presumed as safe in the environment.

Other applications of FCM have focussed on composting processes. Composting is an aerobic biological process that uses the innate microbial community to convert the biodegradable organic fraction present in municipal solid wastes (such as vegetables, fruits or animal rests) into a humus-like product. Interestingly, Fujii and Hiraishi (2009) sorted metabolically active cells from a composting process which were further phylogenetically identified, as an approach for linking both functional and structural aspects of bacterial community studies. In this case, microbial community dynamics with metabolically active bacteria during the start-up operation of a personal fed-batch composting (FBC) reactor were studied. Metabolically active bacteria were monitored by the redox-dye-staining method using CTC, and the fluorescent formazans produced were detected by epifluorescence microscopy and FCM. A drastic population change from ubiquinone-containing members of the Proteobacteria to the Actinobacteria took place during the overall period of operation. DGGE of PCR-amplified 16S rRNA gene fragments showed that the FACS-sorted population had a different community structure from the global population. Data indicated that the metabolically active bacteria, measured by CTC staining, were not always predominant in the FBC process. This may have arisen because the exogenous microorganisms present in the fresh biowaste, with which the FBC reactor was seeded, could temporarily grow faster or show higher levels of activity than the predominant bacteria in the reactor. The authors highlighted the promising utility of FCM sorting followed by PCR-DGGE profiling or other molecular approaches for the rapid detection and phylogenetic identification of metabolically active bacteria in communities of different environments.

FCM has been useful for assessing inoculation of vegetable waste compost with isolated *Geobacillus* strains during the thermophilic stage of composting (Sarkar *et al.*, 2010) to improve bioprocess efficiency. The inoculum contained two amylolytic and three cellulolytic thermophilic bacteria, isolated from a preceding compost pile and were identified as *Geobacillus* species by 16S rRNA gene sequencing. This mixed inoculum was applied to a mixture of market waste, rice straw and cow dung. FCM analysis determined that inoculation increased the total cell count particularly during the thermophilic stage. Concomitantly, a significant rise in microbial metabolism in the compost pile took place, as reflected by the dehydrogenase activity, measured by the method based on the estimation of the rate of reduction of 2,3,5-triphenyl tetrazolium chloride (TTC) to triphenyl formazan (TPF). As a result, the C/N ratio dropped more rapidly in the inoculated mixture than the control without the inoculum. This study suggested that inoculation by thermophilic bacteria may be effective in the thermophilic stage of the composting process.

It is known that microbial production of hydrogen may help to solve not only the problems caused by organic wastes but also create an alternative energy source. Fermentative bio-hydrogen production has been considered as an environmentally friendly process for recovering hydrogen from solid wastes. FCM-FISH has been applied for analysis and isolation of Clostridial strains in an anaerobic semi-solid bio-hydrogen batch producing system by targeting of the hydrogenase gene to improve hydrogen production from brewery yeast waste (Jen *et al.*, 2007). This study also noted that the general environmental engineering approaches for hydrogen production are performed in open systems by mixed microflora showing a high degree of diversity, thus making the system difficult to control. Isolation and characterization of the major hydrogen-producing contributors from each substrate-specific fermentation system would be helpful for operation of the system and elevating hydrogen-producing efficiency. In a previous study by Chang *et al.* (2006), the predominant clostridial hydrogenases that may contribute to bio-hydrogen production in an anaerobic semi-solid fermentation system were detected by gene-targeted PCR and reverse transcriptase PCR (RT-PCR). The results strongly suggested that clostridia possessing two specific hydrogenase genes (a specific *Clostridium pasteurianum*-like hydrogenase and a specific *Clostridium saccharobutylicum*-like hydrogenase) were the major hydrogen producers in the system. Subsequently, by using hydrogenase gene-targeted PCR and RT-PCR, the predominant clostridial hydrogenase contributing to bio-hydrogen production in the anaerobic semi-solid fermentation system was monitored. A *C. pasteurianum*-like hydrogenase gene sequence could be detected by both PCR and RT-PCR, suggesting that the bacterial strain possessing this specific hydrogenase gene was dominant in the population. Hydrogenase gene-targeted FISH–FCM analysis confirmed that only a small percentage of the total eubacterial cells in a hydrogen-producing culture expressed the *C. saccharobutylicum*-like hydrogenase. In contrast, the eubacteria that expressed the *C. pasteurianum*-like hydrogenase constituted a much higher proportion of the population. A clostridial strain possessing the identical nucleotide sequence of the *C. saccharobutylicum*-like hydrogenase gene was then isolated and identified by 16S rRNA sequence as *Clostridium butyricum*. Comparing with the original inoculum of mixed microbiota, using either the isolated *C. butyricum* M1 as single inoculum or co-culturing with a *Bacillus thermoamylovorans* isolate, an improved hydrogen production from brewery yeast waste could be achieved.

Bacterial viability in soils and sediments

Estimates of the number of different species in a soil sample (which obviously varies depending on the method used) include 467 species, 6300 species, 10,000 species or even 500,000 species. Speculating about diversity on a larger scale, it was estimated that although the entire bacterial diversity of the oceans might be unlikely to exceed 2×10^6, one ton of soil could contain 4×10^6 different taxa (for a review see Keller and Zengler, 2004). This immense biodiversity represents a challenge for researchers interested in soil ecology since, despite the emergence of metagenomics, methods to isolate total microorganisms are still affected by selection biases towards fast growing abundant microorganisms (Hugenholtz and Tyson, 2008). Some environmental factors known to influence soil community composition include the presence of heavy metals, organic contaminants or pesticides. Even small changes in the soil environment appear to exert a major influence on the overall diversity of species and community structure. Indeed, because of their sensitivity and rapid response to even minor changes, bacteria can be used as indicators of soil quality.

Even in this complex sampling environment, FCM can be useful since low abundance microorganisms can be enriched by cell sorting. This can be followed by taxonomic identification of selected populations by PCR amplification of 16S rDNA genes, library construction and sequencing of different gene fragments, thereby obtaining more comprehensive phylogenetic data. However, special care should be taken when applying appropriate detachment separation treatments before FCM analysis of microorganisms associated with soil and sediments. Physicochemical treatments are required to concentrate attached cells into a liquid suspension and reduce particle-masking effects. Some studies have evaluated metabolic activities of separated cells following various treatments. Amalfitano *et al.* (2009) used FCM to identify specific subpopulations of attached prokaryotes by applying the NADS assay to gain reliable information at single-cell level on the effect of separation treatments and detected two groups of cells exhibiting different nucleic acid contents. Chemical treatments (including sodium pyrophosphate and Tween 20) and physical detachment was performed by shaking and sonication. However, NADS showed that separation procedures negatively affected membrane integrity, perhaps due to changes in the state of cell membranes, thus raising concerns on the applicability of cytometric metabolic assays on treated sediment samples.

The use of antibodies has been reported for analysis of microbes in environmental soil samples (Porter *et al.*, 1997) but, prior isolation of the target microorganism in pure culture is normally required to generate antibodies against specific antigens (for direct staining techniques). To overcome this limitation, indirect staining methods use non-fluorescent primary antibodies followed by fluorescent secondary antibodies directed against the first ones. For FCM protocols, antibody molecules can be made fluorescent by covalently attaching them to fluorescent organic compounds such as fluorescein isothiocyanate (FITC), the most widely used, but others such as tetramethylrhodamine isothiocyanate (TRITC), Texas Red, phycobiliproteins, and cyanine dyes including the indocarbocyanine dyes (CY3, CY5 and CY7), and ALEXA dyes have been also reported (Vives-Rego *et al.*, 2000). To analyse soil samples Porter *et al.* (1997) reported on a range of extractants evaluated for their ability to disperse soil. The most effective agent was sodium cholate, which gave significantly greater dispersal, yielding a bacterial cell suspension suitable for FCM analysis. To reduce the problems associated with the high background particulate count, lectin-mediated magnetic bead capture was used. The method was tested using added bacterial populations which were detected by immunofluorescence, nucleic acid staining and membrane potential analysis. Selected fluorescent dyes were then used for viability analysis of indigenous bacteria from soil.

For cultivation of mixed communities of recalcitrant microcolony-forming soil bacteria, a soil substrate membrane system (SSMS) was developed and used in combination with cell sorting for the isolation of individual microcolonies or novel filamentous bacterial morphotypes after enrichment (Ferrari *et al.*, 2008; Bergquist *et al.*, 2009). SSMS was developed to mimic the natural terrestrial environment of soil bacteria. In SSMS, microcolony formation occurs on a polycarbonate membrane which is in immediate contact with a slurry of soil substrate.

An alternative FCM strategy to analyse heterogeneity within complex microbial populations, which avoids killing cells for FISH staining, is based on monitoring the physiological or functional aspects of target populations. Kalyuzhnaya *et al.* (2008) reported the use of a novel stain, redox sensor green (RSG), a novel fluorogenic redox indicator dye, to overcome this problem in sediments. This dye yields green fluorescence when modified by bacterial reductases, many of which are components of electron transport systems. Unlike tetrazolium salts, such as 5-cyano-2,3-ditolyl tetrazolium chloride (CTC) or 2-(p-iodophenyl-3-(p-nitrophenyl)-5-phenyltetrazolium chloride (INT), RSG does not suppress cellular metabolism and could be used for real-time detection of cells actively respiring *in situ*. The advantage of combining RSG staining with FCM and cell sorting lies in the potential for downstream applications of populations of live cells which can be separated based on their specific metabolic responses.

As an application related to soil quality, FCM has been used as a method for detecting and tracking the fate of insecticidal toxins in soil without requiring their extraction or purification, using an antibody conjugated with fluorescein isothiocyanate (FITC–Ab) (Tapp and Stotzky, 1997). The accumulation and persistence of insecticidal toxins from *Bacillus thuringiensis* in soil and other habitats may result in environmental hazards, such as toxicity to non-target species and selection of toxin resistant target species. Toxins from *B. thuringiensis* subsp. *tenebrionis* and *kurstaki* bound on clay- or silt-sized particles separated from Kitchawan soil were unmodified (naturally contains predominantly kaolinite) or modified with the clay minerals montmorillonite or kaolinite as controls. All particles that bound to the toxin from *B. thuringiensis* subsp. *tenebrionis* showed a significant shift in the peak of fluorescence to the right on the x-axis compared with the non-specific fluorescence from the control FITC–Ab complexes with particles in the absence of the toxin. This method was more sensitive and rapid than the dot-blot ELISA and many samples could be easily processed.

Another application of FCM has focussed on biotreatment of contaminated soils. This research interest in direct detection of specific strains in contaminated soils, either indigenous or inoculated, is linked to bioremediation objectives. An FCM method described by Thomas *et al.* (2000) was used to detect and enumerate a polycyclic aromatic hydrocarbon (PHA)-degrading bacterial strain, *Sphingomonas* sp. 107, inoculated into a soil sample artificially contaminated with pyrene. To compare the FCM method with colony forming unit assays, a rifampicin-resistant *Sphingomonas* sp. 107 was obtained which could be distinguished from the indigenous microflora, since there was no organism resistant to rifampicin in the soil that could transform indole to indigo (naphthalene dioxygenase activity). By combining light-scattering profiles (morphological properties), ethidium bromide influx (cell wall permeability), and FISH against 16S rRNA specificity, it was possible to enumerate the population of interest from the indigenous microflora and soil debris during biotreatment. FCM detection was improved by adding a blocking reagent to the hybridization buffer to minimize non-specific attachment of the fluorescent probe to soil particles. As another example of this application, Leys *et al.* (2004) described a culture-independent PCR-based detection method using specific primers targeting the *Sphingomonas* 16S rRNA gene combined with DGGE to assess *Sphingomonas* diversity in PAH-contaminated soils. Analysis of *Sphingomonas* population structures of several PAH contaminated soils by a newly developed PCR–DGGE method revealed that soils containing the highest phenanthrene concentrations showed the lowest *Sphingomonas* diversity. Sequence analysis of cloned PCR products amplified from soil DNA revealed new 16S rRNA gene *Sphingomonas* sequences significantly different from sequences obtained from known cultivated isolates.

Another interesting application of FCM is related to the development of GFP-producing whole-cell biosensors *in situ* to study interactions between antibiotic producers and their target organisms in soil (Hansen *et al.*, 2001). A combination of highly specific bacterial biosensors with the resolution power of FACS generated qualitative data regarding oxytetracycline production by *Streptomyces rimosus* within soil microcosms. A plasmid containing a transcriptional fusion between the *tetR*-regulated P_{tet} promoter from Tn*10* and a FACS-optimized *gfp* gene was constructed. When harboured by *E. coli*, this plasmid produced large amounts of green fluorescent protein (GFP) in the presence of tetracycline. This tetracycline biosensor was used to detect the production of oxytetracycline by *S. rimosus* introduced into sterile soil. The tetracycline induced GFP-producing biosensors were detected by FACS, enabling the detection of oxytetracycline encounters by single biosensor cells.

Future trends

Traditional growth-based microbiological assays, considered to be both slow and biased, are continuously being replaced by automated optical detection methods. Innovative methods are especially required to address the complex challenges faced in the environmental field. The potential of nanobiotechnology is increasing rapidly. Recently, it has been shown that the optical properties of the quantum dots (QDs) (inorganic crystalline nanoparticles made of semiconducting materials) can be used as fluorescent labels for biomolecules. Since then, interest in their applications has increased enormously. QDs could prove useful for the study of environmental samples where microorganisms of interest need to be isolated from the background debris, which generally causes high autofluorescence problems. The potential use of QDs in FCM has been reported by Ibáñez-Peral *et al.* (2008), who studied binding of QDs modified with biotin to paramagnetic beads (Dynabeads®). These researchers compared their application as probes targeting specific DNA sequences from microorganisms with coupling of an oligonucleotide labelled with the commonly used dye fluorescein isothiocyanate (FITC). Fluorescent intensity of Dynabeads® labelled with different amounts of QDs or LinkerFITC was analysed by

FCM. Comparisons indicated that the minimum fluorophore concentration required for detection of QDs above autofluorescent background was 100-fold less than for FITC.

Positive cells containing novel catabolic genes can be detected through the expression of green fluorescence protein (GFP) and physically separated for sequence analysis. In the near future, this approach should facilitate isolation of novel biocatalysts from the environment (Bergquist *et al.*, 2009). FCM is also useful in the development of biosensor field-based assays for real time detection. Kramer *et al.* (2007) noted that rapid and effective biology-based technologies were a societal need for the prevention of waterborne diseases worldwide.

Identification of specific metabolic capabilities in the community, understanding synergistic/antagonistic interactions with other species (integrated community ecology and community assembly) remains a key challenge to improve environmental bioprocess efficiency. It should be noted that functional redundancy (i.e. different strains that can carry out the same or similar functions) and metabolic versatility may exist within environmental microbial communities. To face the challenges in environmental biotechnology, it has recently been proposed by Stepanauskas and Sieracki (2007) that single-cell isolation techniques such as FCM may potentially link '-omics' data to phenotypic data. These authors used high-speed FACS, whole-genome multiple displacement amplification (MDA), and subsequent PCR screening to directly detect metabolic genes in uncultured microorganisms, even when the metabolic and phylogenetic markers were located far apart on the chromosome.

It has been reported that relationship between structure and function in a community could only be understood, predicted and engineered through an understanding of the source of diversity from which the community is drawn (Curtis and Sloan, 2004). The robustness and automation of FCM increases the opportunity to tackle these interesting questions. Furthermore, additional advantages may be gained when applying automated techniques such as FCM to the development of kinetic models to control and predict microbial population dynamics in biotreatments (Díaz *et al.*, 2010).

In environmental biotreatment plants, routine measurement of key chemical compounds and parameters are subjected to legislative regulation and is therefore of the greatest importance in plant operation and control. Relatively inexpensive chemical methods are performed routinely by the operating staff. In a similar way, rapid and reliable analysis for the active microbial populations, especially those key microorganisms harbouring functional gene markers responsible for the critical steps of the process, is also a key requirement. This approach should offer information about bioprocesses kinetics and performance, allowing decision making in almost real time conditions regarding biotreatment strategies (Fig. 3.2). In this way, the increased benefit compared with biases of time-consuming methods such as tedious microscopic observations or plate counts could be achieved. The goal should be to add practical microbiological parameters related to the activity of the key physiological groups to the conventional chemical parameters already routinely performed in engineered biotreatments.

This goal requires technological capabilities for almost on-line microbial detection systems, quantification and analysis. Design, commercialization and implementation of automated devices with high sample output should provide information about target microorganisms within a short period of time (minutes or hours). There is an increasing interest in developing a fully functional, networkable system platform capable of rapid detection of multiple species/functional groups of microorganisms under real environmental conditions. Different FCM analysis systems are commercially available. The design of robust flow injection systems coupled to FCM (FIA-FCM) (Zhao *et al.*, 1998) is an interesting outcome. One microsystem, miniaturized and integrated in accordance with recent advances in micro-electromechanical systems technology is the Micro Integrated Flow Cytometer (MIFC) (Gruden *et al.*, 2004). Recently, an integrated microfluidic device (μFlowFISH), capable of performing 16S rRNA FISH followed by FCM detection has been applied (Liu *et al.*, 2011) for identifying microbial communities in agreement with results obtained with conventional FCM-FISH, providing an automated platform for quantitative detection of

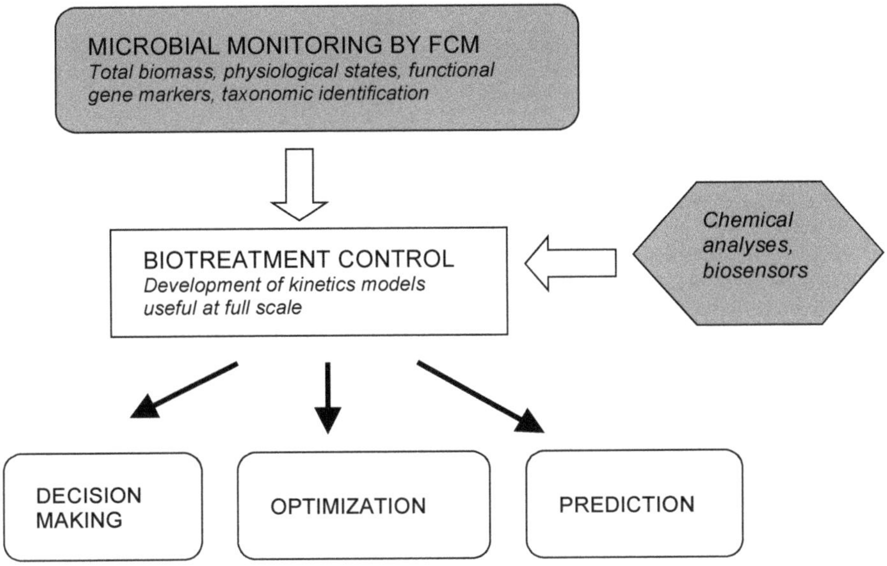

Figure 3.2 Utility of FCM under routine operating conditions in environmental bioprocess optimization and control.

microbial cells from complex samples. The robust outcomes already achieved makes it possible to think that, in the near future, with the efforts of multidisciplinary approaches, microbial consortia in environmental engineered bioprocesses may be controlled under routine operating conditions to gain better performance.

References

Akram, A., and Stuckey, D.C. (2008). Flux and performance improvement in submerged anaerobic membrane bioreactor (SAMBR) using powdered activated carbon (PAC). Process Biochem. *43*, 93–102.

Amalfitano, S., Fazi, S., and Puddu, A. (2009). Flow cytometric analysis of benthic prokaryotes attached to sediment particles. J. Microbiol. Methods *79*, 246–249.

Barber, W.P., and Stuckey, D.C. (1999). The use of the anaerobic baffled reactor (ABR) for wastewater treatment: A review. Water Res. *33*, 1559–1578.

Barbesti, S., Citterio, S., Labra, M., Baroni, M.D., Neri, M.G., and Sgorbati, S. (2000). Two and three-color fluorescence flow cytometric analysis of immunoidentified viable bacteria. Cytometry *40*, 214–218.

Bergquist, P.L., Hardiman, E.M., Ferrari, B.C., and Winsley, T. (2009). Applications of flow cytometry in environmental microbiology and biotechnology. Extremophiles *13*, 389- 401.

Berney, M., Hammes, F., Bosshard, F., Weilenmann, H.-U., and Egli, T. (2007). Assessment and interpretation of bacterial viability by using the LIVE/DEAD BacLight Kit in combination with flow cytometry. Appl. Environ. Microbiol. *73*, 3283–3290.

Berney, M., Vital, M., Hülshoff, I., Weilenmann, H.-U., Egli, T., and Hammes, F.A. (2008). Rapid, cultivation-independent assessment of microbial viability in drinking water. Water Res. *42*, 4010–4018.

Bosshard, F., Berney, M., Scheifele, M., Weilenmann H.-U., and Egli, T. (2009). Solar disinfection (SODIS) and subsequent dark storage of *Salmonella typhimurium* and *Shigella flexneri* monitored by flow cytometry. Microbiology *155*, 1310–1317.

Chang, J.J., Chen, W.E., Shih, S.Y., Yu, S.J., Lay, J.J., Wen, F.S., and Huang, C.C. (2006). Molecular detection of the clostridia in an anaerobic biohydrogen fermentation system by hydrogenase mRNA-targeted reverse transcription-PCR. Appl. Microbiol. Biotechnol. *70*, 598–604.

Curtis, T.P., and Sloan, W.T. (2004). Prokaryotic diversity and its limits: microbial community structure in nature and implications for microbial ecology. Curr. Opin. Microbiol. *7*, 221–226.

Curtis, T.P., Head, I.M., and Graham, D.W. (2003). Theoretical Ecology for engineering biology. Environ. Sci. Technol. *37*, 64A–70A.

Díaz, M., Herrero, M., García, L.A., and Quirós, C. (2010). Application of Flow Cytometry to industrial microbial bioprocesses. Biochem. Eng. J. *48*, 385–407.

Falcioni, T., Manti, A., Boi, P., Canonico, B., Balsamo, M., and Papa, S. (2006). Comparison of disruption procedure for enumeration of activated sludge floc bacteria by flow cytometry. Cytometry Part B, *70B*, 149–153.

Ferrari, B.C., Winsley, T., Gillings, M., and Binnerup, S. (2008). Cultivating previously uncultured soil bacteria using a soil substrate membrane system. Nat. Protoc. *3*, 1261–1269.

Foladori, P., Bruni, L., Tamburini, S., and Ziglio, G. (2010a). Direct quantification of bacterial biomass in influent, effluent, and activated sludge of wastewater treatment plants by using flow cytometry. Water Res. *44*, 3807–3818.

Foladori, P., Tamburini, S., and Bruni, L. (2010b). Bacteria permeabilisation and disruption caused by sludge reduction technologies evaluated by flow cytometry. Water Res. *44*, 4888–4899.

Freese, H., Karsten, U., and Schumannet, R. (2006). Bacterial abundance, activity, and viability in the eutrophic River Warnow, Northeast Germany. Microb. Ecol. *51*, 117–127.

Fujii, Y., and Hiraishi, A. (2009). Combined use of cyanoditolyl tetrazolium staining and flow cytometry for detection of metabolically active bacteria in a fed-batch composting process. Microbes Environ. *24*, 57–63.

van der Gast, C.J., Knowles, C.J., Starkey, M., and Thompson, I.P. (2002). Selection of microbial consortia for treating metal working fluids. J. Ind. Microbiol. Biotechnol. *29*, 20–27.

Günther, S., Trutnau, M., Kleinsteuber, S., Aus, G., Bley, T., Röske, I., Harms, H., and Müller, S. (2009). Dynamics of polyphosphate-accumulating bacteria in wastewater treatment plant microbial communities detected via DAPI (4′,6′-diamidino-2-phenylindole) and tetracycline labelling. Appl. Environ. Microbiol. *75*, 2111–2121.

Gruden, C., Skerlos, S., and Adriaens, P. (2004). Flow cytometry for microbial sensing in environmental sustainability applications: current status and future prospects. FEMS Microbiol. Ecol. *49*, 37–49.

Hansen, L.H., Ferrari, B., Sørensen, A.H., Veal, D., and Sørensen, S.J. (2001). Detection of oxytetracycline production by *Streptomyces rimosus* in soil microcosms by combining whole-cell biosensors and flow cytometry. Appl. Environ. Microbiol. *67*, 239–244.

Hammes, F.A., and Egli, T. (2005). New method for assimilable organic carbon determination using flow-cytometric enumeration and a natural microbial consortium as inoculum. Environ. Sci. Technol. *39*, 3289–3294.

Hammes, F., Berney, M., Wang, Y., Vital, M., Köster, O., and Thomaset, E. (2008). Flow-cytometric total bacterial cell counts as a descriptive microbiological parameter for drinking water treatment processes. Water Res. *42*, 269–277.

Hoefel, D., Grooby, W.L., Monis, P.T., Andrews, S., and Saint, C.P. (2003). Enumeration of water-borne bacteria using viability assays and flow cytometry: a comparison to culture-based techniques. J. Microbiol. Methods *55*, 585–597.

Hoefel, D., Monis, P.T., Grooby, W.L., Andrews, S., and Saint, C.P. (2005). Culture-independent techniques for rapid detection of bacteria associated with loss of chloramine residual in a drinking water system. Appl. Environ. Microbiol. *71*, 6479–6488.

Hugenholtz, P., and Tyson, G.W. (2008). Microbiology: metagenomics. Nature *455*, 481–483.

Ibáñez-Peral, R., Bergquist, P.L., Walter, M., Gibbs, M., Goldys, E.M., and Ferrari, B. (2008). Potential use of quantum dots in flow cytometry. Int. J. Mol. Sci. *9*, 2622–2638.

Jen, C.J., Chou, C.-H., Hsu, P.-C., Yu, S.-J., Chen, W.-E., Lay, J.-J., Huang, C.-C., and Wen, F.-S. (2007). Flow-FISH analysis and isolation of clostridial strains in an anaerobic semi-solid bio-hydrogen producing system by hydrogenase gene target. Appl Microbiol. Biotechnol. *74*, 1126–1134.

Jolivet-Gougeon, A., Sauvager, F., Bonnaure-Malle, M., Colwell, R.R., and Cormier, M. (2006). Virulence of viable but nonculturable *S. typhimurium* LT2 after peracetic acid treatment. Int. J. Food Microbiol. *112*, 147–152.

Kalyuzhnaya, M.G., Lidstrom, M.E., and Chistoserdova, L. (2008). Real-time detection of actively metabolizing microbes by redox sensing as applied to methylotroph populations in Lake Washington. ISME J. *2*, 696–706.

Keller, M., and Zengler, K. (2004). Tapping into microbial diversity. Nat. Rev. Microbiol. *2*, 141–150.

Kramer, M.F., Vesey, G., Look, N.L., Herbert, B.R., Simpson-Stroot, J.M., and Lim, D.V. (2007). Development of a Cryptosporidium oocyst assay using an automated fiber optic-based biosensor. J. Biol. Eng. *1*, 3.

Kumar, A., Pandey, A.K., Singh, S.S., Shanker, R., and Dhawan, A. (2011). Cellular uptake and mutagenic potential of metal oxide nanoparticles in bacterial cells. Chemosphere *83*, 1124–1132.

Leys, N.M.E.J., Ryngaert, A., Bastiaens, L., Verstraete, W., Top, E.M., and Springael, D. (2004). Occurrence and phylogenetic diversity of *Sphingomonas* strains in soils contaminated with polycyclic aromatic hydrocarbons. Appl. Environ. Microbiol. *70*, 1944–1955.

Liu, P., Meagher, R.J., Light, Y.K., Yilmaz, S., Chakraborty, R., Arkin, A.P., Hazen, T.C., and Singh, A.K. (2011). Microfluidic fluorescence *in situ* hybridization and flow cytometry (μFlowFISH). Lab Chip *11*, 2673–2679.

Miyauchi, R., Oki, K., Aoi, Y., and Tsuneda, S. (2007). Diversity of nitrite reductase genes in "*Candidatus* Accumulibacter phosphatis"-dominated cultures enriched by flow-cytometric sorting. Appl. Environ. Microbiol. *73*, 5331–5337.

Nebe-von-Caron, G., Stephens, P.J., Hewitt, C.J., Powell, J.R., and Badley, R.A. (2000). Analysis of bacterial function by multicolour fluorescence flow cytometry and single cell sorting. J. Microbiol. Methods *42*, 97–114.

Park, H.S., Schumacher, R., and Kilbane, J.J. (2005). New method to characterize microbial diversity using flow cytometry. J. Ind. Microbiol. Biotechnol. *32*, 94–102.

Phe, M.H., Dossot, M., and Block, J.C. (2004). Chlorination effect on the fluorescence of nucleic acid staining dyes. Water Res. *38*, 3729–3737.

Phe, M.H., Dossot, M., Guilloteau, H., and Block, J.C. (2007). Highly chlorinated *Escherichia coli* cannot be stained by propidium iodide. Can. J. Microbiol. *53*, 664–670.

Porter, J., Pickup, R., and Edwards, C. (1997). Evaluation of flow cytometric methods for the detection and viability assessment of bacteria from soil. Soil Biol. Biochem. *29*, 91–100.

Reis, A., Lopes da Silva, T., Kent, C.A., Kosseva, M., Roseiroa, J.C., and Hewitt, C.J. (2005). Monitoring population dynamics of the thermophilic *Bacillus licheniformis* CCMI 1034 in batch and continuous cultures using multi-parameter flow cytometry. J. Biotech. *115*, 199–210.

Resina-Pelfort, O., Comas-Riu, J., and Vives-Rego, J. (2001). Effects of deflected droplet electrostatic cell sorting on the viability and exoproteolytic activity of bacterial cultures and marine bacterioplankton. Syst. Appl. Microbiol. *24*, 31–36.

Sarkar, S., Banerjee, R., Chanda, S., Das, P., Ganguly, S., and Pal, S. (2010). Effectiveness of inoculation with isolated *Geobacillus* strains in the thermophilic stage of vegetable waste composting. Bioresour. Technol. *101*, 2892–2895.

Singer, A.C., van der Gast, C.J., and Thompson, I.P. (2005). Perspectives and vision for strain selection in bioaugmentation. Trends Biotechnol. *23*, 74–77.

Sloan, W.T., Woodcock, S., Lunn, M., Head, I., and Curtis, T.P. (2007). Modeling taxa-abundance distributions in microbial communities using environmental sequence data. Microb. Ecol. *53*, 443–455.

Stenuit, B., Eyers, L., Schuler, L., Agathos, S.N., and George, I. (2008). Emerging high-throughput approaches to analyze bioremediation of sites contaminated with hazardous and/or recalcitrant wastes. Biotechnol Adv. *26*, 561–575.

Stepanauskas, R., and Sieracki, M.E. (2007). Matching phylogeny and metabolism in the uncultured marine bacteria, one cell at a time. Proc. Natl. Acad. Sci. U.S.A. *104*, 9052–9057.

Tapp, H., and Stotzky, G. (1997). Monitoring the insecticidal toxins from *Bacillus thuringiensis* in soil with flow cytometry. Can. J. Microbiol. *43*, 1074–1078.

Thomas, J.-C., St-Pierre, Y., Beaudet, R., and Villemur, R. (2000). Monitoring by laser-flow-cytometry of the polycyclic aromatic hydrocarbon-degrading *Sphingomonas* sp. strain 107 during biotreatment of a contaminated soil. Can. J. Microbiol. *46*, 433–440.

Ullrich, S., Karrasch, B., Hoppe, H., Jeskulke, K., and Mehrens, M. (1996). Toxic effects on bacterial metabolism of the redox dye 5-cyano-2, 3- itolyl tetrazolium chloride. Appl. Environ. Microbiol. *62*, 4587–4593.

Vives-Rego, J., Lebaron, P., and Nebe-von Caron, G. (2000). Current and future applications of flow cytometry in aquatic microbiology. FEMS Microbiol. Rev. *24*, 429–448.

Volsch, A., Nader, W.F., Geiss, H.K., Nebe, G., and Birr, C. (1990). Detection and analysis of two serotypes of ammonia-oxidizing bacteria in sewage plants by flow cytometry. Appl. Environ. Microbiol. *56*, 2430–2435.

Wang, Y., Hammes, F., Boon, N., Chami, M., and Egli, T. (2009). Isolation and characterization of low nucleic acid (LNA)-content bacteria. ISME J. *3*, 889–902.

Wang, Y., Hammes, F., De Roy, K., Verstraete, W., and Boon, N. (2010). Past, present and future applications of flow cytometry in aquatic microbiology. Trends Biotechnol. *28*, 416–424.

Woodcock, S., van der Gast, C.J., Bell, T., Lunn, M., Curtis, T.P., Head, I.M., and Sloan, W.T. (2007). Neutral assembly of bacterial communities. FEMS Microbiol. Ecol. *62*, 171–180.

Xia, S., Li, J., Wang, R., Li, J., and Zhang, Z. (2010). Tracking composition and dynamics of nitrification and denitrification microbial community in a biofilm reactor by PCR-DGGE and combining FISH with flow cytometry. Biochem. Eng. J. *49*, 370–378.

Zimmermann, S.G., Wittenwiler, M., Hollender, J., Krauss, M., Ort, C., Siegrist, H., and Gunten, U. (2011). Kinetic assessment and modeling of an ozonation step for full-scale municipal wastewater treatment: micropollutant oxidation, by-product formation and disinfection. Water Res. *45*, 605–617.

Zhao, R., Natajaran, A., and Srienc, F. (1998). A flow injection flow cytometry system for on-line monitoring of bioreactors, Biotechnol. Bioeng. *62*, 609–617.

Flow Cytometry and Microbial Community Fingerprinting

4

Susann Müller and Kathleen Schleinitz

Abstract

Flow cytometry (FCM) can be used to reveal community patterns by analysing certain cell parameters that reflect microbial community structures. The method is valuable for separating a microbial community into sub-sets of individuals with shared physiological properties, such as proliferation activity and scatter characteristics. Fingerprinting tools elucidating community composition, i.e. describing the phylogenetic affiliation of the single community members can also complement FCM. These techniques are introduced with a short overview. Their applications for microbial community composition and structure analyses will be compared. The chapter then describes how fingerprinting techniques can be applied to characterize the FCM-separated subsets. This chapter does not provide detailed protocols for applying FCM to microbial communities, as these protocols will vary depending on the community and physiological parameters of interest. Instead, this chapter highlights the general methodology and approach and discusses potential future advances in the field.

What are natural communities?

Microorganisms comprise the highest amount of biomass and the highest genetic information on earth. They inhabit all spheres such as water or sediment, the subsurface or the open surface. They can also be found in air, or living on or within other organisms such as plants, animals and human beings. In nature they do not occur as pure cultures but always as mixed communities. Their preferred life style is as a biofilm because such a conglomerate ensures sharing of metabolites and protection against predation and abiotic stress factors such as weather pattern changes or toxic assaults.

Microorganisms have been used by mankind for thousands of years in the production of food. Only within the last 100 years have people started to use pure cultures to generate processed nutrients as, for example, in case of beer, yoghurt or cheese. In addition, white biotechnology (that is the chemical and pharmaceutical industry) relies on pure culture exploration. Research in biology and medical sciences also depends on pure cultures due to necessary knowledge gained by studies on metabolic pathways, carbon fluxes and their regulation.

However, in, for example, textile wastewater cleaning processes, certain microbes are used that act as specific degraders of distinct dyes (Novotny *et al.*, 2011). Additionally, biotechnologists plan to use artificially mixed cultures to facilitate complicated chemical reactions. It is quite common to engage known mixed cultures in the nutritional industry – e.g. for the production of yoghurt or wine (Breeuwer and Abee, 2000). However, despite their widespread commercial use, the nature of relationships between community members is frequently not understood.

This lack of knowledge is even more important for natural communities. Natural communities are part of the human body, are the main drivers of wastewater treatment plants or key players in detoxification processes of contaminated landscapes. Therefore, they are responsible for many processes deeply connected to human life. Yet, knowledge on community structure, composition, dynamics and functioning is often scarce. A key reason for this is the high degree of microbial complexity, since communities consist of thousands of different organisms. Also, intra-community dynamics are generally unknown, since communities quickly react to varying abiotic conditions and the generation time of certain members can be very short. Therefore, the composition of a community can change within minutes. A second reason is that most of the entire organisms within a community cannot be cultivated. As a result their phylogenetic affiliations, their metabolism and their main regulation pathways remain unknown.

Owing to such severe limitations there is no *modus operandi* available enabling the generation of enhanced knowledge on how these communities work. However, in the past 10 years a number of techniques have emerged which may provide feasible solutions towards this problem.

In this chapter we concentrate on two aspects: how to obtain information on the composition of a natural community by using phylogenetic tools and how to obtain information on structure of a community by using single cell tools.

Community structure analysis using flow cytometry

To analyse microbial communities at the single cell level they need to be extracted from the environment. This is simple in case of taking samples from groundwater and even simpler when wastewater treatment plants are the focus of research. The samples just require to be stored in clean sterile plastic tubes. Since cells adhere easily to plastic surfaces and may not be removed during the washing steps glass vials can be a better solution for sample storage. If soils or sediments are sampled these cells are often attached to sand or stone surfaces or captured in capillaries of the investigated material. For single cell analytics the detachment of cells is a precondition and can prove to be a challenging undertaking. As many cells as possible should be removed and they should be in a healthy state after performing this procedure, meaning that their cell membrane and cell wall should at least be intact. The procedures for detachment vary and should be tested for every single micro-environmental system. The procedures include gentle shaking (manual or on a shaker at slow rpm values) in phosphate-buffered saline (PBS) or pyrophosphate buffer up to long term treatment using sonication. An optimal procedure will try to avoid destroying cells (by too harsh treatment) or losing cells (undetached cells). To check the number of cells in solution after treatment, a classic protein determination can be undertaken. Another possibility is to use the inorganic material under investigation, add distinct cell numbers of a live pure strain known to be able to survive on this material, incubate it and afterwards, perform the detachment procedure under consideration, and estimate how many cells are finally recovered. Some workers have used a Nycodenz – gradient to separate soil material from the cells. The washing and isolation steps often require centrifugation. Therefore it is generally advisable to work at low centrifugation values (like 3000 *g* for 5 minutes) since especially groundwater and sediment organisms can be very fragile in their structure and disrupt at higher velocities. On the other hand the same habitats often comprise of very small bacteria which do not sediment at these velocities and must be centrifuged at higher speed and/or for longer durations. All these methods have disadvantages but should be tested to obtain as many cells as possible from the respective micro-environment.

When the cells are extracted they need to be stabilized or fixed for further investigation. Usually living cells are not investigated by flow cytometry since, until now, no technique exists which provides reliable information on overall viability when investigating communities with unknown members. The classical fixation procedures rely on para-formaldehyde, glutaraldehyde or ethanol. All of them increase autofluorescence and agglutination of the cells. Para-formaldehyde stabilizes the membrane by linking the surface proteins whereas ethanol extracts lipids from the membrane. Fixatives can also be combined in their application. A good approach is to first treat cells with para-formaldehyde for only 30 min to stabilize the cell membrane and afterward treat with 70% ethanol to make them susceptible to larger probes used afterwards. Such cells can be stored at −20°C for an extended period (depending on community for either days or weeks) if the cell number per ml is sufficiently low (not exceeding 10^9 cells per ml, Wallner *et al.*, 1993). All procedures need a certain time to act which should considered as around one hour.

Much quicker is the use of sodium azide at high concentrations (10%) since it immediately blocks the energy flux due to inhibition of the respiratory chain and prevents aggregation. The disadvantage is that this method can only be applied to aerobic systems. Therefore, heavy metal ions can be employed for this method and also for communities extracted from micro-aerobic or anaerobic environments. Under sulfate-reducing conditions molybdenum can be applied (Vogt *et al.*, 2005), whereas barium and nickel were successfully applied in wastewater treatment plants (Günther *et al.*, 2008). Barium, however, can cause problems when the nucleic acid content of a cell is to be investigated after fixation, since barium is thought to disturb RNA structure (Stöhr and Petrova, 1975). Freezing generally leads to cell membrane and cell wall disruption and is therefore useful for nucleic acid isolation but limited in its application for single cell analytics.

The fixed or conserved communities can now be further investigated for their members. This chapter concentrates on phylogeny and will therefore not discuss techniques to enable information on cell function on a single cell level to be obtained.

When aiming at phylogenetic classification of community members, there are two main approaches to specifically stain cells with fluorescent probes. These are application of oligonucleotide probes or antibodies. When oligonucleotide probes are used an rRNA sequence needs to be available at least. Many sequences are already present in data bases which allow the possibility to construct probes very easily. These probes are oligomers of between 15 and 30 bases and can be labelled by attaching one or two fluorescent dyes, the latter preferably at the ends of the probe. The labelling procedure has given the technique its name: fluorescence *in situ* hybridization or FISH and many fluorescent dyes are available for this purpose. Some of them show a high quantum yield like cyanine dye Cy3, others are specifically developed for laser excitation such as the Alexa or Bodipy dye groups. The labelled oligomers specifically bind to the respective 16S rRNA sequences of the target organisms. However, several drawbacks may occur. In nature, the rRNA content of an organism can be low if optimal growth conditions are not present. Since at least one thousand of the target rRNA molecules need to be attached to be visible under a microscope, or by laser interrogation, many organisms simply cannot be detected since the fluorescence intensity is too low. Secondly, for optimal annealing of the probe, optimal conditions need to be determined. That is impossible for natural communities since the annealing conditions will be different for each different species. Thirdly, the secondary structure of the rRNA inhibits an equally high-affinity binding to every sequence section. This is well described by Fuchs *et al.* (1998) for *E. coli*, which can be seen as only one representative member within a microbial community.

To circumvent some of the problems previously described, probes were developed for amplifying fluorescent marker characteristics relying on, for example, catalysed reporter deposition (CARD–FISH) or tyramide signal amplification (TSA–FISH). The first method uses oligomer-bound enzymes to convert fluorogenic into fluorescent dyes which accumulate at the oligomer site. The signal is not quantitative but generates high signal intensity. TSA–FISH relies on the reaction of highly reactive fluorescent tyramide radicals with each other after their activation by enzymes such as horseradish peroxidase (HRP). An insoluble precipitate in the vicinity of the HRP location is formed thus increasing fluorescence intensity even further. The major problem for the combination using flow cytometry analysis is that the introduction of large molecules like HRP into cells requires disruption of cell wall and cell membrane. This presents no problem for microscopic applications since disrupted cells are still located on the same position on a microscopic slide. However, flow cytometry requires a strong trigger signal to distinguish cells from debris. This is especially true when environmental samples are to be investigated. Therefore, only a few protocols are available which demonstrate CARD–FISH labelling for flow cytometric application in bacterial cells (Sekar *et al.*, 2004). Apart from

FISH-technologies, labelled antibodies or lectins can also be used to follow abundances of certain species. Both molecule types can only be used when directed to the surface of the microorganisms. Otherwise the same problem as described above will hinder any quantitative measurements. Lectins bind relatively non-specifically to sugar moieties, therefore they can be used to differentiate Gram-positive from Gram-negative cells. To specifically target cells with antibodies, particular antibodies need to be readily available. Since most of the cells in natural communities are not cultivable it is difficult and/or expensive to produce specific antibody molecules against them. Therefore, antibodies are often used to detect specific cell types such as known pathogens. Antibodies and lectins can be easily fluorescently labelled with high intensity probes like the above mentioned HRP. Several amplification techniques are available for this purpose. In cases where several antibodies or lectins are applied at the same time to a sample, one needs to be aware that cross reactions are quite common. Precise control strategies are therefore a precondition for reliable determination of target cell types.

Using FISH, antibodies or lectins enables the determination of certain cell abundances within microbial communities. However, these techniques do not allow a study of the dynamics of the whole community. This can only be done if dyes are used which stain every cell. Several dye classes are available for this purpose. Lipophilic dyes such as Nile red label every cell but they are not very stable in their binding behaviour: due to their highly lipophilic features they quickly form crystals on the outside of cells. Protein dyes often have a high quantum yield but the protein contents of bacteria are not disparate enough to use this feature as a community characteristic. Therefore, at the moment, the nucleic acid content is the most interesting cell parameter to target. Nucleic acids can be stained using SYBR or SYTO dyes. In the past, several groups used them to create plots with so called high nucleic acid (HNA) and low nucleic acid (LNA) content bacterial subcommunities. HNAs were related to highly active bacteria although the LNA were also shown to comprise actively growing bacteria (Wang *et al.*, 2009). Another dye which is a promising probe is 4′,6-diamidino 2-phenylindole (DAPI) which binds solely to AT-rich regions of the DNA. Since bacteria may contain several chromosomes of different length and information, as well as several copies of them, DAPI staining gives a surprisingly clear discriminative signal. The only disadvantage of this dye is that a UV-laser needs to be used as excitation source. In wastewater samples DAPI was shown to discriminate up to 20 sub-communities (Günther *et al.*, 2009). In this way, natural community behaviour can be studied by measuring DNA content and forward scatter behaviour over time. In many applications, the origin of, and the dynamics within, a microbial community may be recognizable due to its specific dot plot pattern.

The idea has developed to use these patterns as a kind of a fingerprint and to denote it as 'community structure' as opposed to 'community composition' which refers to the phylogenetic classification (see 'Community composition analysis using fingerprinting techniques', below). Upcoming subclusters of cells can be gated using cytometric software and abundances of cells in gates can be determined. The following information can be determined: numbers of clusters, positions of clusters in dot plots and abundances of cells in clusters. This data set is informative enough to recognize a specific community and to monitor changes in cell abundances within certain gates.

Without knowing the phylogenetic affiliation (and therefore the potential metabolic capacities these organisms bear in relation to their genetic information) flow cytometry measurements can give quantitative information on the dynamics within a biological system. The cells can be seen as biosensors since subclusters may vanish or appear in reaction to certain micro-environmental conditions (Günther *et al.*, 2012). In the following paragraph an example is given on how information on natural community structure may be obtained using single cell analytics.

In this example, flow cytometry and fingerprinting techniques were used to follow the dynamics of an m-xylene degrading enrichment culture over a time period of approximately 2000 hours. The community investigated was obtained from a contaminated groundwater aquifer and was grown statically, under strictly anaerobic conditions, in the laboratory as an enrichment culture. From these enrichments, the cultures investigated were set up in serum bottles (56 ml, 50 parallels) and grown at 12°C, which was the temperature of the original aquifer. The substrate m-xylene was supplied as the sole source of carbon and energy. Xylene belongs to the BTEX (benzene, toluene, ethylbenzene, xylene) group of contaminants. In the aquifers around Leipzig/ Germany, BTEX-compounds are priority pollutants introduced by accidental spillage from the chemical industry decades ago. The concentration of benzene for example can be as high as 1 g/ml, whereas xylene concentrations are usually much lower. As m-xylene is toxic and has a low water solubility, a 1% (v/v) stock solution in the organic carrier phase 2,2,4,4,6,8,8-heptamethylnonane (HMN) was prepared. The medium of each cultivation bottle was overlaid with 1.5 ml of this stock solution, to ensure a constant supply of low concentrations of the substrate. The flasks were inoculated with 10 ml of 3-week-old inocula. In the experiment, sulfate was used as an electron acceptor to copy the anaerobic conditions of the natural aquifer. Control flasks without m-xylene were prepared. The control flasks were also inoculated with the enrichment culture and harvested at the same times as the m-xylene grown community.

For cytometric analysis cells were harvested and fixed in 0.1% sodium chloride containing 3.8 mM sodium azide, 15 mM sodium molybdate and 100 mM hydrogen peroxide and stored at 4°C. Since the performance of the whole community rather than specific community members was the focus of the experiment all cells were stained with DAPI to follow dynamics over an extended time period. For staining the cells were centrifuged at 10,300 *g* for 30 minutes at 4°C. The resulting cells were adjusted to an optical density (O.D.) of 0.04 (d = 0.5mm, 700 nm). The cells were treated with 2 ml stock A (2.1 g citric acid, 0.5 g Tween 20 in 100 ml distilled water) for 10 minutes. They were re-centrifuged and stained with 0.68 μM DAPI in 0.4 M disodium hydrogen phosphate at pH 7 for 60 minutes in the dark.

A MoFlo cell sorter was used to analyse the cells. Forward and side scatter were measured using the 488 nm line of an argon-ion laser, side scatter was used as trigger. DAPI was measured using an argon-ion MLUV-laser (Coherent, USA). Alignment was performed using blue and green fluorescent beads of 1 and 2 μm, respectively. Also, an internal biological standard was used to ensure correct staining. The dot plots were evaluated using the SUMMIT V3.1 Software (DakoCytomation, USA). Using these parameters information was obtained on DNA contents of cells (blue fluorescence), scatter related cell size (forward scatter: FSC) and cell density (side scatter: SSC).

Additionally, cell number was measured by inclusion of a defined number of fluorescent beads as well as protein concentration and sulfide production as bulk parameters. All methods are fully described by Vogt *et al.* (2005).

In these experiments, sulfide production served as a marker for sulfate reduction and it increased to values over 1 mM sulfide per litre medium within 2000 hours of cultivation. The control, without carbon or energy source, showed no increase in sulfide concentration. In contrast, there were nearly parallel increases in protein contents and cell numbers in the substrate-containing flasks as well as in the controls without substrate. The protein content varied over time but did not exceed double values and ended up with the same quantities as the inoculum after 2000 hours cultivation. For cell number a doubling in population was shown. These data indicated (slow) growth, independent of the m-xylene supply, in both m-xylene and control cultures under sulfate reducing conditions. The community dynamics are presented in Fig. 4.1. Over an extended time greater than 2000 hours several cultivation flasks were harvested and analysed by flow cytometry. Fig. 4.1a presents a number of dot plots of the community grown on m-xylene. Alignment beads are shown for comparison (circle at 0 hours). During the first 200 hours the structure of the community remained nearly constant. The two main dominating clusters are marked within squares at 216 hours (Fig. 4.1a). Since both of the clusters always occurred as double sub-sets it can be assumed that the cells belonged to the same phylotypes and underwent cell cycling. In the following time beyond 216 hours, however, the community structure showed varying patterns.

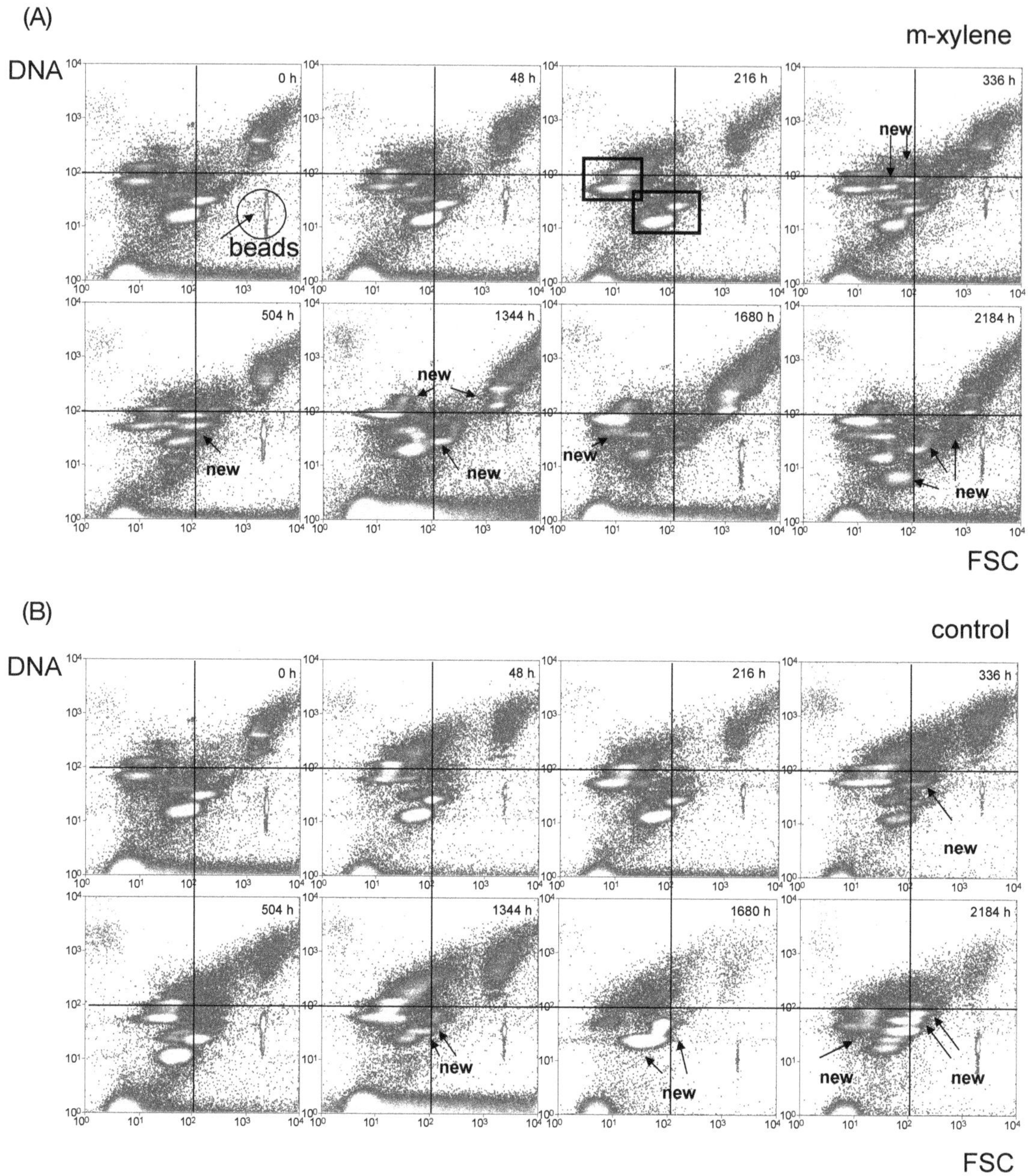

Figure 4.1 Community pattern of samples harvested from enrichment cultures grown on m-xylene (A), and without any carbon and energy source, (B). The x-axis represents the forward scatter which is related to cell size and the y-axis represents the DNA content of single cells. Beads (circle) were used to align the measurements to each other. Squares are used to mark dominant sub-communities. Arrows highlight newly emerging sub-communities.

New subcommunities appeared whereas others disappeared. The arrows show a few obvious examples for newly emerging clusters. The behaviour of this community was surprising in the first place. The m-xylene is provided steadily via the HMN layer on the top of the medium. Therefore it should be expected that the community pattern remains constant from the start for the whole period of cultivation. The appearance of new subclusters may be explained by a variation in the micro-environments. Owing to sulfate reduction an increase in the sulfide concentration was noted during the cultivation time and may have influenced conditions in the closed flask systems and may also explain why other organisms were increasing in their abundances. Fig. 4.2b presents the control community cultivated without any substrate or energy source. It was interesting to observe that this community also changed its structure, though in a different way compared with the community grown on m-xylene. A relatively high similarity in the structure was observed during the first 500 hours. A major change was visible after 1344 hours which became even stronger after 1680 and 2184 hours of cultivation. Huge subclusters of cells disappeared and a completely new community structure evolved. The variation in community structure in the control may be related to fermenting activities. Since not only the whole protein content (from 7 μg/ml in the inoculum to about 16 μg/ml after 1500 hours, declining to the original 7 μg/ml level at the end of cultivation) but also the cell number increased (from about 2×10^5 to about 5×10^5 cells/ml in maximum over 2000 hours, which was similar to the m-xylene grown community) the changes in the structure of the community may be explained by organisms which can live on each other's metabolic intermediates and on cell debris. A general shift was observed from cells in clusters with high scatter and high DNA contents to those characterized by lower FSC and DNA fluorescence.

It should be mentioned that during analysis of natural communities abiotic particles can also be determined by flow cytometry. Particles may originate from soil, dust or water constituents but may also emerge during the cytometric measurement itself as is not unusual for samples to contain sulfide. Fig. 4.2 shows a community harvested after 1680 hours grown on m-xylene with sulfate as the electron acceptor. The scatter plot shows sulfur granules (in yellow, FSC vs. SSC) which emerged after sulfide reacted with oxygen during cytometric measurement. The phenomenon is not unknown and has already been described by Hübschmann *et al.* (2007). The fluorescence plots showed the sulfur granules as unstained events (DAPI vs. SSC and DAPI vs. FSC) which precipitated with the cells due to high centrifugation velocities used for the sample preparation. Therefore, when investigating samples isolated from natural environments the researcher needs to be aware of abiotic particles in the sample which may also form distinct clusters in the scatter plots. To differentiate those non-biotic clusters from the biotic ones, a staining step should be used to provide clarification.

The advantage of cytometric community structure analysis is that every cell in a community is detected and that dynamics can be measured

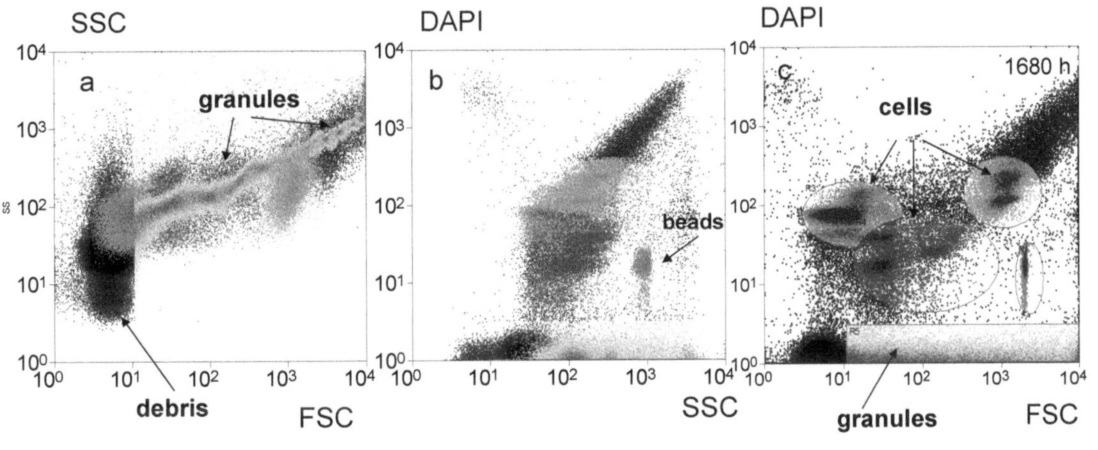

Figure 4.2 Community pattern of one sample shown after staining with DAPI. Scatter characteristics (a), DNA versus side scatter (b) and DNA versus forward scatter (c) distributions are shown. Sulfur granules developed when keeping the sample in air before sample treatment (seen in (a)). The community structure is best presented in (c).

quickly every few minutes with low laboratory inputs. The methods are easy to perform and evaluation can be undertaken without much effort. It is a low cost approach as long as a suitable cytometer is at hand which can analyse DAPI/ scatter characteristics of a community.

Community composition analysis using fingerprinting techniques

T-RFLP (*terminal restriction fragment length polymorphism*) analysis has been found to be very useful in the analysis of microbial communities from a wide range of habitats, freshwater and marine, sediment, soil and sewage sludge. It involves DNA extraction and quantification, PCR amplification, enzymatic restriction of PCR products and separation of the resulting fragments by capillary electrophoresis. These steps will be briefly described later. As other fingerprinting techniques such as denaturing gradient gel electrophoresis (DGGE), temperature gradient gel electrophoresis (TGGE), single-strand conformation polymorphism (SSCP), amplified rDNA restriction analysis (ARDRA) and amplified ribosomal intergenic spacer analysis (ARISA) are also PCR based, problems and limitations connected to DNA extraction and PCR amplification are as relevant for these techniques as they are for T-RFLP.

Similar to flow cytometry, extraction of the information used for analysis – in this case DNA – is crucial and must be optimized for each sample type. DNA extraction from environmental samples and complex microbial communities can be hampered by a number of obstacles which include DNA yield, molecular size and PCR amplifiability.

Insufficient cell lysis is one problem that frequently concerns certain groups of microorganisms (such as Gram-positives) and can lead to distorted results when these groups are excluded from further analysis. Numerous cell lysis protocols exist that apply chemical extractants [e.g. sodium dodecyl sulfate (SDS), chloroform, phenol, guanidinium isothiocyanate], physical lysis techniques (sonication, bead beating, freeze–thaw cycles, microwave), enzymatic lysis (e.g. proteinase K, lysozyme), or combinations thereof. Most of these agents and procedures are also used in commercially available DNA extraction kits which are sometimes specialized in the extraction of DNA from a certain habitat (e.g. soil). Insufficient cell lysis will result in the exclusion of (certain groups of) microorganisms. On the other hand, too harsh a lysis, can result in highly fragmented DNA that gives a suboptimal template for PCR amplification.

Another problem concerns the co-extraction of PCR inhibiting substances alongside the target DNA. The co-extraction of humic acids, which in their physico-chemical characteristics closely resemble DNA is a problem inherent in DNA extractions from soil, and has been addressed in a number of specialized commercial extraction kits. Co-elution of divalent cations that severely inhibit PCR amplification can be prevented by washing of the sample or chelating these ions prior to cell lysis.

Low or varying DNA yield can be a problem for T-RFLP analysis as the same amount of DNA has to be added to each PCR reaction in order to allow for quantitative analysis and comparison of different samples. In the case of water samples, cells can be enriched by filtering prior to DNA extraction. However, this is not an option for sediment samples. There, DNA from parallel extractions can be pooled and concentrated by precipitation.

After extraction DNA has to be quantified. This can be done using electrophoresis which compares the concentration of the extracted DNA with a standard of known concentrations or by staining with fluorescent dyes. Determination of absorbance at 260 nm is also a widely used method. However, it has major disadvantages including: detection of single-stranded DNA and nucleotides alongside double-stranded DNA, inference by contaminants often found in DNA extracts from environmental samples, inability to distinguish DNA from RNA or an inability to reliably detect small amounts of DNA. Therefore this method is not recommended for use in T-RFLP analyses. Generally, quantification methods, when compared to each other, can give differing results. Therefore it is of importance to use one and the same method throughout the experiment.

In PCRs aiming at the elucidation of community composition the 16S rRNA gene is the gene most frequently targeted, when aiming at describing the metabolic potential functional marker genes chosen. PCR amplification is known to be a major source of bias in T-RFLP profiles (Hartmann and Widmer, 2008, and literature therein). The bias is caused by preferential, inhibited or unspecific amplification, generation of chimeric sequences, template re-annealing, formation of heteroduplexes and nucleotide misincorporation. Careful optimization of PCR conditions will reduce bias. As for all PCRs, a no template control is of utmost importance to detect DNA contaminations of reagents.

Enzymatic restriction is another major source of bias in T-RFLP profiles (Hartmann and Widmer, 2008), due to residual polymerase activity, partially single stranded fragments giving rise to artefact peaks, and incomplete or unspecific enzymatic restrictions. The latter can largely be circumvented by adding sufficient enzyme and optimizing restriction conditions. Each enzyme should resolve community composition as well as possible. Highly complex communities cannot be resolved using T-RFLP, and other techniques such as amplicon sequencing (see following passage) should be considered to characterize community composition in such habitats.

Detection of T-RF (terminal restriction fragment) abundances during capillary electrophoresis can be compromised by reduced efficiency in injection and reduced peak heights due to broader peaks, which occur at larger T-RF sizes. When identifying phylotypes on the basis of their T-RFs it should be kept in mind that detected T-RFs lengths can differ by up to 4 bp (−4 or +4) from the theoretical T-RFs lengths. Considering the high numbers of 16S rRNA sequences in the databases it is virtually impossible to determine community composition merely by comparison of T-RF lengths with the theoretical T-RF lengths of sequences in databases. Therefore, a clone library or clone libraries should be constructed from the samples analysed. Screening (e.g. by ARDRA [amplified rDNA restriction analysis] or T-RFLP) coupled with subsequent sequencing of members of each clade gives a qualitative overview of community compositions. T-RF lengths for each phylotype can be determined by performing T-RFLP on single clones. This approach also allows for testing of the suitability of restriction enzymes for use in T-RFLP. Statistical analysis of T-RFLP data are carried out using either peak areas or peak heights. Normalization prior to analysis should be performed.

Generally T-RFLP is well suited to the analysis of microbial communities displaying low or medium complexity. The more complex communities become, the more often different, unrelated members will have T-RFs of the same length, making it impossible to resolve a significant proportion of the community. Although the biases introduced during the single steps of the technique preclude detection of community compositions with high resolution, T-RFLP profiles have been shown to reliably reflect relative changes in community composition and diversity, e.g. changes over time, or under different environmental conditions. A major advantage of T-RFLP over gel based fingerprinting techniques like DGGE or TGGE lies in its suitability for high throughput analysis.

Besides T-RFLP, DGGE and TGGE have been widely used in microbial ecology (for a review see Muyzer, 1999), as a consequence their potential and pitfalls are well documented. Both provide a reasonable means to investigate microbial communities, especially with regard to strong shifts in community composition, thereby resembling T-RFLP. The main pitfall lies in the reproducibility between gels. Standardized procedures for gel preparation can reduce, but not eliminate this problem. The accuracy of comparison is heavily dependent on the use of good internal standards. As DGGE and TGGE separate fragments of the same length, but different sequence, the use of degenerate primers is limited as their use can result in multiple bands leading to an overestimation of diversity. On the other hand, even the most sensitive staining methods are unable to visualize the whole diversity in a sample, especially with regard to phylotypes of low abundance. Another problem is related to fragment length: shorter fragments that usually result in better resolution contain less sequence information.

For the analysis of highly diverse communities the resolution of the above mentioned fingerprinting techniques is insufficient. Therefore, the use of next generation sequencing techniques are the methods of choice. Pyrosequencing allows the massive parallel sequencing of 16S rDNA amplicons, resulting in a significantly higher sampling depth and provides orders of magnitude greater sequence information than traditional techniques (Sogin *et al.*, 2006). However, similar to other PCR-based approaches, the choice of primers and PCR conditions can severely affect the result, and sequences obtained should undergo a rigid quality assessment prior to being used for further analysis (Engelbrektson *et al.*, 2010; Kunin *et al.* 2010).

As an example, the same aquifer community is investigated as in the section on 'Community structure analysis using flow cytometry'. Every T-RFLP-pattern shown in Fig. 4.3 was obtained from one serum bottle (see page 81). T-RFLP patterns indicated that two abundant phylotypes were detected throughout the course of the experiment. The dominating one was affiliated to the Epsilonproteobacteria. As this class of proteobacteria is not known to be involved in the degradation of contaminants this phylotype is presumed not to be directly involved in the breakdown of m-xylene, but rather to metabolize a central intermediate like acetate. The second most abundant T-RF belonged to the Deltaproteobacteria and is the presumed m-xylene degrader. Apart from that a number of less abundant T-RFs were detected that could not be identified on the basis of the clone library constructed from the consortium. This could be caused by two reasons: the clone library (96 clones) was too small to cover all of the community members or some of the T-RFs detected may be artefacts from PCR or restriction errors such as those described above.

Nevertheless it can be seen that the presumed degrader was clearly more abundant in the culture supplied with m-xylene (Fig. 4.3a) compared with the cultures lacking a carbon and energy source (Fig. 4.3b). Furthermore, it can be observed that some changes occurred in the composition of the less abundant community members. In the m-xylene degrading culture the OTU (operational taxonomic unit) producing a T-RF of 159.65 bp was mainly present during the first half of the experiment. On the contrary, another OTU with a T-RF of 86.21 bp only emerged after 1344 hours.

When comparing the cytometric community structure (Fig. 4.1) with the T-RFLP patterns (= community composition, Fig. 4.3) it is obvious that both data sets showed the domination of two main organisms or two main subclusters (structure information given in Fig. 4.1a, 216 hours). The most dramatic change in community structure (control, 1680 hours, Fig. 4.1b) was also found for the composition using the T-RFLP fingerprint (Fig. 4.3b). Such data can be related to each other by using n-MDS similarity analyses which provide additional insights into community development trends (Bombach *et al.*, 2011).

The application of a fingerprinting technique such as T-RFLP is able to provide information on potential metabolic capacities but is more elaborate and cost intensive in comparison to cytometry. It gives a good resolution of community composition; however, identification of community members is also dependent on the size of the clone library. T-RFLP can only be applied to follow community composition via selective sampling (and gives something like a snapshot of the community) and is therefore not advantageous to follow quick dynamic changes.

Combination of community fingerprinting and single cell analysis by cell sorting

The two approaches outlined in this chapter are very helpful in understanding natural community behaviour. Cytometric and phylogenetic techniques were combined by our group within several approaches (for an overview see Müller and Nebe-von-Caron, 2010). Using these techniques, enrichment cultures from contaminated soil, bacteria grown on carbon beads in ground water aquifer wells and wastewater communities were characterized. Even a pure culture on an anode emerging out of a natural community within a microbial fuel cell was phylogenetically affiliated after cytometric measurement and application of a sequencing approach (Harnisch *et al.*, 2011).

However, the advantage with these techniques is even greater when combined with cell sorting. Certain cells in certain subclusters can be gated and sorted, afterwards being phylogenetically investigated by using T-RFLP for example. This enables quantification of cell numbers in gates

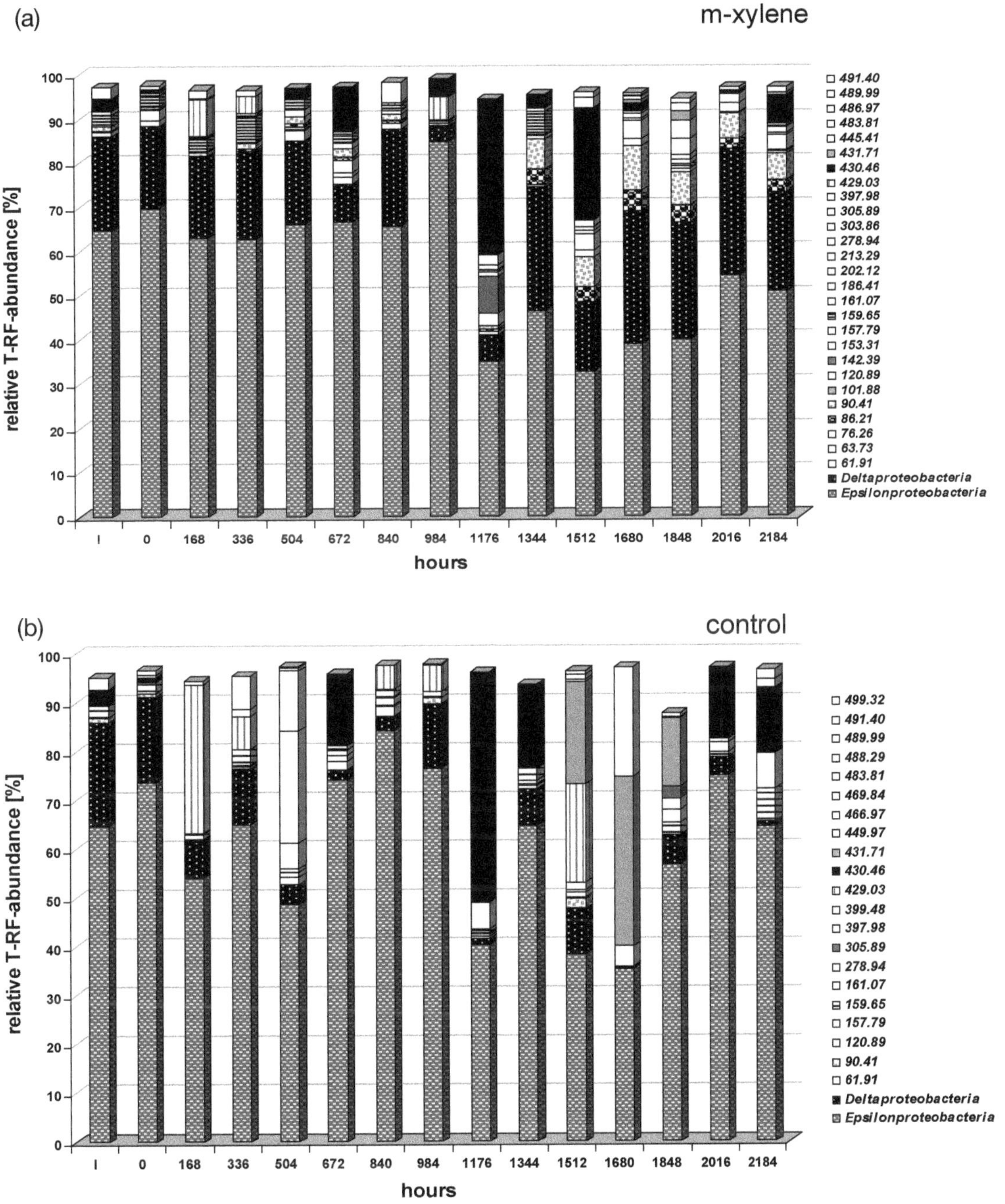

Figure 4.3 T-RFLP pattern of samples harvested from enrichment cultures grown on m-xylene (a) and without any carbon and energy source (b, control). The highly abundant microorganisms and T-RFs were denoted with various markings. Low abundant T-RFs were all marked in white.

with a certain phylogenetic affiliation. The main advantage is that quantification is combined with metabolic capacity information. These approaches are described by (Kleinsteuber *et al.*, 2006; Günther *et al.*, 2009; Bombach *et al.*, 2011; Günther *et al.*, 2012).

If a cell sorting step is involved in community analysis, several points need to be regarded. At least 10^5 cells should be sorted to obtain sufficient DNA for subsequent phylogenetic analysis. Firstly, sort gates should be chosen for highly abundant subclusters. Clusters with low cell abundances demand a longer sorting time depending on the speed of the respective cell sorter. Sorting speed (about 1000 cells/s) should be as low as possible to avoid coincidences. The machine needs to be cleaned and rinsed for several hours. Before starting the sorting process, tubes and reagents must be free from contaminating cells or DNA which is done by performing an elaborate washing step or using one-way materials. The sample needs to be centrifuged immediately and the cells can then be frozen at minus 20°C.

When certain clusters dominate a community their organisms' phylogenetic affiliation may provide a hint as to its metabolic capacities and thus also information on the main role within the community. Also, low abundant organisms may come into focus since after sorting they can clearly be described. In a whole community approach, percentages below 1% are often not detected by common molecular biology tools like T-RFLP. Next generation sequencing techniques are able to find these rare organisms but these are, to date, too expensive and still require extensive time for evaluation and interpretation. Nevertheless, a sorting step lowers the number of data which need to be interpreted. Efforts spent on data evaluation decreases considerably in such a case. Therefore, cytometric community structure analyses combined with cell sorting and common phylogenetic tools like T-RFLP or new ones like next generation sequencing approaches are future strategies to resolve natural communities and to obtain information on their dynamics and potential capacities. It should be pointed out that if the phylogenetic affiliations of cells in clusters are known, these do not need to be determined regularly. If the cytometric analyses are aligned the respective sub-set of cells will appear at the same position in the dot plot. Therefore, most of the analyses can be undertaken using cytometric measurements only. This enables potential applications in process monitoring or even process prediction since the organisms in the sub-sets serve as biosensors for the state of an industrial process such as wastewater. Also developments in natural environments like a contaminated aquifer can be quickly followed without an immediate necessity to include phylogenetic techniques. Therefore, community fingerprinting techniques and cell sorting will only come in place if the biological system changes, e.g. by intrinsic (cell related) or extrinsic parameters (abiotic conditions) which are easily indicated by changes in community structure as measured by flow cytometry.

Acknowledgements

We would like in particular to thank Thomas Hübschmann, who analysed the shown samples on the MoFlo and Helga Engewald for cytometric sample handling. Furthermore, we want to thank Christin Dathe and Ute Lohse for cultivation and performing the T-RFLP analysis of the community samples. The research programmes SAFIRA II and CITE are acknowledged for their financial support.

References

Bombach, P., Hübschmann, T., Fetzer, I., Kleinsteuber, S., Geyer, R., Harms, H., and Müller, S. (2011). Resolution of natural microbial community dynamics by community fingerprinting, flow cytometry and trend interpretation analysis. Adv. Biochem. Engin. Biotechnol. *124*, 151–181.

Breeuwer, P., and Abee, T. (2000). Assessment of viability of microorganisms employing fluorescence techniques. Int. J. Food Microbiol. *55*(1–3), 193–200.

Engelbrektson, A., Kunin, V., Wrighton, K.C., Zvenigorodsky, N., Chen, F., Ochman, H., and Hugenholtz, P. (2010). Experimental factors affecting PCR-based estimates of microbial species richness and evenness. ISME J. *4*, 642–647.

Fuchs, B.M., Wallner, G., Beisker, W., Schwippl, I., Ludwig, W., and Amann, R. (1998). Flow cytometric analysis of the *in situ* accessibility of *Escherichia coli* 16S rRNA for fluorescently labeled oligonucleotide probes. Appl. Environm. Microbiol. *64*, 4973–4982.

Günther, S., Hübschmann, T., Rudolf, M., Eschenhagen, M., Röske, I., Harms, H., and Müller, S. (2008). Fixation procedures for flow cytometric analysis of environmental bacteria. J. Microbiol. Methods. *75*, 127–134.

Günther, S., Trutnau, M., Kleinsteuber, S., Hause, G., Bley, T., Röske, I., Harms, H., and Müller, S. (2009). Dynamics of polyphosphate-accumulating bacteria in wastewater treatment plant microbial communities detected via DAPI (4′,6′-diamidino-2-phenylindole) and tetracycline labeling. Appl. Environ. Microbiol. *75*, 2111–2121.

Günther, S., Koch, C., Hübschmann, T., Röske, I., Müller, R.A., Bley, T., Harms, H., and Müller, S. (2012). Correlation of community dynamics and process parameters as a tool for the prediction of the stability of wastewater treatment. Environ. Sci. Technol. *46*, 84–92.

Harnisch, F., Koch, C., Patil, S.A., Hübschmann, T., Müller, S., and Schröder, U. (2011). Revealing the electrochemically driven selection in natural community derived microbial biofilms using flow-cytometry. Energy Environ. Sci. *4*, 1265–1267.

Hartmann, M., and Widmer, F. (2007). Reliability for detecting composition and changes of microbial communities by T-RFLP genetic profiling, FEMS Microbiol. Ecol. *63*, 249–260.

Hübschmann, T., Vogt, C., Till, S., Rohwerder, T., Sand, W., Harms, H., and Müller, S. (2007). Detection of sulfur microparticles in bacterial cultures by flow cytometry. Eng. Life Sci. *7*, 403–407.

Kleinsteuber, S., Riis, V., Fetzer, I., Harms, H., and Müller, S. (2006). Population dynamics of a microbial consortium during growth on diesel fuel in saline environments. Appl. Environ. Microbiol. *72*, 3531–3542.

Kunin, V., Engelbrektson, A., Ochman, H., and Hugenholtz, P. (2010). Wrinkles in the rare biosphere: pyrosequencing errors can lead to artificial inflation of diversity estimates. Environ. Microbiol. *12*, 118–123.

LaMontagne, M.G., Michel, F.C. Jr., Holden, P.A., and Reddy, C.A. (2002). Evaluation of extraction and purification methods for obtaining PCR-amplifiable DNA from compost for microbial community analysis, J. Microbiol. Meth. *49*, 255–264.

Miller, D.N., Bryant, J.E., Madsen, E.L., and Ghiorse, W.C. (1999). Evaluation and optimization of DNA extraction and purification procedures for soil and sediment samples. Appl. Environ. Microbiol. *65*, 4715–4724.

Müller, S., and Nebe-von-Caron, G. (2010). Functional single-cell analyses – flow cytometry and cell sorting of microbial populations and communities. FEMS Microbiol. Rev. *34*, 554–587.

Muyzer, G. (1999). DGGE/TGGE a method for identifying genes from natural ecosystems. Curr. Opin. Microbiol. *2*, 317–322.

Nocker, A., Burr, M., and Camper, A.K. (2007). Genotypic microbial community profiling: a critical technical review. Microbiol. Ecol. *54*, 276–289.

Novotny, C., Svobodova, K., Benada, O., Kofronova, O., Heissenberger, A., and Fuchs, W. (2011). Potential of combined fungal and bacterial treatment for color removal in textile wastewater. Bioresource Technol. *102*, 879–888.

Schütte, U.M.E., Abdo, Z., Bent, S.J., Shyu, C., Williams, C.J., Pierson, J.D., and Fornay, L.J. (2008). Advances in the use of terminal restriction fragment length polymorphism (T-RFLP) analysis of 16S rRNA genes to characterize microbial communities. Appl. Microbiol. Biotechnol. *80*, 365–380.

Sekar, R., Fuchs, B.M., Amann, R., and Pernthaler, J. (2004). Flow sorting of marine bacterioplankton after fluorescence *in situ* hybridization. Appl. Environ. Microbiol. *70*, 6210–6219.

Sogin, M.L., Morrison, H.G., Huber, J.A., Mark Welch, D., Husse, S.M., Neal, P.R., Arrieta, J.M., and Herndl, G.J. (2006). Microbial diversity in the deep sea and the underexplored "rare biosphere". Proc. Natl. Acad. Sci. U.S.A. *103*, 12115–1210.

Stöhr, M., and Petrova, L. (1975). The alkaline hydrolysis of nucleic acid for removal of RNA associatedfluorescence in phenantridium related flow through cytofluorometry. Histochemistry *4*, 95–99.

Vogt, C., Lösche, A., Kleinsteuber, S., and Müller, S. (2005). Population profiles of a binary bacterial culture grown with toluene under sulphate reducing conditions. Cytometry Part A *66*, 91–102.

Wallner, G., Amann, R., and Beisker, W. (1993). Optimizing fluorescent *in situ* hybridization with rRNA-targeted oligonucleotide probes for flow cytometric identification of microorganisms. Cytometry *14*, 136–143.

Wang, Y., Hammes, F., Boon, N., Chami, M., and Egli, T. (2009). Isolation and characterization of low nucleic acid (LNA)-content bacteria. ISME J. *3*, 889–902.

Application of Flow Cytometry to the Detection of Pathogenic Bacteria

5

Deirdre Kennedy and Martin G. Wilkinson

Abstract

Outbreaks of infections have emphasized the necessity for rapid and economic detection methods for pathogens in samples ranging from those of clinical origin to food products during production and retail storage, and increasingly, in environmental samples. Flow cytometry (FCM) allows the rapid acquisition of multi-parametric data regarding cell populations within fluidised samples. However, the application of FCM to pathogen detection depends on the availability of specific fluorescent probes such as antibodies and RNA probes capable of detecting and isolating pathogens from these diverse samples. A particular issue for FCM methodology is the ability to recover and discriminate bacteria from the sample matrix which may pose a major technical hurdle towards accurate and sensitive analysis. This chapter focuses on detection of pathogens using FCM in samples originating from food, water, environmental and clinical sources and outlines the current state of the art and potential future applications.

Microbial pathogenesis

Understanding the biology of microbial pathogens and the pathways to disease are of major importance to medical microbiology and form the cornerstone for the development of preventative or therapeutic approaches to such conditions. Disease can result from the introduction of a single factor into the host such as a pathogenic agent e.g. bacteria or virus, or may require a series of complex interactions to occur between the host and the pathogen e.g. requiring a specific mode of transmission or an altered status of the host immune system. The diversity of microbial end products e.g. toxins generated in contaminated food products may account for the spectrum of human diseases (Stein, 1998). In recent decades, infectious diseases have become a vital and resurgent field of research in which FCM is becoming a highly useful analytical technique. Traditional approaches used to study and detect bacterial, fungal or parasitic pathogens may involve a combination of culturing on non-selective and selective media, microscopic examination, followed by a range of biochemical, immunological, and genetic confirmatory tests. Infectious pathogens of interest for FCM analysis include viruses such as human papillomavirus (HPV) and rubella virus, bacteria such as *Escherichia coli*, *Listeria monocytogenes*, *Staphylococcus aureus*, and a range of spore formers including *Bacillus* and *Clostridia* species, fungi such as *Candida* and *Aspergillus*, protozoa such as *Plasmodium* spp., and multicellular parasites such as *Trichinella* and *Schistosoma*. Each of these agents presents a significantly different challenge related to their specific detection. Detection teams who wish to identify these potential pathogens will face analysis of complex samples including those originating from foods, human or animal tissue or clinical fluids. These samples may contain a mixture of endemic microorganisms, the target pathogen(s), spores, and/or inorganic debris which complicate the identification procedure for the target pathogen. From such diverse materials, one needs to develop FCM based assays for rapid and specific detection of the harmful pathogens. Such FCM assays should be rapid, sensitive, credible by validation, suitable for simultaneous agent detection (multiplex analysis) and applicable to a variety of complex samples (Zahavy *et al.*, 2012). More recently, because of concerns regarding bioterrorism, accurate analytical techniques for pathogenic agents are needed which can be deployed at public transport hubs or scenes of terrorist incidents. Many of the current detection kits depend on specific binding of antibodies to the surface antigenic components of pathogenic bacteria or viruses. Immunological detection mechanisms have both advantages and disadvantages; however, their combination with differential FCM staining and multi-parametric analysis may allow significant improvements in assay sensitivity, selectivity and data analysis (Robinson, 2001).

Advantages of flow cytometry for pathogen detection

From the point of view of a microbiologist, a key advantage of FCM methodology is the ability to simultaneously collect multiple data outputs for an individual cell (Winson and Davey, 2000). This analysis typically generates data regarding cell viability, cell vitality, structural integrity, physiological status or stage of growth cycle. Flow cytometers offer the advantage of being able to analyse thousands of cells/events per second, depending on sample type and cell concentration in the sample. While some flow cytometers may require significant monetary investment with specialized training for operators, more cost-effective and user friendly machines are becoming increasingly available. It is clear from the increasing numbers of publications and other technical data that FCM is emerging as an alternative rapid method for microbial detection, enumeration and population profiling. Indeed a number of studies have reported its use for the study of a range of pathogenic microorganisms (Allegra *et al.*, 2008; Berney *et al.*, 2006; Cronin and Wilkinson, 2007; Jarzembowski *et al.*, 2008; Nocker *et al.*, 2011; Ryumae *et al.*, 2010; Ueckert *et al.*, 1995). The techniques currently used to identify microbial pathogens rely upon well-established conventional clinical microbiology approaches but suffer from a number of limitations. While standard culture and susceptibility tests permit pathogen identification they are laborious, time-consuming, expensive, and may require the use of labile natural products such as egg yolk supplements and other ingredients with a short shelf life (see Fig. 5.1). Detection and counting is carried out using optical methods,

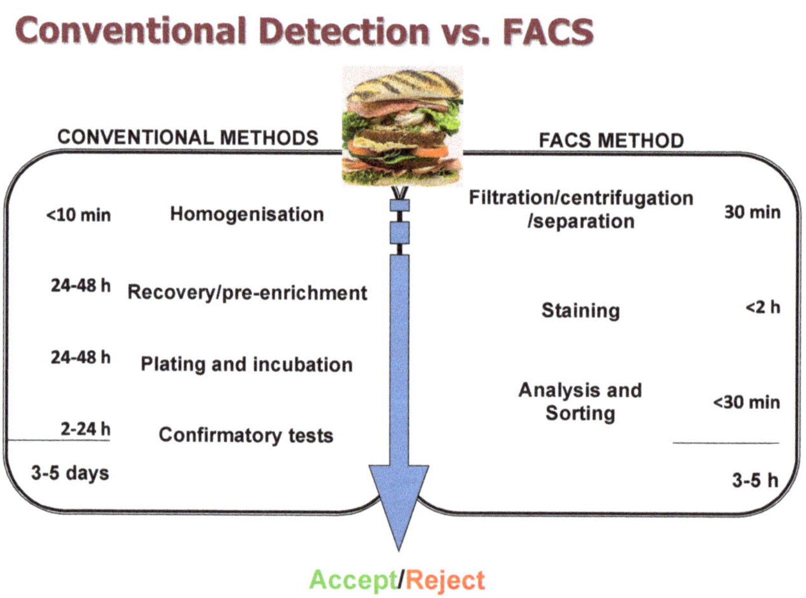

Figure 5.1 Time comparison of typical conventional microbial detection and enumeration methods with potential flow cytometric methods for detection of target bacteria.

mainly by ocular inspection (Lazcka *et al.*, 2007) and detection of typical colonies is often followed by sub-culturing and use of a variety of tests for species identification (Entis *et al.*, 2001). For the modern medical or public health laboratory there is clearly a gap between these slower and well accepted traditional detection methods and the requirement to provide real time data to enable decisions to be made regarding medical treatment options, or implementation of risk management programmes.

Molecular biology and immunology has provided a broad array of materials available for labelling of specific pathogens including some commercially available antibodies and a more extensive sequence database of information for designing molecular probes. These two approaches have advanced the field with the development of commercial rapid methods for pathogen detection (Harkins and Harrigan, 2004). However if combined with FCM, developments in both molecular biology and immunology should potentially enable further advances such as on/at line detection of target cells present at low levels within mixed populations of non-target cells with almost instantaneous results. This is also of particular benefit to allied healthcare, pharmaceutical and food industries where product release can only be carried out after full testing results are obtained and interpreted.

Challenges in pathogen detection

Most detection and enumeration of pathogenic microbes is derived from cultures growing in pathogen-specific enrichment and/or selective media. However, this methodology requires that pathogenic bacteria are both viable and culturable in order to be detected which limits detection to a specific physiological state (Harkins and Harrigan, 2004). If a pathogen is not capable of growing under defined conditions these methods may lead to false-negative results or underestimate the population within a given sample. More importantly, tests that are routinely utilized for pathogen identification do not directly characterize virulence factors and do not provide the necessary information regarding the potential pathogenicity or virulence of the identified organisms (Vora *et al.*, 2004). In terms of laboratory work load and sample throughput, conventional microbiological techniques do not lend themselves well to analysis of large numbers of environmental or clinical samples. To quickly determine the presence of pathogen(s), microbiologists need reliable and accurate tools which can respond to the increasing pressures to find faster, more accurate analytical techniques for identifying infectious agents. Such efforts have led to the development of rapid FCM-based methods which can potentially rapidly detect low concentrations of pathogens in water, food and clinical samples (Hammes *et al.*, 2008; Lautenschlager *et al.*, 2013; Riyaz-Ul-Hassan *et al.*, 2008). The main experimental difficulty in analysing bacteria using FCM is that many of their biological characteristics (including size, shape and DNA content) vary depending upon the growth conditions used, or the sources from which the organism were obtained. For example, exponentially growing cells are larger than dormant or starved cells and may also contain higher levels of nucleic acids (Sincock and Robinson, 2001). Therefore, strict reproducibility of conditions is required in order to produce consistent data and all new methods require validation before widespread acceptance. In addition, the capital cost involved in flow cytometry analysis, especially where cell sorting is involved, is quite high which can restrict its widespread analytical usage (Ivnitski *et al.*, 1999). An area of potential concern to the laboratory analyst using cell sorting of biosafety level two pathogens is the generation of an aerosol of bacteria and some studies have reported the use of a killing step during sample preparation for this reason. Clearly this killing step is not practical for methods where cell sorting, viability testing or culturing is required. However, most flow cytometers capable of sorting have aerosol control such as filters but the biosafety levels required will vary with the type of sorting being carried out (Schmid *et al.*, 2001) especially where they are also routinely used for sorting of cells from infectious human blood samples (Raybourne and Tortorello, 2003). However, it is possible to house cell sorters within laminar flow hoods as is the case in the author's own laboratory for use in sorting of pathogens such as *E. coli*, *Staphylococcus aureus* and *Listeria monocytogenes* (Kennedy *et al.*, 2011).

Instrument set up

As bacteria are much smaller than mammalian cells, the detection and analysis of pathogenic microorganisms poses particular challenges for FCM. If the instrument is not specifically designed for microbial analysis one must first determine whether it is capable of such measurements and then optimize its performance (Harkins and Harrigan, 2004). Generally flow cytometers used for such applications are fitted with in-line filters to prevent contamination from microorganisms that may be present in sheath fluid or sheath fluid tanks. Therefore in the initial machine set up, the operator must validate the ability of the machine to detect and enumerate a specific pathogen(s). A range of fluorescent beads are commercially available for calibration and counting and are used as standards to provide a reference point allowing the operator to optimize instrument settings for detection of a particular pathogen. These beads allow detection at the size level required for analysis of small particles such as bacteria and are available in the range of 0.1 μm upwards. Thereafter determination of threshold values and photomultiplier tube (PMT) voltages are undertaken to enable optimization of a range of parameters for analysis of pathogens. Fluorescence and side-scatter or forward scatter ratio information allows the operator to anticipate where, for example, labelled pathogens are located on a two-parameter histogram. The concept of using FCM to analyse bacterial suspensions is not new, however the successful commercial development or modification of flow cytometers devoted entirely to the detection of a specific bacterial pathogen is still relatively uncommon.

A RAPID-B flow cytometer (LITMUS RAPID-B now Vivione BioSciences, AR, USA) described by Wilkes *et al.* (2012) is purported to have superior optical and physical characteristics for microbial applications including 130 nm resolution, a large cross-section flow cell channel, and syringe-controlled sample introduction for detecting and quantifying bacterial sized particles in complex matrices. The RAPID-B approach utilizes fluorescein-conjugated polyclonal antibodies that tag the target pathogen while a second detergent based reagent exposes bacterial cell surface epitopes. Thereafter the labelled pathogen is detected in a flow cytometer using the modified detection cell. Currently, this company claims to have developed rapid immuno-FCM tests for pathogens such as *E. coli* O157, and *Staphylococcus aureus*. Shortened enrichment times (4.5 hours) enables data to be generated rapidly with simultaneous information on total counts and viability based on Propidium Iodide (PI) staining. In total, sample preparation was reported as ~30 minutes, followed by enrichment of 4–6 hours with analysis times of 3–4 minutes. Hence the assay could deliver data within a typical manufacturing process cycle and was claimed to have a limit of detection (LOD) of 1 viable cell per 25 g of a contaminated spinach sample. Therefore the system may be used for process quality control and/or bio-mapping of a food production facility.

When developing methodology for microbial detection and enumeration by FCM it is important to first prepare and analyse pure cultures of the pathogen of interest for setting up the instrumentation and for initial validation of the labelling and enumeration method. This also sets the control profiles for the particular microbial species in terms of typical live, dead or damaged cells and where they may be expected to be located within a typical cytograph. It is important to note that species, sub-species and strain-related differences in FCM profiles can be expected and the analyst, in many cases, will have to develop a database of these profiles for each pathogen of interest in order to recognize and differentiate target from non-target bacteria and to interpret data. Sample preparation protocols may be quite extensive in order to remove debris and to allow a sample meet the requirements of a reliable assay. Signal filtering techniques and instrument set up are important as are validation of counts as described by Subires *et al.* (2014). These workers used a polyclonal antibody conjugated to R-phycoerythrin (RPE) along with viability staining by PI and SYBR Green 1 for detection of *E. coli* O157:H7. This pathogen is the causative agent of human haemorrhagic enteritis which may lead to severe kidney failure in specific population cohorts. Signal filtering was based on evaluation of SSC events from 50–200, with optimal signal to noise ratio obtained with an SSC trigger at 150 events for threshold value. This allowed a lower limit of detection (LOD) to be established from 10^3 CFU/ml while a limit of quantification (LOQ) was established at 10^4 CFU/ml. Using signal triggering at 150 SSC the authors reported very good correlation between FCM data and viable agar plate counts. Some assay formats will require the exact enumeration of pathogens within a specific sample volume, e.g. a 1 ml analyte volume derived from 25 g of sample in the case of pathogens in foods. These detection levels are significantly lower than those frequently presented in FCM research studies where optimal levels of 10^6 cells/ml are analysed as pure cultures obtained from dilute media without the presence of significant interfering substances. Therefore, sensitivity and clearly defined limits of detection need to be considered and optimized when setting up assays for low levels of target pathogens present in food, clinical or environmental samples.

Sample preparation

Generally, most microbial testing is performed on the final product. Both the food and pharmaceutical industries have implemented a hazard analysis at critical control points (HACCP) system which also includes raw-material and in-process testing for pathogens. Therefore methods for rapid screening of a variety of sample types are of significant interest to these industries. Samples of various origins to be screened for pathogens often require a clean up/pretreatment step for successful analysis by FCM. This is necessary as substances within a sample matrix can affect the accuracy of flow cytometric counts (King, 2007), especially if particles are auto-fluorescent and are within the same size range as the bacteria of interest. These issues cause high background fluorescence and interfere with light scatter profiles. Samples may also contain high background levels of non-pathogenic microflora which will require to be accounted for, most likely by selective labelling of target bacteria and/or discarding of other microorganisms. Finally, all bacteria to be analysed by FCM must be separated in a free form without the presence of adhering matrix particles. Sample preparation may involve homogenization of solid samples with a stomacher, with or without serial dilution, followed by separation and concentration steps such as centrifugation and filtration. Large particulates in samples can be removed via a series of filtration steps that may include initial filtration through a larger mesh size such as 2000 μm and subsequently through a smaller filter (e.g. 3 μm) (Harkins and Harrigan, 2004). Centrifugation may be carried out in conjunction with filtration as it assists in sedimenting large particles leaving bacterial cells in suspension in the supernatant. Addition of special reagents may also be involved such as use of a clearing solution that removes micelles from milk or egg samples (McClelland and Pinder, 1994). Growth in enrichment media can increase the number of target organisms to detectable levels and allow recovery of stressed cells. This step may be necessary as the infectious dose of certain pathogens in samples may be lower than can be potentially detected by FCM. However, data gathered using an enrichment step is usually accepted as being qualitative rather than quantitative for the target microorganism.

Food or pharmaceutical samples are typically diluted into a defined volume of enrichment medium for a specific pathogen following clear guidelines (e.g. 65 g beef per 585 ml of pathogen-specific enrichment medium for testing of *E. coli* O157:H7), followed by a culture period ranging from 18 to 72 hours depending on the target microorganism. For many samples, it may be necessary to both concentrate cells and remove matrices from the sample. Concentration may involve immunomagnetic separation (IMS) which is used as a pre-treatment and/or pre-concentration step to capture and extract the target pathogen from a bacterial suspension using antibody-coated magnetic beads (Lazcka *et al.*, 2007). Many studies have shown IMS to be a useful concentration and isolation method for bacteria (Favrin *et al.*, 2003; Hibi *et al.*, 2006; Power *et al.*, 2003). Hibi *et al.*, (2006) showed that IMS combined with FCM could detect and enumerate *L. monocytogenes* within one minute in the range of 10^2–10^8 CFU/ml after a 2 hour preparation time with excellent agreement with standard plate counts ($r = 0.97$). Without IMS, sensitivity decreased with high R values noted at cell concentrations of $> 8 \times 10^4$ CFU/ml. Power *et al.* (2003) combined IMS with FCM for the sensitive detection of *Cryptosporidium* in faecal samples. The IMS/FCM technique was found to be approximately 50-fold more sensitive than formol-ether concentration, commonly used for *Cryptosporidium* epidemiological investigations. Tortorello *et al.* (1998) compared six rapid methods of enriching and detecting *E. coli* O157:H7 in apple juice samples. At the lowest inoculum level (0.1 CFU/ml), FCM required a 24-hour total enrichment and preparation time (and ranked third after ab-DEFT and IMS-PCR) but had the shortest analysis time of just 1 minute per sample.

Rymae *et al.* (2010) reported a study that used high gradient magnetic separation (HGMS) with flow cytometry to detect *Flavobacterium psychrophilum*. This technique uses a strengthened magnetic force by introducing a magnetic gradient between the magnetic filter and nearby column. This technique resulted in a much more sensitive assay with cell numbers determined by FCM correlating with those obtained using the colony counting method in the range of approximately 10^1–10^5 CFU/ml.

Immunolabelling

A combination of FCM with monoclonal antibodies against specific cellular antigens can enable FCM to identify specific cell populations within the background microflora of a sample matrix. Light scatter is first utilized to identify the cell population(s) of interest, while measurement of fluorescence intensity provides specific information about each individual cell or target labelled cells which will fluoresce in the channel of interest. Both polyclonal and monoclonal antibodies against a variety of different microbes can be sourced commercially from a range of suppliers. Polyclonal and monoclonal antibodies are becoming increasingly available against a variety of different microbes such as *Salmonella* Typhimurium, *Listeria monocytogenes, Escherichia coli O157, Staphylococcus aureus* (Guttikonda *et al.*, 2007; Jung *et al.*, 2003; McCarthy and Culloty, 2011; McClelland and Pinder, 1994). Polyclonal antibodies are generally noted as giving a more adequate and sufficient signal for FCM detection. Antibodies may be purchased in pure unconjugated form, as enzyme conjugates for ELISA assays, biotin tags, or fluorescent markers, depending upon the vendor and their intended use. For FCM analysis many researchers use pure unconjugated primary antibodies to label target molecules. Thereafter, a range of commercially available secondary antibodies conjugated to a range of markers are used to reveal the labelled complex (primary antibody-antigen). Hence the fluorescence intensity of the target microbial population is generally measured by using a primary/secondary antibody labelling approach. For example, the use of 635-nm (Red) excitation will require the analyst to perform a conjugation step with either allophycocyanin (APC), or Alexa Fluor 647, or to employ an indirect labelling method using a commercially available secondary antibody conjugate (Harkins and Harrigan, 2004). The use of 488 nm (blue) laser excitation allows the use of common fluorochromes such as Phycoerythrin (PE), fluorescein isothiocyanate (FITC), and peridinin chlorophyll protein (PerCP). Because many combinations of lasers, detectors, filters and fluorochromes are possible for multicolour analysis, precautions need to be taken (i.e. bandpass filters, dichroic mirrors, longpass filters, etc.) by the operator to ensure each fluorochrome is being detected by only one detector including selection and set up of bandpass filters, dichroic mirrors, longpass filters, etc. When bacteria are labelled using antibodies, a variety of factors can affect surface antigen density (growth condition, species, serovars, and strain variation).

Schellenberg *et al.* (2006) described an immuno-staining procedure for *Staphylococcus aureus* in a mixed microbial population containing six various strains of lactic acid bacteria. Cells were first diluted in PBS, centrifuged, and the resultant pellet re-suspended in phosphate-buffered saline (PBS) and 1% Tween to prevent agglomeration. Samples were then heat fixed and cooled prior to staining using an *S. aureus*-specific fluorescently conjugated (FITC) antibody. These workers counterstained samples with PI for the purpose of analysis by FCM and oil immersion fluorescent microscopy. In this study, a combination of the specificity and sensitivity of immunofluorescent probes with cytometric analysis of thousands of bacterial cells was claimed to be a powerful analytical approach allowing a new insight into bacterial ecology. Significant correlation was found between estimated mean cell-unit concentrations as determined by the FCM and plate counting methods ($r^2 = 0.732$, $P = 0.0004$) (Schellenberg *et al.*, 2006). FCM analysis and sorting studies using monoclonal antibodies against defined surface markers on normal or neoplastic cell populations has created the basis for routine clinical diagnostic assays, for example leukaemia classification or monitoring CD4 T-cell loss during HIV disease progression (Herzenberg *et al.*, 2000). These diagnostic applications of FCM are based on the availability of highly specific labelling antibodies. FCM detection of *Bacillus anthracis* spores (Dang *et al.*, 2001) demonstrated the utility of antibody-based labelling techniques for monitoring of environmental pathogens. Spores were treated with irradiation or by autoclaving and then analysed using a number of methods including FCM immunolabelling by three types of primary antibodies. The study illustrated the necessity to optimize and choose suitable primary antibodies for the spores of interest as some epitopes were destroyed after inactivation treatments and hence the staining ability of these antibodies was lost. Despite the obvious advantages in time saving that microbial immuno-FCM can offer, antibodies are not always available for a particular application and overall few antibodies are commercially available which target whole cells of pathogenic bacteria or other specific bacterial groups of interest.

Fluorescence *in situ* hybridisation (FISH)

Historically, FISH and other *in situ* hybridization data played a primary role in mapping genes on human chromosomes. In microbial analysis, FISH is used to label specific nucleic acid sequences inside intact viable cells and to enable identification of bacterial species present in various samples that may contain mixed microbial populations. The first step in the procedure involves preparation of short sequences of single-stranded DNA probes that hybridize to a portion of the target gene. The next step is to label these probes by attaching one of a number of colours of fluorescent dye. Depending on the desired level of specificity, the oligonucleotides complementary to these sequences may function as group or species-specific probes for microbial cells in a sample. As the probe is fluorescent it is possible to detect the site of hybridization directly. In general, the small subunit ribosomal RNA (rRNA) provides an abundant intracellular target that is commonly used as a phylogenetic marker of bacterial classification. This abundance allows generation of a signal strong enough for detection. FISH with rRNA-targeted oligonucleotide probes has been shown to be a rapid method for identification of human pathogenic bacteria and yeast in samples such as blood cultures (Banerjee *et al.*, 2002; Kempf *et al.*, 2000; Moter and Göbel, 2000). Most FISH detection and analysis is carried out using fluorescence microscopy. However, a combination of FISH and FCM (FC-FISH or FLOW-FISH) has been shown in certain cases to be both a rapid and reliable technique for bacterial identification (Rigottier-Gois *et al.*, 2003). These workers successfully identified *Clostridium coccoides*, *Fusobacterium prausnitzii*, *Bifidobacterium* spp. and Enterobacteria in human faecal samples with data obtained not statistically different from standard counting methods. Probe retention/binding can be quantified by measuring median fluorescence in a population of cells, using a fluorescence channel of interest on the cytometer. The primary advantage of this technique is that it eliminates the time required for preparation and subsequent analysis of slides. One of the most intriguing potential applications of FCM is the possibility for direct extraction of specific subpopulations from samples, omitting a cultivation step, followed by genetic or possibly genomic characterization (Kalyuzhnaya *et al.*, 2006).

FISH procedures for human pathogen detection using peptide nucleic-acid (PNA) probes in place of nucleic-acid probes have also been reported (Kenny *et al.*, 2008; Stender *et al.*, 2002). PNA is a DNA mimic, where the entire negatively charged sugar phosphate backbone has been replaced with a neutral one consisting of repeated N-(2-aminoethyl) glycine units linked by peptide bonds and which is stable both chemically and biologically. Peptide bonds can enable PNA to be readily conjugated with peptides, fluorescent dyes, and other useful biomolecules. Because PNA probes have a neutral backbone, the kinetics of binding and the stability of the hybridization are significantly better than that obtained using standard nucleic-acid probes (Harkins and Harrigan, 2004). These PNA probes can also be used in conjunction with FCM as the same principles apply if they are conjugated to a fluorescent marker. Hartmann *et al.* (2005) used a combination of PNA FISH probes in conjunction with FCM to rapidly identify *Staphylococcus aureus* in blood cultures. Results obtained with PNA FISH using FCM did not reveal any discrepancies in sensitivity or specificity compared to standard laboratory methods and offered a much quicker alternative to standard methods as results were obtained within 4 hours. The detection of growing pathogens (e.g. *Stenotrophomonas maltophilia*) in blood cultures was achieved more rapidly by FC-FISH compared with standard agar plate detection methods (Kempf *et al.*, 2005). The target organisms grown on agar plates subsequently had to be identified with standard laboratory methods including: cytochrome oxidase test, API 20E strip, API 20NE strip, ID32 C, Vitek II system. Using the FC- FISH method required a one hour hybridization followed by washing and analysis on a FACS Calibur cytometer. Rapid identification of pathogens was achieved using specific FISH probes, Ckru-1453: specific for *C. krusei*, Sma-633: specific for *S. maltophilia*. Hybridization probes with FCM identified *C. albicans* and Enterobacteriaceae directly from blood culture bottles. *C. albicans* hybridized with the universal yeast probe PF-2 (green) and with a *C. albicans*-specific probe. Enterobacteriaceae, hybridized with the universal bacterial probe and also with a Enterobacteriaceae specific probe. Results were then confirmed with microscopy and species identified further with standard microbiology tests. For example, blood cultures were artificially spiked with approximately 10–50 CFU/ml *S. maltophilia*. BacTec9240® blood culture system indicated bacterial growth 31 hours after inoculation, where using FC–FISH, bacteria were detectable after 23 hours.

Green fluorescent protein (GFP)

Several bioluminescent coelenterates use a secondary fluorescent protein, green fluorescent protein (GFP), in an energy transfer reaction to generate green light. The most studied of these proteins have been GFPs from the jellyfish *Aequorea victoria* and the sea pansy *Renilla reniformis*. Proteins from these organisms have a chromophore, which is derived from the primary amino acid sequence of GFP (Chalfie, 1995). The GFP from *A. victoria* has a major excitation peak at a wavelength of 395 nm and a minor one at 475 nm. Its emission peak is at 509 nm, which is in the lower green portion of the visible spectrum. GFP from *Renilla reniformis* has a single major excitation peak at 498 nm. The gene for GFP has been isolated and has become a widely used and useful tool for generating chimeric proteins of GFP where it is linked to other proteins and functions as a fluorescent tag. As a non-invasive fluorescent marker in living cells, it has a wide range of applications as a cell lineage tracer, a reporter of gene expression, or as an indicator of protein–protein interactions (Mocz, 2012). Variants with more intense fluorescence or alterations in the excitation and emission spectra have been produced and FCM has proven to be a useful method with which to study infectivity and host–pathogen interactions.

Ornithogalum dubium (commonly known as the Star of Bethlehem) is a flowering plant and is a natural host of the soft rot pathogen *Pectobacterium carotovorum* subsp. *carotovorum* (Pcc). Golan *et al.* (2010) developed a quantitative assay for Pcc expressing a GFP reporter gene, using fluorescent activated cell sorting to quantify bacterial cell proliferation in *O. dubium* plantlets inoculated with Pcc. The cell sorting approach for quantification of Pcc was found to be accurate, reproducible and reduced analytical time. Douesnard-Malo and Daigle (2011) used FCM to demonstrate that *Salmonella* Typhi can persist for extended periods within environments containing a suitable host organism. This study investigated interactions between *Salmonella* Typhi and *Acanthamoeba castellanii* (a host Amoeba) by using co-cultures. Growth of both organisms was estimated by cell count, viable count, FCM and fluorescence microscopy. These workers tracked the growth and infectivity of *Salmonella* Typhi by monitoring auto fluorescence from the Amoebae and green fluorescent protein signal from the salmonellae. Cell population analyses by FCM confirmed a decline in the bacterial population, clearly seen by the loss of fluorescence over time when bacteria were grown in pure culture. It was only in the presence of amoebae that fluorescence was observed after 14 days. Results confirmed what earlier co-cultures have shown by viable plate counts, i.e. survival of *Salmonella* Typhi for 14 days. This FCM method could provide a means to monitor the interaction between a pathogen and its host, allowing a better understanding of pathogen transmissibility.

Quantum dot labelling

Commonly used cell sorting systems are based upon single or dual excitation as excitation source both for light scatter parameters and for several fluorescence detectors. Hence, for multi-labelling detection, there is a requirement for fluorophores having a broad excitation wavelength and sharp emission bands. Such fluorophores should have high fluorescence efficiency, stability, and be available for conjugation with other bio-molecules. Quantum dots (QDs) provide practical features which meet many of these criteria (Zahavy *et al.*, 2012) as they are inorganic crystalline nanoparticles made of semiconducting materials. QDs used for biological applications typically are composed of three different layers starting with a core, normally cadmium selenide (CdSe) coated with a semi-conductor outer shell of zinc sulphide (ZnS). The core-shell ranges in size from 3 to 10 nm and the core size defines the fluorescence emission of the QDs. A third layer is composed of a polymer which solubilizes the QDs while incorporating specific functional groups such as proteins or chemical compounds which allow specific binding to the desired target. QDs are currently being used as a novel fluorophores as their physical and chemical properties confer significant advantages over traditional dyes, including brighter fluorescence and resistance to photobleaching (Ibanez-Peral *et al.*, 2008). In comparison to organic fluorophores, a single excitation source can simultaneously excite several QDs with distinctive emission bands which can be easily separated by optical filters (Zahavy *et al.*, 2012). QDs have applications for bacterial and pathogen detection in combination with FCM as well as for bacteriophages (Edgar *et al.*, 2006). A QD method has combined *in vitro* biotinylation of engineered host specific bacteriophage and conjugation of the phage to streptavidin-coated QDs. This phage-based assay reduces the amplification to a shorter time (20–45 minutes from infection to lysis) because each infected bacterium can result in a release of 10–1000 phage particles that can be readily detected by QDs. Ibanez-Peral *et al.* (2008) investigated the potential of QDs for FCM based assays by labelling beads in various ways. These workers indicated that the minimum fluorophore concentration required for detection of QDs above an autofluorescent background was 100-fold less than for FITC. The fluorescent signals of Dynabeads®, magnetic separation beads from Invitrogen, labelled with both fluorophores (FITC and QD655) were analysed by FCM and the MFI values collected. When QDs were used as fluorophores to label beads, they still exhibited lower detection limits in FCM in comparison to FITC. The effects of coupling an oligonucleotide labelled with FITC and QDs modified with biotin to paramagnetic beads (Dynabeads®) as probes targeting specific DNA sequences from microorganisms may be of interest. Dynabeads are of a larger size than bacteria and can be distinguished readily by FCM. However, there are few reports of specific nucleotide probes being used for microbial identification using FCM. Most reports have described QDs conjugated to specific antibodies for this purpose where they have also been used in immuno-phenotyping applications of human cells (Chattopadhyay *et al.*, 2006; Summers *et al.*, 2010). Conjugates need to maintain the immunospecific properties of the antibodies with the fluorescence properties of the QDs. However, QDs have been found to be unsuitable in certain applications such as environmental sampling for detection of *Cryptosporidium* (Ferrari and Bergquist, 2007). These workers noted the superior spectral characteristics of QDs; however, they noted that oocysts stained with QD-conjugates exhibited significantly lower fluorescence intensity than organic conjugates. Additionally, the level of non-specific binding by QD-conjugates to other particles present in the water concentrate was significantly higher that of the organic conjugates. Overall, it is important when developing accurate FCM methods for water analysis to recognize that there may be high levels of auto-fluorescent and background particles in the samples. Water analysis and FCM will be dealt with extensively in another chapter in this text.

FCM analysis of food samples

The extent of growth, survival and biochemical activity of microorganisms in foods is the result of stress reactions. These reactions are a response to the changing physical and chemical conditions within the food microenvironment, e.g. pH, oxygen, water activity, salt, and temperature and the presence of other microorganisms within foods. The above factors greatly influence the ability of bacteria to colonize the food matrix and to grow and subsequently be detected using traditional plating on solid selective media (Fleet, 1999). Legislation pertaining to food safety is particularly stringent, as failure to detect infection/intoxication in food products may have serious consequences for the consumer including illness and/or death. Hence, accurate and early detection of pathogenic bacteria is key to prevention and identification of problems related to food and public health. Despite the obvious need for obtaining analytical results in the shortest possible time, the continued widespread use of traditional and standard bacterial enumeration and/or identification methods can take up to 7 or 8 days to yield useful data. This timescale is clearly insufficient for products with a short shelf life including chilled ready meals. Therefore, researchers have focused their efforts towards the development of FCM-based rapid methods to replace or augment these methodologies (Lazcka *et al.*, 2007). This is quite technically challenging for FCM methodology, as certain pathogens have a low infectious dose and hence low limits of detection (LOD) and limit of quantification (LOQ) are required. The infectious dose of *E. coli* 0157:H7 is of the order of 10–100 cells and so a highly sensitive detection system is necessary for its detection in foods (Subires *et al.*, 2014; Sunwoo *et al.*, 2006). The ingestion of low numbers (< 10) of pathogens such as *Salmonella* Typhimurium may lead to food poisoning; thus, preparation methods are required to concentrate low levels of the target microorganisms (Rotariu *et al.*, 2005). Compared with chemical and physical hazards the occurrence or emergence of microbiological hazards depends on a variety of factors which may occur during food production, processing and storage. Therefore implementation of well-structured hygiene and food safety management systems are essential (Untermann, 1998).

FCM is a commonly used technique for quantifying bacteria in milk (Gunasekera *et al.*, 2003; Holm *et al.*, 2004), milk powder (Flint *et al.*, 2006) and to monitor dairy starters during cheese ripening (Bunthof and Abee, 2002). Quantification of bacteria in these matrices are undertaken by several commercial instruments such as the BactoCount (Bentley Instruments Inc., Chaska, USA). A range of FCM based technologies are also available from under the Chemunex® brand, (Biomerieux, France) including the BactiFlow®, BactiFlowALS® and D-Count® for rapid viable cell analysis of bacteria, yeast and moulds in food processing environments. In addition, FCM-based methods have also been used for Gram-typing bacteria in bulk milk by labelling with various fluorochromes (Holm and Jespersen, 2003; Langerhuus *et al.*, 2013). The same principle could potentially be used for Gram-typing of mastitis-associated bacteria. Langerhuus *et al.* (2013) proposed an assay which allowed rapid screening of milk samples for mastitis-causing bacteria. Results indicated that it was possible to detect and determine the Gram-type of pathogenic bacteria in mastitis affected milk samples based on bacterial biotin-conjugated wheat germ agglutinin and Acridine Orange (AO). Wheat germ agglutinin binds to N-acetyl-glucosamine in the peptidoglycan layer, which is prevented in Gram-negative bacteria, as wheat germ agglutinin cannot penetrate the outer lipopolysaccharide layer. In addition, Gram-negative bacteria also had less uptake of AO. The authors admitted that the method cannot compete with the traditional culturing techniques or PCR technology with respect to bacterial identification. However, they claim it could potentially provide the dairy farmer or the attending veterinarian with on-site information on bacterial detection and Gram-type in cases of clinical mastitis (Langerhuus *et al.*, 2013).

FCM analysis of clinical samples

Life-threatening and harmful infections that require prompt antimicrobial therapy need rapid and accurate diagnostic tests in order to be effective in treating such diseases caused by specific pathogens. FCM is a promising and emerging technique with enormous potential for clinical microbiology applications. Alvarez-Barrientos *et al.* (2000) published an excellent review entitled the 'Applications of flow cytometry to clinical microbiology' outlining the advances made using FCM. This section will now outline some of the more recent advances in this area. FCM has been shown to detect viral antigens either on the surface of, or within, infected cells using antibodies that specifically recognize surface or internal antigens. FCM analysis of virus-infected cells is best suited to samples of blood, bronchoalveolar lavage fluid, and urine. However, it is also possible to analyse cells from tissues that have previously been pretreated for other clinical tests (Alvarez-Barrientos *et al.*, 2000).

As far back as 1990, McSharry *et al.* (1990) developed a method to detect and quantify human immunodeficiency virus (HIV)-infected peripheral blood mononuclear cells obtained from HIV-seropositive patients. Intracellular HIV antigens were detected by indirect immunofluorescence using monoclonal antibodies against HIV antigens as the primary antibody and fluorescein isothiocyanate-conjugated goat anti-mouse immunoglobulin G F(ab')2 antibody as the secondary antibody. After overnight amplification of clinical samples suspected to contain HIV, FCM enabled detection of the virus 1 to 3 days prior to cytopathic effects were evident in cell culture. Honda *et al.* (1997) used FCM to detect cytomegalovirus (CMV)-infected cell populations in peripheral blood lymphocytes from CMV-infected patients. Using monoclonal antibodies directed against immediate-early CMV antigen or against several cell membrane markers to phenotype-infected peripheral blood cells from bone marrow transplant recipients; these authors developed a rapid and quantitative FCM method for the detection of immediate-early CMV antigen.

Dixon *et al.* (2005) evaluated an FCM method for the detection and enumeration of *Cyclospora* oocysts in human faecal specimens which had previously been associated with food-borne outbreaks of cyclosporiasis in Ontario, Canada. In the absence of commercially available anti-*Cyclospora* monoclonal antibodies, the FCM method was based on the autofluorescence of *Cyclospora* oocysts. FCM data were generally very comparable to the original microscopy results for these specimens, both in terms of the presence or absence of oocysts in samples and their relative concentrations.

Rapid and accurate methods to quantify bacterial adhesion and invasion to epithelial and various other types of cells has been evaluated using FCM (Grootaert *et al.*, 2011; Pils *et al.*, 2006; Trouillet *et al.*, 2011). Variants of *Neisseria gonorrhoeae* expressing distinct colony opacity-associated (Opa) proteins bind to receptors of the carcinoembyronic antigen-related cell adhesion molecule (CEACAM) family to direct their uptake by human cells. Pils *et al.* (2006) suggested that quantitative analysis of bacterial invasion *in vitro* may be key to understanding the molecular basis of the various internalization processes. The analytical approach was based on direct fluorescein labelling of bacteria prior to infection followed by FCM based detection of cell-associated fluorescence as an estimate of the number of internalized bacteria. FCM analysis not only determined the percentage of infected cells, but also provided data on the relative amount of internalized bacteria based on the strength of the fluorescence signal. These workers demonstrated that this protocol could correctly quantify intracellular bacteria when compared with a conventional antibiotic protection assay or microscopic techniques.

Flow cytometers are routinely used to measure specific cell subpopulations in blood, bone marrow aspirates, body fluids and tissues. Septicaemia or blood poisoning is one of the leading causes of death in hospital patients and in the USA is the tenth leading cause of death (Melamed and Sorvillo, 2009). Rapid identification of sepsis-causing pathogens is crucial as appropriate antimicrobial agents can be selected earlier, avoiding unnecessary/ineffective treatments with a potential for improved patient prognosis. Definitive identification of pathogens grown from blood cultures by traditional methods involves sub-culturing and biochemical analysis requiring at least 24 hours for data generation after the first indication of growth (Kempf *et al.*, 2005). As described earlier, the rapid identification of bacterial pathogens by FC-FISH provides a potential method to overcome this problem. Schellenberg *et al.* (2008) used FCM as a tool to quantify total bacterial cells in vaginal specimens from adolescents. Typically a reduction in *Lactobacillus* populations and an overgrowth by anaerobic and Gram-negative species defines a frequently asymptomatic clinical condition known as bacterial vaginosis. FCM was used by these workers as a simple method to quantify total bacterial cells in vaginal swab samples. Bacterial cells could easily be distinguished from eukaryotic cells on the basis of size and shape and the study provided data regarding patients over a six week period of sampling. Kadkhoda *et al.* (2011) identified the requirement for a rapid, easy and reliable screening method for urinary tract infections (UTI), which would obviate the requirement to culture samples that do not contain pathogens. These workers used a UF-1000i™ flow cytometer which stains cell components with a fluorescent dye and rapidly measures cells in the urine such as leucocytes, bacteria, and erythrocytes. The method rapidly determines both the white blood cell and bacterial counts providing useful information regarding which samples may or may not need to be forwarded for traditional culture methods. Hussein *et al.* (2002) used FCM and immuno-labelling to detect the presence of *E. coli* 0157:H7 in bovine rumen fluid and faeces. Concentrations of 10^4 CFU per ml were accurately identified and enumerated, however lower concentrations required enrichment.

FCM has found applications in other medical fields such as the study of bacteria which may cause periodontal disease. Spirochetes such as *Treponema denticola* are highly fastidious organisms and are extremely difficult to culture and enumerate on solid agar media. The current method can take between 7 and 14 days to yield results. Orth *et al.* (2009) described an FCM based approach using the LIVE/DEAD® BacLight™ Bacterial Viability and Counting Kit (Invitrogen). A standard curve was first prepared using known dilutions of cells and cell count estimates were then derived from this. The percentage of bacteria that were SYTO 9 and PI positive, i.e. dead cells, were excluded from analysis as viable counts were compared with a standard method. In enumerating cells, a known concentration of beads was added to each sample which could be easily identified from bacterial whole cells and cell debris by SYTO 9 staining and forward scatter. The actual viable cell numbers in the sample could be determined based on the ratio of the number of viable bacteria and the number of events in the gate containing the fluorescent counting beads. Using FCM methods *T. denticola* could be detected and enumerated by comparison with a known number of counting beads within a positive linear correlation coefficient (R^2) of 0.9874.

Physiology of microbial pathogens and FCM analysis

The ability to distinguish different physiological states is important in order to assess the survival and growth of pathogenic microorganisms in various environments such as aquatic environments (Allegra *et al.*, 2008; Berney *et al.*, 2007) or to assess the effects of processing treatments and potential antibacterial treatments on pathogen response (Cronin and Wilkinson, 2008; Kennedy *et al.*, 2011). Bacterial responses and physiological states are of particular interest to the food industry as bacteria in various physiological states can be incorporated into food matrices where they are exposed to rapid and dynamic changes in environmental conditions which can allow recovery and outgrowth (Cronin and Wilkinson, 2009; Doherty *et al.*, 2010). Pianetti *et al.* (2008) used FCM to assess the survival and viability of a pathogenic strain of *Aeromonas hydrophilia* at various salt concentrations and temperatures over an extended period (188 days). SYBR Green and PI were used to stain the cells prior to FCM analysis for total cell count and viability studies. Membrane effects were assessed by adding 3,3′-diethyloxadicarbocyanine ($DiOC_2$). Data obtained by FCM were generally comparable to those obtained by the plate count method. However, in the presence of 6% NaCl differences were noted between the numbers of viable cells detected by FCM and by agar plate counts. These workers also noted that trends resulting in populations with high membrane potential values, indicative of membrane integrity, were strongly correlated with the viable cells. This suggested that $DiOC_2$ may be a useful fluorochrome for detecting changes in the membrane potential of *Aeromonas* cells and changes in their viability.

The application of staining probes for physiological evaluation of the target microorganism for certain samples also requires a detailed knowledge of the effect of the dye itself on the cell itself as well as the physiology and structure of the microorganism (Kennedy *et al.*, 2011; Shi *et al.*, 2007). Bacterial endospores of *Bacillus* and *Paenibacillus* were characterized into live, dead, and germinating, based on a number of physiological properties monitored by staining with cytometric dyes (Comas-Riu and Vives-Rego, 2002; Cronin and Wilkinson, 2007). Using simultaneous staining protocols with different fluorochromes, FCM can reveal previously unknown levels of heterogeneity within a culture arising as a result of various degrees of cell injury and differing physiological states (Berney *et al.*, 2007; Davey, 2011; Want *et al.*, 2011). These studies tracked physiological states of bacteria subjected to different environments or growth limiting factors using stains such as PI and SYTO9, 3,3-dihexylocarbocyanine iodide-$DiOC_6(3)$. An increasing range of functional probes exist, mainly for assessing microbial membrane potential, intracellular pH and metabolic activity within bacterial cells. Viability is typically classified as the ability of cells to form a colony on an agar plate (Davey, 2011; Davey and Kell, 1996). However, the use of FCM as a microbiological tool casts new light on this definition. Using FCM, it is not necessary to culture an organism to determine its viability, at least in terms of metabolic state. Plate counts, although usually considered to be the 'gold standard' of viability measurement indicates only how many of the cells can replicate under the specific conditions provided for growth. FCM viability staining meanwhile provides data on how many of the cells can exclude, become permeable to, or utilize a particular stain or combinations of stains (Davey, 2011). Excellent descriptions of the range of dyes that are available to the microbiologist interested in FCM are published in a number of reviews (Alvarez-Barrientos *et al.*, 2000; Joux and Lebaron, 2000; Nebe-von-Caron *et al.*, 2000; Shapiro, 2003; Sincock and Robinson, 2001; Tracy *et al.*, 2010).

Much work has been reported regarding the suitability of various probes and combinations of probes for different assays, generating varied outcomes and recommendations. FCM methodology has also allowed development of quantitative procedures to assess antimicrobial susceptibility and drug cytotoxicity using rapid, accurate, and highly reproducible assays (Gauthier *et al.*, 2002; Mason *et al.*, 1994; Pore, 1994). The importance of using a number of viability indicators to characterize the physiological state of stressed bacteria has been emphasized by Berney *et al.* (2006) Hewitt and Nebe-von-Caron (2004) and Nocker (2011) as each parameter reflects differing levels of cellular integrity and functionality. Therefore the sum of multiple parameters (including culturability) may be employed to generate a wider perspective on the overall physiological state of bacteria. The use of cell sorting to evaluate culturability has also led to advances in understanding and interpretation of physiological states and staining methods. (Kennedy *et al.*, 2011; Nebe-von-Caron *et al.*, 2000; Want *et al.*, 2011). In these cases it has been used to establish the relationship between the culturable state of the bacteria on agar that had been classified by FCM as permeabilized/injured or dead.

In summary, the presence of a range of cellular features are considered as prerequisites for a 'living' cell and many of these can be rapidly and simultaneously measured by FCM. The viability-related parameters measurable by FCM includes: enzyme activity such as intracellular enzymes responsible for key reactions in metabolic pathways, membrane polarization reflecting the maintenance of the cellular energy status, membrane integrity to indicate the presence of an intact functional barrier between the extracellular environment and the cytoplasm. These parameters which are easily measurable by FCM are of great relevance when studying the physiological state of pathogens originating from clinical, food or environmental samples.

Cell separation/cell sorting

Sorting of bacteria by fluorescence activated cell sorting (FACS) has tremendous potential for sensitive and selective analysis and subsequent study of particular bacterial subpopulations in samples. This ability to detect, identify, and enumerate cells in a single device based on phenotype has proven to be invaluable in the fields of medicine, food science, military applications, and basic and clinical research (Pappas and Wang, 2007). However, while FACS is well established as a reliable method for sorting cells, it can be expensive, voluminous, and requires dedicated highly trained staff to operate the equipment (Pappas and Wang, 2007). Using cell sorters, cells with a specific characteristic or indeed a combination of characteristics can be separated from the sample for further analysis or growth (Winson and Davey, 2000). Many cell separation methods use immunofluorescence, previously described in this chapter, as the triggering signal for detecting and isolating pathogens of interest (Lazcka *et al.*, 2007; Pappas and Wang, 2007). The use of cell sorting has many novel applications in the analysis and study of pathogens. FACS has been used as a novel enrichment technique to isolate herpes simplex viruses (HSV) for further study (Kolb and Brandt, 2004). FCM and FACS have been employed to study host–pathogen interactions. For example, FACS methods have been used to compare the infectivity and other properties of host cells, e.g. weakly or highly *Salmonella*-infected cells (Thöne *et al.*, 2007). Spleen cells that differed in *Salmonella* infectivity were sorted and further analysed for the presence of cell surface markers to better understand the relevance of different host cell populations in *Salmonella* infections. Martinon *et al.* (2011) demonstrated that sorting of single cells using FACS was useful to generate defined population standards for quantification of *S. aureus* by PCR. This allowed improved standard curves for PCR quantification of *S. aureus* to be generated based on the fact that single cells could be sorted to give highly defined populations. FCM has been highlighted as potentially useful in studying antibiotic resistance of pathogenic microorganisms (Jarzembowski *et al.*, 2008). In this study, the heterogeneity of populations treated with antibiotics was revealed using fluorescent penicillin. Isolated strains of *Enterococcus faecalis* and *S. aureus* were compared with known penicillin-resistant and penicillin-sensitive strains. Fluorescent intensities were lower in penicillin resistant and suspected penicillin resistant strains. The presence of dormant endospores in certain foods is a likely cause of food-borne poisoning. Bacterial endospores of *Bacillus* and *Paenibacillus* have been characterized into live, dead, and germinating, based on a number of physiological properties achieved by staining with cytometric dyes (Comas-Riu and Vives-Rego, 2002; Cronin and Wilkinson, 2007)

Conclusion

This review has focused on the application and potential of FCM as a reliable technique for use in pathogen detection in a range of different sample types. It could be argued that if such methods were to become automated they would remove costs associated with highly trained microbiologists and continuous on-site operation would speed up detection. However, there is a huge challenge for microbiologists in standardizing FCM assays particularly from complex samples from various food/clinical/environmental origins. It also remains to be seen how quickly regulatory authorities will assess and allow FCM based methods to become part of the methodology for routine analysis of pathogens. Overall, it is clear that FCM offers a variety of new possibilities for rapid species and strain-specific labelling of target pathogens.

References

Allegra, S., Berger, F., Berthelot, P., Grattard, F., Pozzetto, B., and Riffard, S. (2008). Use of flow cytometry to monitor *Legionella* Viability. Appl. Environ. Microbiol. *74*, 7813–7816.

Alvarez-Barrientos, A., Arroyo, J., Canton, R., Nombela, C., and Sanchez-Perez, M. (2000). Applications of flow cytometry to clinical microbiology. Clin. Microbiol. Rev. *13*, 167–195.

Banerjee, A., Yasseri, M., and Munson, M. (2002). A method for the detection and quantification of bacteria in human carious dentine using fluorescent *in situ* hybridisation. J. Dent. *30*, 359–363.

Berney, M., Weilenmann, H.-U., and Egli, T. (2006). Flow-cytometric study of vital cellular functions in *Escherichia coli* during solar disinfection (SODIS). Microbiology *152*, 1719–1729.

Berney, M., Hammes, F., Bosshard, F., Weilenmann, H.-U., and Egli, T. (2007). Assessment and interpretation of bacterial viability by using the LIVE/DEAD BacLight Kit in combination with flow cytometry. Appl. Environ. Microbiol. *73*, 3283–3290.

Bunthof, C.J., and Abee, T. (2002). Development of a flow cytometric method to analyze subpopulations of bacteria in probiotic products and dairy starters. Appl. Environ. Microbiol. *68*, 2934–2942.

Chalfie, M. (1995). Green fluorescent protein. Photochem. Photobiol. *62*, 651–656.

Chattopadhyay, P.K., Price, D.A., Harper, T.F., Betts, M.R., Yu, J., Gostick, E., Perfetto, S.P., Goepfert, P., Koup, R.A., De Rosa, S.C., *et al.* (2006). Quantum dot semiconductor nanocrystals for immunophenotyping by polychromatic flow cytometry. Nat. Med. *12*, 972–977.

Comas-Riu, J., and Vives-Rego, J. (2002). Cytometric monitoring of growth, sporogenesis and spore cell sorting in *Paenibacillus polymyxa* (formerly *Bacillus polymyxa*). J. Appl. Microbiol. *92*, 475–481.

Cronin, U.P., and Wilkinson, M.G. (2007). The use of flow cytometry to study the germination of *Bacillus cereus* endospores. Cytometry A *71*, 143–153.

Cronin, U.P., and Wilkinson, M.G. (2008). Physiological response of *Bacillus cereus* vegetative cells to simulated food processing treatments. J. Food Protect. *71*, 2168–2176.

Cronin, U.P., and Wilkinson, M.G. (2009). The growth, physiology and toxigenic potential of *Bacillus cereus* in cooked rice during storage temperature abuse. Food Control *20*, 822–828.

Davey, H.M. (2011). Life, death, and in-between: meanings and methods in microbiology. Appl. Environ. Microbiol. *77*, 5571–5576.

Davey, H.M., and Kell, D.B. (1996). Flow cytometry and cell sorting of heterogeneous microbial populations: the importance of single-cell analyses. Microbiol. Rev. *60*, 641–696.

Dixon, B.R., Bussey, J.M., Parrington, L.J., and Parenteau, M. (2005). Detection of *Cyclospora cayetanensis* oocysts in human fecal specimens by flow cytometry. J. Clin. Microbiol. *43*, 2375–2379.

Doherty, S.B., Wang, L., Ross, R.P., Stanton, C., Fitzgerald, G.F., and Brodkorb, A. (2010). Use of viability staining in combination with flow cytometry for rapid viability assessment of *Lactobacillus rhamnosus* GG in complex protein matrices. J. Microbiol. Methods *82*, 301–310.

Douesnard-Malo, F., and Daigle, F. (2011). Increased persistence of *Salmonella enterica* serovar Typhi in presence of *Acanthamoeba castellanii*. Appl. Environ. Microbiol. *77*, 7640–7646.

Edgar, R., McKinstry, M., Hwang, J., Oppenheim, A.B., Fekete, R.A., Giulian, G., Merril, C., Nagashima, K., and Adhya, S. (2006). High-sensitivity bacterial detection using biotin-tagged phage and quantum-dot nanocomplexes. Proc. Natl. Acad. Sci. U.S.A. *103*, 4841–4845.

Entis, P., Fung, D., Griffiths, M., McIntyre, L., Russell, S., Sharpe, A.N., and Tortorello, M.L. (2001). Rapid methods for detection, identification, and enumeration. In Compendium of Methods for the Microbiological Examination of Foods, F.P. Downes, and K. Ito, eds. (American Public Health Association, Washington, DC, USA), pp. 89–126.

Favrin, S.J., Jassim, S.A., and Griffiths, M.W. (2003). Application of a novel immunomagnetic separation-bacteriophage assay for the detection of Salmonella enteritidis and Escherichia coli O157:H7 in food. Int. J. Food Microbiol. *85*, 63–71.

Ferrari, B.C., and Bergquist, P.L. (2007). Quantum Dots as alternatives to organic fluorophores for *Cryptosporidium* detection using conventional flow cytometry and specific monoclonal antibodies: lessons learned. Cytometry A *71*, 265–271.

Fleet, G.H. (1999). Microorganisms in food ecosystems. Int. J. Food Microbiol. *50*, 101–117.

Flint, S., Drocourt, J.-L., Walker, K., Stevenson, B., Dwyer, M., Clarke, I., and McGill, D. (2006). A rapid, two-hour method for the enumeration of total viable bacteria in samples from commercial milk powder and whey protein concentrate powder manufacturing plants. Int. Dairy J. *16*, 379–384.

Gauthier, C., St-Pierre, Y., and Villemur, R. (2002). Rapid antimicrobial susceptibility testing of urinary tract isolates and samples by flow cytometry. J. Med. Microbiol. *51*, 192–200.

Golan, A., Kerem, Z., Tun, O.M., Luzzatto, T., Lipsky, A., and Yedidia, I. (2010). Combining flow cytometry and *gfp* reporter gene for quantitative evaluation of *Pectpbacterium carotovorum* ssp. *carotovorum* in *Ornithogalum dubium* plantlets. J. Appl. Microbiol. *108*, 1136–1144.

Grootaert, C., Boon, N., Zeka, F., Vanhoecke, B., Bracke, M., Verstraete, W., and Van de Wiele, T. (2011). Adherence and viability of intestinal bacteria to differentiated Caco-2 cells quantified by flow cytometry. J. Microbiol. Methods *86*, 33–41.

Gunasekera, T.S., Veal, D.A., and Attfield, P.V. (2003). Potential for broad applications of flow cytometry and fluorescence techniques in microbiological and somatic cell analyses of milk. Int. J. Food Microbiol. *85*, 269–279.

Guttikonda, S., Tang, X.L., Yang, B.M., Armstrong, G.D., and Suresh, M.R. (2007). Monospecific and bispecific antibodies against *E. coli* O157 for diagnostics. J. Immunol. Methods *327*, 1–9.

Hammes, F., Berney, M., Wang, Y., Vital, M., Köster, O., and Egli, T. (2008). Flow-cytometric total bacterial cell counts as a descriptive microbiological parameter for drinking water treatment processes. Water Res. *42*, 269–277.

Harkins, K.R., and Harrigan, K. (2004). Labeling of bacterial pathogens for flow cytometric detection and enumeration. Curr. Protoc. Cytom. *11*, 11.17.

Hartmann, H., Stender, H., Schäfer, A., Autenrieth, I.B., and Kempf, V.A.J. (2005). Rapid identification of *Staphylococcus aureus* in blood cultures by a combination of fluorescence *in situ* hybridization using peptide nucleic acid probes and flow cytometry. J. Clin. Microbiol. *43*, 4855–4857.

Herzenberg, L.A., De Rosa, S.C., and Herzenberg, L.A. (2000). Monoclonal antibodies and the FACS: complementary tools for immunobiology and medicine. Immunol. Today *21*, 383–390.

Hewitt, C.J., and Nebe von Caron, G. (2004). The application of multi-parameter flow cytometry to monitor individual microbial cell physiological state. Adv. Biochem. Eng. Biotechnol. *89*, 197–223.

Hibi, K., Abe, A., Ohashi, E., Mitsubayashi, K., Ushio, H., Hayashi, T., Ren, H., and Endo, H. (2006). Combination of immunomagnetic separation with flow cytometry for detection of *Listeria monocytogenes*. Anal. Chim. Acta *573–574*, 158–163.

Holm, C., and Jespersen, L. (2003). A flow-cytometric Gram-staining technique for milk-associated bacteria. Appl. Environ. Microbiol. *69*, 2857–2863.

Holm, C., Mathiasen, T., and Jespersen, L. (2004). A flow cytometric technique for quantification and differentiation of bacteria in bulk tank milk. J. Appl. Microbiol. *97*, 935–941.

Honda, J., Okubo, Y., Imamura, Y., Kusaba, M., Saruwatari, N., and Oizumi, K. (1997). Flow cytometric detection of cytomegalovirus antigen in peripheral blood cells after bone marrow transplantation. Br. J. Haematol. *99*, 415–417.

Hussein, H.S., Thran, B.H., and Redelman, D. (2002). Detection of *Escherichia coli* O157:H7 in bovine rumen fluid and feces by flow cytometry. Food Control *13*, 387–391.

Ibanez-Peral, R., Bergquist, P.L., Walter, M.R., Gibbs, M., Goldys, E.M., and Ferrari, B. (2008). Potential use of quantum dots in flow cytometry. Int. J. Mol. Sci. *9*, 2622–2638.

Ivnitski, D., Abdel-Hamid, I., Atanasov, P., and Wilkins, E. (1999). Biosensors for detection of pathogenic bacteria. Bios. Bioelectron. *14*, 599–624.

Jarzembowski, T., Wiśniewska, K., Józwik, A., Bryl, E., and Witkowski, J. (2008). Flow cytometry as a rapid test for detection of penicillin resistance directly in bacterial cells in *Enterococcus faecalis* and *Staphylococcus aureus*. Curr. Microbiol. *57*, 167–169.

Joux, F., and Lebaron, P. (2000). Use of fluorescent probes to assess physiological functions of bacteria at single-cell level. Microbes Infect. *2*, 1523–1535.

Jung, Y.S., Frank, J.F., and Brackett, R.E. (2003). Evaluation of antibodies for immunomagnetic separation combined with flow cytometry detection of *Listeria monocytogenes*. J. Food Protect. *66*, 1283–1287.

Kadkhoda, K., Manickam, K., Degagne, P., Sokolowski, P., Pang, P., Kontzie, N., and Alfa, M. (2011). UF-1000i flow cytometry is an effective screening method for urine specimens. Diagn. Microbiol. Infect. Dis. *69*, 130–136.

Kalyuzhnaya, M.G., Zabinsky, R., Bowerman, S., Baker, D.R., Lidstrom, M.E., and Chistoserdova, L. (2006). Fluorescence *in situ* hybridization-flow cytometry-cell sorting-based method for separation and enrichment of type I and type II methanotroph populations. Appl. Environ. Microbiol. *72*, 4293–4301.

Kempf, V.A.J., Trebesius, K., and Autenrieth, I.B. (2000). Fluorescent *in situ* hybridization allows rapid identification of microorganisms in blood cultures. J. Clin. Microbiol. *38*, 830–838.

Kempf, V.A.J., Mändle, T., Schumacher, U., Schäfer, A., and Autenrieth, I.B. (2005). Rapid detection and identification of pathogens in blood cultures by fluorescence *in situ* hybridization and flow cytometry. Int. J. Med. Microbiol. *295*, 47–55.

Kennedy, D., Cronin, U.P., and Wilkinson, M.G. (2011). The response of *Escherichia coli*, *Listeria monocytogenes* and *Staphylococcus aureus* to simulated food processing treatments using fluorescence activated cell sorting and plate counting. Appl. Environ. Microbiol. *77*, 4657–4668.

Kenny, J.H., Zhou, Y., Schriefer, M.E., and Bearden, S.W. (2008). Detection of viable *Yersinia pestis* by fluorescence *in situ* hybridization using peptide nucleic acid probes. J. Microbiol. Methods *75*, 293–301.

King, D.N. (2007). A critical evaluation of a flow cytometer used for detecting enterococci in recreational waters. J. Water Health *5*, 295–305.

Kolb, A.W., and Brandt, C.R. (2004). Enhanced isolation of low frequency herpes simplex virus recombinants using green-fluorescent protein and FACS. J. Virol. Methods *115*, 73–81.

Langerhuus, S.N., Ingvartsen, K.L., Bennedsgaard, T.W., and Røntved, C.M. (2013). Gram-typing of mastitis bacteria in milk samples using flow cytometry. J. Dairy Sci. *96*, 267–277.

Lautenschlager, K., Hwang, C., Liu, W.-T., Boon, N., Köster, O., Vrouwenvelder, H., Egli, T., and Hammes, F. (2013). A microbiology-based multi-parametric approach towards assessing biological stability in drinking water distribution networks. Water Res. *47*, 3015–3025.

Lazcka, O., Campo, F.J.D., and Munoz, F.X. (2007). Pathogen detection: A perspective of traditional methods and biosensors. Bios. Bioelectron. *22*, 1205–1217.

McCarthy, M., and Culloty, S. (2011). Optimization of two immunofluorescent antibodies for the detection of *Escherichia coli* using immunofluorescent microscopy and flow cytometry. Curr. Microbiol. *62*, 402–408.

McClelland, R.G., and Pinder, A.C. (1994). Detection of *Salmonella typhimurium* in dairy products with flow cytometry and monoclonal antibodies. Appl. Environ. Microbiol. *60*, 4255–4262.

McSharry, J.J., Costantino, R., Robbiano, E., Echols, R., Stevens, R., and Lehman, J.M. (1990). Detection and quantitation of human immunodeficiency virus-infected peripheral blood mononuclear cells by flow cytometry. J. Clin. Microbiol. *28*, 724–733.

Martinon, A., Cronin, U., and Wilkinson, M. (2011). Development of defined microbial population standards using fluorescence activated cell sorting for the absolute quantification of *S. aureus* using real-time PCR. Mol. Biotechnol. *50*, 1–10.

Mason, D.J., Allman, R., Stark, J.M., and Lloyd, D. (1994). Rapid estimation of bacterial antibiotic susceptibility with flow cytometry. J. Microsc. *176*, 8–16.

Melamed, A., and Sorvillo, F.J. (2009). The burden of sepsis-associated mortality in the United States from 1999 to 2005: an analysis of multiple-cause-of-death data. Critical Care (London, England) *13*, R28.

Mocz, G. (2012). The fluorophore of green fluorescent protein. In The Use of Fluorescence in Research into Protein Structures. Available online: http://dwb.unl.edu/Teacher/NSF/C08/C08Links/pps99.cryst.bbk.ac.uk/projects/gmocz/gfp.htm. Accessed 2 April 2015.

Moter, A., and Göbel, U.B. (2000). Fluorescence *in situ* hybridization (FISH) for direct visualization of microorganisms. J. Microbiol. Methods *41*, 85–112.

Nebe-von-Caron, G., Stephens, P.J., Hewitt, C.J., Powell, J.R., and Badley, R.A. (2000). Analysis of bacterial function by multi-colour fluorescence flow cytometry and single cell sorting. J. Microbiol. Methods *42*, 97–114.

Nocker, A., Caspers, M., Esveld-Amanatidou, A., van der Vossen, J., Schuren, F., Montijn, R., and Kort, R. (2011). A multiparameter viability assay for stress profiling applied to the food pathogen *Listeria monocytogenes* F2365. Appl. Environ. Microbiol. *77*, 6433–6440.

Orth, R., O'Brien-Simpson, N., Dashper, S., Walsh, K., and Reynolds, E. (2009). An efficient method for enumerating oral spirochetes using flow cytometry. J. Microbiol. Methods *80*, 123–128.

Pappas, D., and Wang, K. (2007). Cellular separations: A review of new challenges in analytical chemistry. Anal. Chim. Acta *601*, 26–35.

Pianetti, A., Manti, A., Boi, P., Citterio, B., Sabatini, L., Papa, S., Rocchi, M.B.L., and Bruscolini, F. (2008). Determination of viability of *Aeromonas hydrophila* in increasing concentrations of sodium chloride at different temperatures by flow cytometry and plate count technique. Int. J. Food Microbiol. *127*, 252–260.

Pils, S., Schmitter, T., Neske, F., and Hauck, C.R. (2006). Quantification of bacterial invasion into adherent cells by flow cytometry. J. Microbiol. Methods *65*, 301–310.

Pore, R.S. (1994). Antibiotic susceptibility testing by flow cytometry. J. Antimicrob. Chemother. *34*, 613–627.

Power, M.L., Shanker, S.R., Sangster, N.C., and Veal, D.A. (2003). Evaluation of a combined immunomagnetic separation/flow cytometry technique for epidemiological investigations of *Cryptosporidium* in domestic and Australian native animals. Vet. Parasitol. *112*, 21–31.

Raybourne, R.B., and Tortorello, M.L. (2003). Microscopy techniques: DEFT and flow cytometry. In Detecting Pathogens in Food, T.A. McMeekin, ed. (Woodhead Publishing, Cambridge, UK), pp. 186–216.

Rigottier-Gois, L., Rochet, V., Garrec, N., Suau, A., and Doré, J. (2003). Enumeration of bacteroides species in human faeces by fluorescent *in situ* hybridisation combined with flow cytometry using 16S rRNA probes. Syst. Appl. Microbiol. *26*, 110–118.

Riyaz-Ul-Hassan, S., Verma, V., and Qazi, G.N. (2008). Evaluation of three different molecular markers for the detection of *Staphylococcus aureus* by polymerase chain reaction. Food Microbiol. *25*, 452–459.

Robinson, J.P. (2004). Overview of flow cytometry and microbiology. Curr. Protoc. Cytom. *30* (Unit 11.1), 11.1.1–11.1.4.

Rotariu, O., Ogden, I.D., MacRae, M., Badescu, V., and Strachan, N.J.C. (2005). An immunomagnetic separator for concentration of pathogenic micro-organisms from large volume samples. J. Magn. Magn. Mater. *293*, 589–596.

Ryumae, U., Hibi, K., Yoshiura, Y., Ren, H., and Endo, H. (2010). Rapid and highly sensitive detection of *Flavobacterium psychrophilum* using high gradient immunomagnetic separation with flow cytometry. Aquaculture *309*, 125–130.

Schellenberg, J., Smoragiewicz, W., and Karska-Wysocki, B. (2006). A rapid method combining immunofluorescence and flow cytometry for improved understanding of competitive interactions between lactic acid bacteria (LAB) and methicillin-resistant *S. aureus* (MRSA) in mixed culture. J. Microbiol. Methods *65*, 1–9.

Schellenberg, J., Blake Ball, T., Lane, M., Cheang, M., and Plummer, F. (2008). Flow cytometric quantification of bacteria in vaginal swab samples self-collected by adolescents attending a gynecology clinic. J. Microbiol. Methods *73*, 216–226.

Schmid, I., Lambert, C., Ambrozak, D., and Perfetto, S.P. (2007). Standard safety practices for sorting of unfixed cells. Curr. Protoc. Cytom. *39* (Unit 3.6), 3.6.1–3.6.20.

Shapiro, H.M. (2003). Practical Flow Cytometry, 4th edn (Wiley-Liss, Hoboken, NJ, USA).

Shi, L., Günther, S., Hübschmann, T., Wick, L.Y., Harms, H., and Müller, S. (2007). Limits of propidium iodide as a cell viability indicator for environmental bacteria. Cytometry A *71*, 592–598.

Sincock, S.A., and Robinson, J.P. (2001). Flow cytometric analysis of microorganisms. Methods Cell Biol. *64*, 511–537.

Stein, D.C. (1998). Chapter 7 – Principles of bacterial pathogenesis. Prin. Med. Biol. *9*, 85–97.

Stender, H., Fiandaca, M., Hyldig-Nielsen, J.J., and Coull, J. (2002). PNA for rapid microbiology. J. Microbiol. Methods *48*, 1–17.

Subires, A., Yuste, J., and Capellas, M. (2014). Flow cytometry immunodetection and membrane integrity assessment of Escherichia coli O157:H7 in ready-to-eat pasta salad during refrigerated storage. Int. J. Food Microbiol. *168–169*, 47–56.

Summers, H.D., Holton, M.D., Rees, P., Williams, P.M., and Thornton, C.A. (2010). Analysis of quantum dot fluorescence stability in primary blood mononuclear cells. Cytometry A *77*, 933–939.

Sunwoo, H.H., Wang, W.W., and Sim, J.S. (2006). Detection of *Escherichia coli* O157:H7 using chicken immunoglobulin Y. Immunol. Lett. *106*, 191–193.

Thöne, F., Schwanhäusser, B., Becker, D., Ballmaier, M., and Bumann, D. (2007). FACS-isolation of *Salmonella*-infected cells with defined bacterial load from mouse spleen. J. Microbiol. Methods *71*, 220–224.

Tortorello, M.L., Reineke, K.F., Stewart, D.S., and Raybourne, R.B. (1998). Comparison of methods for determining the presence of *Escherichia coli* O157:H7 in apple juice. J. Food Protect. *61*, 1425–1430.

Tracy, B.P., Gaida, S.M., and Papoutsakis, E.T. (2010). Flow cytometry for bacteria: enabling metabolic engineering, synthetic biology and the elucidation of complex phenotypes. Curr. Opin. Biotechnol. *21*, 85–99.

Trouillet, S., Rasigade, J.-P., Lhoste, Y., Ferry, T., Vandenesch, F., Etienne, J., and Laurent, F. (2011). A novel flow cytometry-based assay for the quantification of *Staphylococcus aureus* adhesion to and invasion of eukaryotic cells. J. Microbiol. Methods *86*, 145–149.

Ueckert, J., Breeuwer, P., Abee, T., Stephens, P., von Caron, G.N., and ter Steeg, P.F. (1995). Flow cytometry applications in physiological study and detection of foodborne microorganisms. Int. J. Food Microbiol. *28*, 317–326.

Untermann, F. (1998). Microbial hazards in food. Food Control 9, 119–126.

Vora, G.J., Meador, C.E., Stenger, D.A., and Andreadis, J.D. (2004). Nucleic acid amplification strategies for DNA microarray-based pathogen detection. Appl. Environ. Microbiol. *70*, 3047–3054.

Want, A., Hancocks, H., Thomas, C., Stocks, S., Nebe-von-Caron, G., and Hewitt, C. (2011). Multi-parameter flow cytometry and cell sorting reveal extensive physiological heterogeneity in *Bacillus cereus* batch cultures. Biotechnol. Lett. 33, 1395–1405.

Wilkes, J.G., Tucker, R.K., Montgomery, J.A., Cooper, W.M., Sutherland, J.B., and Buzatu, D.A. (2012). Reduction of food matrix interference by a combination of sample preparation and multi-dimensional gating techniques to facilitate rapid, high sensitivity analysis for Escherichia coli serotype O157 by flow cytometry. Food Microbiol. *30*, 281–288.

Winson, M.K., and Davey, H.M. (2000). Flow cytometric analysis of microorganisms. Methods Enzymol. *21*, 231–240.

Zahavy, E., Ber, R., Gur, D., Abramovich, H., Freeman, E., Maoz, S., and Yitzhaki, S. (2012). Application of nanoparticles for the detection and sorting of pathogenic bacteria by flow-cytometry. Adv. Exp. Med. Biol. *733*, 23–36.

The Use of Flow Cytometry to Study Sporeforming Bacteria

6

Ultan Cronin

Abstract

Spore formers comprise a diverse group of important bacteria and possess two unique traits – the ability to form dormant and resistant spores and their ability to restart metabolism and germinate. Flow cytometry (FCM) offers many possibilities for the study of spore formers, bringing advantages in terms of speed, novelty and the ability to measure heterogeneity within a population. In this chapter, five categories of FCM studies involving spore formers (mainly involving species within the genera, *Bacillus* and *Clostridium*) have been identified: those based on autofluorescence or light scatter; those involving antibodies or fluorescence *in situ* hybridization probes; those using reporter proteins in the context of genetic transformation; those involving permeability dyes; those utilizing physiological dyes. In the past, instrumentation was expensive, difficult to operate and the range of fluorescent dyes and antibodies available to the microbiologist was limited. In the last half a decade, however, the number of low-cost, easy-to-run and powerful (two or more lasers, six or more detectors) digital instruments has mushroomed. In a similar vein, the variety of monoclonal antibodies and dyes available to the microbiologist is rapidly increasing. Both developments allow the researcher interested in spore formers to design more information-rich and daring experiments than ever before. Experiments combining the use of fluorescent reporter proteins, physiological dyes, a permeability marker and a monoclonal antibody are now possible. Using antibodies or fluorescent dyes specific to particular structural components of endospores such as the exosporium or protein coat, FCM may become a very useful tool in the real-time analysis of sporulation or resistance mechanisms. FISH probes could be used for environmental studies. The growing variety of reporter proteins such as GFP or mCherry allows the possibility of simultaneously studying the expression of a number of genes, an experimental approach which may help unravel the sequence of events leading to sporulation. The use of multiplex bead arrays could prove a powerful tool in characterizing a species' repertoire of secreted proteins or toxins. Of prime importance in implementing new methods for the study of spore formers is their comparison with existing gold standard techniques. Even a research tool as powerful as FCM must be shown to yield data as robust and reliable as conventional methods.

Introduction

The study of bacterial spores using flow cytometry (FCM) is almost as venerable as the field of microbial FCM itself, with a series of seminal studies on the detection of *Bacillus* spores among some of the earliest studies in the field (e.g. Phillips and Martin, 1983). In the almost 30 years since scientists first applied fluorescent antibodies to suspensions of bacterial spores and collected the fluorescence emitted by each individual spore as it passed through a flow cell, the general field of cytometry has seen quantum advances in terms of instrumentation and the availability of new reagents such as antibodies and physiological dyes. While immunologists and oncologists have taken full advantage of these improvements, microbiologists have been late adaptors, rarely going beyond the use of single or dual staining or harnessing the potential of FCM as a high throughput analytical technology. It is the author's intention in this chapter to succinctly review past flow cytometry studies of microbial spores, present the current state of the art and offer some suggestions as to how full advantage may be taken of the unique resolving power and sensitivity of modern flow cytometers and sorters such that FCM can be fully exploited in the illumination of matters of great importance to microbiologists interested in spore biology.

The importance of spore formers

Bacterial spores are an outstanding example of prokaryotic differentiation and their formation is a survival mechanism *par excellence* (Errington, 2010; Hornstra *et al.*, 2009). There exists a huge variety of spore formers, ranging from familiar and well-studied members of the genera, *Bacillus* or *Clostridium*, to less well known (but no less deserving of study) members of the Actinomycetes to members of genera such as *Myxococcus*, which possess complex and intriguing life cycles (Berleman and Kirby, 2009; Kimura *et al.*, 2010). All branches of microbiology, with the notable of exception of marine microbiology, contain important spore-forming species and, more and more, these species are the focus of flow cytometric studies. Of importance to medical microbiologists are members of the genera *Bacillus* and *Clostridium*. *B. anthracis* is the notorious and feared causative agent of anthrax, while *B. cereus* and other members of the genus have been implicated in cases of food poisoning and opportunistic and nosocomial infections. *C. botulinum* is responsible for botulism, *C. tetani* for tetanus, *C. difficile* for infectious diarrhoea and *C. perfringens* for food infections (Burns *et al.*, 2010; Weese *et al.*, 2010). Food microbiologists study spoilage and food poisoning organisms from both the aforementioned genera, as well as emerging organisms from other genera such as *Anoxybacillus* and *Geobacillus* (Burgess *et al.*, 2010). While environmental and plant microbiologists are especially interested in members of the Actinomycetes, whose role in the soil and rhizosphere is becoming better understood each day, there are a number of other spore-forming genera such as *Sporosarcina* and those referred to above that have important roles in these ecosystems. Being highly efficient secretors of secondary metabolites, members of the genus *Bacillus* are of great value to industrial microbiologists (Errington, 2010). Additionally, Actinomycetes, with their broad and varied metabolic pathways, offer novel approaches to microbiologists in the field of bioremediation or industrial fermentation (Cheng *et al.*, 2009).

Two unique aspects of spore formers are their ability to form spores – sporulation – and the capacity for these spores to later revert from the dormant to the vegetative state – germination.

Both of these processes occur in response to extremes of nutrient availability; sporulation occurs when nutrients are limited and germination when these are plentiful (Higgins and Dworkin, 2012). Additionally, both of these processes are under tight genetic control and reach completion following a highly regulated cascade of gene expression and silencing (Paredes-Sabja *et al.*, 2011). Very briefly, during sporulation, asymmetric cell division gives rise to a large mother and a small daughter cell. The former engulfs and subsequently controls the development of the latter, with the eventual lysis of the former to produce a compact, robust, dehydrated and metabolically inactive spore (Burns and Minton, 2011). During germination, the endospore undergoes a very rapid series of changes, including the loss of Ca^{2+} ions and dipicolinic acid from the core, the breakdown of the cortex and the re-hydration of the core, which subsequently re-initiates metabolism, begins to outgrow and resemble a normal vegetative cell.

Spore formers and flow cytometry: a review of past studies

The majority of flow cytometric studies of spore formers has involved species within the genera, *Bacillus* and *Clostridium*. Outside of these genera, very few examples of the use of FCM for the study of spore formers exist. A comprehensive review of the potential of FCM in the study of *B. cereus* by the current author and editor (Cronin and Wilkinson,

2010) detailed the experimental approaches involving FCM which have been employed in the past and those which could be theoretically applied in the future to the study of this species. Accepting the large differences between spore formers from different genera, in terms of life cycle and structure, flow cytometric strategies which have proved successful for *B. cereus* and other members of the genus *Bacillus* have wide application in the study of other spore formers. Therefore, a summary of flow cytometric studies of species within the genus *Bacillus* is presented in Table 6.1. For the purposes of convenience, flow cytometric studies involving spore formers may be divided into five categories: those harnessing the natural autofluorescence or light scatter of spores or vegetative cells; those involving the use of antibodies or fluorescent *in situ* hybridization (FISH) probes; those using reporter proteins such as green fluorescent protein (GFP) in the context of genetic transformation; those involving permeability dyes such as propidium iodide (PI) or SYTOX Green; those utilizing physiological dyes such as carboxyfluorescein diacetate (CFDA) which yield information on some aspect of vegetative cell or spore metabolism.

Studies involving the use of autofluorescence or scatter

The simplest and least costly FCM studies, performed without the application of expensive probes or dyes to cells or spores, rely on scatter and autofluorescence. Theoretically, two particles as distinct in refractive index as the spore and vegetative cell should have very different forward and side scatter profiles when analysed by FCM.

Table 6.1 Examples of flow cytometric studies carried out on spore formers within the genus, *Bacillus*, in recent years

Topic of study	Species	Stains used	Authors
Bacterial surface display	*B. subtilis*	GFP	Kim *et al.* (2007)
Genetic control of sporulation	*B. subtilis*	GFP	Lulko *et al.* (2007), Veening *et al.* (2005, 2006)
Factors influencing germination	*B. cereus*	CFDA	Cronin and Wilkinson (2007)
Gene expression during germination	*B. subtilis*	GFP	Smits *et al.* (2005), Veening *et al.* (2005, 2006)
Spore structure	*B. anthracis*	Fl Ab	Thompson *et al.* (2007)
Spore structure	*B. cereus*	PI/SYTO 9; Hoechst 33342/CFDA	Cronin and Wilkinson (2008b)
Spore resistance	*B. cereus*	PI/SYTO 9; Hoechst 33342/CFDA	Cronin and Wilkinson (2008a)
Spore resistance	*B. licheniformis*	PI/SYTO 16	Mathys *et al.* (2007)
Spore resistance	*B. subtilis*	Autofluorescence	Laflamme *et al.* (2006)
Spore resistance	*B. licheniformis*	PI	Reis *et al.* (2005)
Spore resistance	*B. subtilis*	$DiOC_6(3)$	Laflamme *et al.* (2005)
Vegetative cell resistance	*B. cereus*	PI/SYTO 9; Hoechst 33342/CFDA; SYTOX Green/C_{12}-resazurin	Cronin and Wilkinson (2008d)
Growth	*B. subtilis*	YFP	Vlamakis *et al.* (2008)
Growth	*B. licheniformis*	PI/$DiOC_6(3)$	Reis *et al.* (2005), Lopes da Silva *et al.* (2005)
Growth	*B. cereus*	CFSE	Cronin and Wilkinson (2008c)
Growth	*B. subtilis*, *B. licheniformis*	SYTO 13	Leser *et al.* (2008)
Detection	*B. anthracis*	Fl Ab	Schumacher *et al.* (2008)

CFDA, carboxyfluorescein diacetate; CFSE, carboxyfluorescein diacetate succinimidyl ester; Fl Ab, fluorescent antibody; GFP, green fluorescent protein; PI, propidium iodide; YFP, yellow fluorescent protein.

Therefore, it has been proposed that, just as in conventional microbiology, the difference in refractive index between spores and vegetative cells may be harnessed in order to discriminate between these two stages in life cycle (Stopa, 2000). These differences may be used to detect sporulation, which, as it begins within the mother cell, new compounds are produced and the process of differentiation starts, the refractile qualities of the mother cell alter (Gould, 1999; Stevenson, 2001). While such applications have proven difficult in the past, owing to the unsuitability of the majority of commercial instruments for the analysis of microbes, the latest cytometers, with increased laser powers and improved optical set-ups allowing for increased forward and side scatter sensitivity, should facilitate the analysis of samples of spore formers based on scatter alone. Similarly, with cytometers now commonly available on the market possessing five or more lasers, it should be possible to differentiate between spores and vegetative cells using autofluorescence alone, given the large differences in structure and protein content. This was shown in a study where viable and non-viable *B. subtilis* endospores were discriminated based on measuring green autofluorescence from UV-excited endospores (Laflamme *et al.*, 2006). As an aside, a very promising technology has been reported where, through a combination of the recording of elastically scattered light and machine learning called 'forward scatter phenotyping', strains of the same species can be reliably differentiated (Rajwa *et al.*, 2010), demonstrating that the use of scatter in the study of microorganisms is far from redundant.

Studies involving antibodies or fluorescence *in situ* hybridisation probes

It has long been a goal of microbiologists to use FCM as a technique for the non-growth based detection of an organism of interest. Given the difficulty and time involved in conventional plate-based detection methods, where the time from sampling to the identification and confirmation of positive colonies can be a number of days, a method involving specific tagging of a sample, followed by immediate flow cytometric analysis taking one to two hours has clear advantages. These specific tags can be either antibodies recognizing a component on the surface of cells or endospores or nucleic acid probes, that specifically bind to a unique sequence such as that of a 16S rRNA gene (Zwirglmaier, 2005). When what is desired is the detection of both vegetative cells and endospores, then two fluorescent antibodies are probably required, as the surfaces of vegetative cells and endospores are very different (Bennet and Belay, 2001). A single nucleic acid probe should be capable of binding to a target sequence in both vegetative cells and endospores, although permeabilization and hydration steps would likely be required for endospores in order to allow access and binding of the probe (Leuschner and Lillford, 2000; Setlow, 2006). What disappoints and frustrates microbial researchers with an interest in FCM is the lack of availability of quality monoclonal antibodies for all but the most commonly studied species. This unavailability probably goes a long way towards explaining the scarcity of studies on spore formers using antibodies. One notable study was that of Schumacher *et al.* (2008), who reported the development of a two-colour FCM assay for simultaneous detection and virulence determination (via an anti-protective antigen fluorescent antibody) of *B. anthracis* endospores. Thompson *et al.* (2007) measured structural differences in the exosporium of *B. anthracis* mutants lacking the *bclB* gene by using fluorescent antibodies raised against the exosporial proteins, BclA and BclB.

Studies involving the genetic transformation of spore formers

The days when green fluorescent protein (GFP) was the only option for researchers interested in studying the expression of a gene of interest are long behind us. Just as the number of fluorochromes that can be used to mark antibodies has exploded in recent years, there is now a bewildering variety of reporter proteins ranging from those that can be excited by the common 488 nm blue laser to red fluorescent protein (RFP) derivatives such as mCherry, excitable by a yellow-green line, to far red proteins such as Katushka, which was designed to avoid problems arising from tissue autofluorescence (Diéguez-Hurtado *et al.*, 2011). GFP-transformed strains in combination with

FCM analysis have been successfully employed to study the genetic control of sporulation where real time visualization of the expression of the genes involved in the differentiation of the mother cell resulting in the production of the endospore was achieved (Veening *et al.*, 2005, 2006). An approach by Vlamakis *et al.* (2008), using a number of YFP reporters under the control of promoters for motility, matrix production and sporulation enabled the spatiotemporal regulation of gene expression within *B. subtilis* biofilms to be studied. This approach allowed almost real-time quantification of the degree of heterogeneity in biofilms over the course of their development. De Jong *et al.* (2010) utilized promoter–GFP fusions to the main phosphorelay genes and altered their expression levels in order to investigate the origins of sporulation heterogeneity. If what is required are simply fluorescently tagged cells or spores for the tracking of how infectious agents interact with the immune system, an alternative to generating GFP-expressing cells is to stain with a tracker dye such as Alexa Fluor 488 as used by Stojkovic *et al.* (2008) in the study of *B. anthracis*.

Studies involving the use of permeability dyes

Just as is the case for non-spore formers, the prototypic flow cytometric study involves the use of a permeability dye with or without a counterstain. The most commonly used permeability dye is PI, with the counterstain normally being SYTO 9. So often used and well characterized is this pair of dyes that it forms the basis of a much-used commercial kit – Invitrogen's LIVE/DEAD® BacLight™ Bacterial Viability Kit (Invitrogen, Oregon, USA; Berney *et al.*, 2007). In the case of vegetative cells, it is generally believed that the large, highly charged molecules of permeability dyes are excluded by the intact membranes of live cells, whereas these molecules readily pass through the membranes of damaged or dead cells. If a counterstain such as SYTO 9 is used, live cells will fluoresce the colour of the counterstain, while dead cells will fluoresce the colour of the permeability dye. Therefore in the case of the PI/SYTO 9 combination, live cells will fluoresce green and dead cells, red. As more e.g. PI enters already green cells, they take on an orange colour, with cells eventually losing all their green colour as the SYTO 9 is displaced from the DNA. Unfortunately, the staining of bacteria with permeability dyes is not as clear-cut as presented above; their use becomes controversial when intermediate populations are observed and researchers struggle to present convincing cases as to the status or 'meaning' of these populations – a problem highlighted by Straüber and Müller (2010). That is why the use of correct controls in the setting up of an FCM-based experiment is so important and the use of cell sorting of various populations with an examination of viability and physiology using traditional techniques is highly recommended (Kennedy *et al.*, 2011).

The current author and editor have employed the permeability stains, PI and Hoechst 33342, to detect the numbers of damaged or dead *B. cereus* vegetative cells in cooked rice (Cronin and Wilkinson, 2009), during continuous culture (Cronin and Wilkinson, 2008c) and in response to simulated food processing treatments (Cronin and Wilkinson, 2008d). Broadly speaking, strong correlations were found between reductions in viability as estimated using plate counts and the percentages of permeabilized cells as detected using FCM. Other authors have also employed permeability stains in the study of the viability of vegetative cells of spore formers and noted that FCM data mirrored plate counts and other data generated using conventional microbiological techniques (Lopes da Silva *et al.*, 2005; Reis *et al.*, 2005).

If the staining of the vegetative cells of spore formers with permeability dyes and their subsequent analysis using FCM can sometimes yield data that is difficult to interpret, then the staining of spores with a combination of dyes such as PI and SYTO 9 can prove even more so. Up to a half-dozen discrete populations could be identified when PI and SYTO 9 were applied to samples of endospores of *B. cereus* subjected to a number of treatments designed to remove or damage various spore components (Cronin and Wilkinson, 2007, 2008b). These authors and others such as Leser *et al.* (2008) have shown that FCM in conjunction with the staining of samples with permeant nucleic acid dyes such as those from the SYTO range can be used to differentiate germinating spores which, because of their rapid rehydration and loss of components such as dipicolinic acid, stain brightly with these stains, in contrast to the spores, which are minimally stained [see also Smelt *et al.* (2008)]. It goes without saying that a dye permeable to vegetative cells and which is excluded by spores can be employed to discriminate between these two forms. Cronin and Wilkinson, (2008a) have shown that highly damaged and non-viable spores tend to be permeable to dyes such as PI and this conclusion is in agreement with the findings of other authors (Laflamme *et al.*, 2005; Reis *et al.*, 2005). It would appear that, as components of spores that contribute to their resistance are damaged or stripped away by treatments such as heat or detergents, they become more permeable to compounds such as PI or Hoechst 33342. For example Mathys *et al.* (2007) used a combination of PI and SYTO 16 to study the effect of pressure and thermal processing on *B. licheniformis* and were able to identify damaged spores based on their increased permeability to PI.

Studies involving the use of physiological dyes

A wide variety of dyes exists for the purposes of investigating the physiology of vegetative cells or germinating spores. Excellent reviews of these compounds are available and the author would recommend in particular those of Alvarez-Barrientos *et al.* (2005) and Shapiro and Nebe-von-Caron (2004). Briefly, dyes are available which allow for the measurement of enzyme activity (general and specific), redox activity, membrane potential, oxidative damage to DNA, intracellular Ca^{2+} and pH. Additionally, cell DNA and protein content can be estimated by staining of cells with fluorochromes which specifically bind to these compounds. Rates of cell division during growth may be measured using a tracking dye such as carboxyfluorescein diacetate succinimidyl ester (CFSE; Cronin and Wilkinson 2008c). In combination with conventional plate counting data or by sorting of populations observed during flow cytometric analysis, the use of these dyes can reveal very useful information on cell physiology in response to challenges such as heat or antibiotics or during processes occurring throughout germination or sporulation. However, without the correct use of carefully selected controls or the backup of complementary techniques, FCM data may be misinterpreted (see Kennedy *et al.* [2011] as an example of how sorting of presumably 'live' cells with membrane potential avoided the drawing of erroneous conclusions). As an example of how the use of physiological dyes in conjunction with FCM can illuminate results obtained using conventional means, the current author and editor found strong correlations between the reductions in viability as measured using plate counting and decreases in the percentages of cells showing redox activity and intracellular esterase activity when investigating the effects of simulated food processing or sanitization treatments on the survival and physiology of *B. cereus* vegetative cells (Cronin and Wilkinson, 2008d).

In any review of bacterial FCM it cannot be stressed enough that the outstanding feature of the technique is its demonstration and measurement of heterogeneity within populations. These data on population heterogeneity are otherwise only collectible using microscopy, but at a much slower and more laborious rate. Together with cell sorting of discreet subpopulations, the application of physiological dyes can be used to give new insights into e.g. resistance mechanisms, gene expression, cell structure and division. The current authors investigated the effects of simulated food processing or sanitization treatments on the survival and physiology of *B. cereus* vegetative cells using a number of the dyes described above (Cronin and Wilkinson, 2008d). Good correlations were found between the reductions in viability as measured using plate counting and decreases in the percentages of cells showing redox or esterase activities.

Potential applications of flow cytometry to the study of spore formers

The field of FCM has advanced greatly in the past decade. We are now firmly in the era of digital cytometers and sorters and the new instruments available to the researcher nowadays bear little resemblance to the rigid and difficult-to-operate machines of the 1990s. Easy-to-operate and maintain bench top cytometers with intuitive, powerful software and possessing up to three lasers and eight detectors are now within the financial reach of most laboratories in the developed world. Whereas previously, entry into the field of FCM involved a steep learning curve with the mastering of a large body of theoretical knowledge needed to successfully operate a cytometer, the new generation of machines on the market are as easy to operate as e.g. plate readers, and any scientist capable of reading an instructions manual and preparing well-stained single-cell suspensions with adequate controls can generate reliable and valuable flow cytometric data. Additionally, there are more dyes available to the researcher than ever before, meaning that, while in the past a researcher was forced into choosing sub-optimal combinations of dyes, he/she can now rationally design a large panel of colours to suit his/her particular instrument. How does a microbiologist with an interest in performing FCM on spore formers take advantage of this brave new world?

Firstly, it must never be forgotten that one of the principal advantages of FCM is its ability to record multiparametric data from each of the particles that trigger a signal above the set threshold. Therefore, even with a modest bench top cytometer, the possibility exists for the collection of up to eight data points from each of 10,000 or more cells analysed. An ambitious microbiologist, anxious to learn as much about his/her cells or spores as possible should take full advantage of the multiparametric nature of FCM and attempt the design of experiments and staining regimes that include as many fluorescent probes as is technically feasible. Microbiologists are urged to follow the well-worn path of the immunologist and learn to routinely work with panels of six to eight colours. With the new wave of instruments which possess multiple lines of powerful and stable laser diodes, detectors in the far red and optical set-ups less prone to spillage and background, previous problems of excessive spillovers between channels, high autofluorescence and difficult compensations are a thing of the past.

The case has been made for the ease of multicolour experiments, but what types of experiments should be carried out by the microbiologist interested in spore formers? Like no other time in the past, experiments combining the use of fluorescent reporter proteins, physiological dyes, a permeability marker and, perhaps, a monoclonal antibody are now possible. This approach in terms of the detection of spore formers in, for example, foodstuffs or soil may allow additional information on the physiological status of detected vegetative cells or endospores to be generated. Through the use of nucleic acid probes, e.g. for toxin genes, one could conceivably detect subsets of strains among the detected vegetative cells and spores.

In the case of the scientist interested in the events leading up to and during sporulation, a multiparameter approach could yield rich and useful information. Using antibodies or fluorescent dyes specific to various structural components of endospores such as the exosporium or protein coat [e.g. the lectin used to bind chitin in the case of *Streptomyces lunalinharesii* spores (Gomes *et al.*, 2008)], FCM may become a very useful tool in the real-time analysis of sporulation – a process that takes up to six hours to complete and involves a very specific sequence of events (Dricks, 2002). Sporulating cultures could be differentially stained at various time-points during the course of sporulation in order to determine the exact timing of the appearance of e.g. cortex. FISH probes could be used in tandem with release-capture strategies for environmental studies of spore formers, along with physiological dyes such as CFSE or any of the recently available CellTracker™ dyes (Invitrogen). As mentioned previously, the large variety of reporter proteins differing from each other in excitation and emission wavelengths allows the possibility of simultaneously studying the expression of a number of genes, and a system similar to the red/green Fucci system, which allows the measurement of different phases of the mammalian cell cycle (Sakaue-Sawano *et al.*, 2008), could be envisaged for the study of sporulation.

In a previous review of the application of FCM to the study of *B. cereus* (Cronin and Wilkinson, 2010), a case was presented for the uptake of multiplex bead array assays for the study of this species. These assays are capable of the simultaneous detection of dozens of soluble proteins within a sample and could have numerous applications to the study of spore formers, from the detection of toxins to the measurement of secreted compounds or signalling proteins.

Conclusions

FCM is a technique that is advancing continuously at a rapid rate. Even compared with 5 years ago there have been large improvements in instrumentation, software and reagents. Entry into the field grows less costly by the day, with the prices of compact bench-top analysers in particular seeing surprising reductions of late. Instruments are also simpler to use and facilitate greater combinations of fluorochromes than previously. Therefore, barriers for the entry of microbiologists in general, and those interested in the study of spores in particular, into the world of FCM are lower than they have ever been. Microbiologists with no experience of FCM should be encouraged to enter the field with the aim of surpassing the one and two colour experiments of the past decade and be bolder in their approaches. With careful choice of fluorochromes highly data-rich experiments are possible, with the simultaneous taking of data at the genetic, physiological and structural level.

References

Alvarez-Barrientos, A., Arroyo, J., Canton, R., Nombela, C., and Sanchez-Perez, M. (2000). Applications of flow cytometry to clinical microbiology. Clin. Microbiol. Rev. *13*, 167–195.

Bennet, R.W., and Belay, N. (2001). *Bacillus cereus*. In Microbiological Examination of Foods, Downes, F.P., and Ito, K., eds. (American Public Health Association, Washington, DC, USA), pp. 311–316.

Berleman, J.E., and Kirby, J.R. (2009). Deciphering the hunting strategy of a bacterial wolfpack. FEMS Microbiol. Rev. 33(5), 942–957.

Berney, M., Hammes, F., Bosshard, F., Weilenmann, H.-U., and Egli, T. (2007). Assessment and interpretation of bacterial viability by using the Live/Dead BacLight Kit in combination with flow cytometry. Appl. Environ. Microbiol. 73(10), 3283–3290.

Burgess, S.A., Lindsay, D., and Flint, S.H. (2010). Thermophilic bacilli and their importance in dairy processing. Int. J. Food Microbiol. *144*, 215–225.

Burns, D.A., and Minton, N.P. (2011). Sporulation studies in *Clostridium difficile*. J. Microbiol. Methods. *87*, 133–138.

Burns, D.A., Heap, J.T., and Minton, N.P. (2010). SleC is essential for germination of *Clostridium difficile* spores in nutrient-rich medium supplemented with the bile salt taurocholate. J. Bacteriol. *192*(3), 657–664.

Cheng, G., Zhao, P., Tang, X.-F., and Tang, B. (2009). Identification and characterization of a novel spore-associated subtilase from *Thermoactinomyces* sp. CDF. Microbiology *155*, 3661–3672.

Cronin, U.P., and Wilkinson, M.G. (2007). The use of flow cytometry to study the germination of *Bacillus cereus* endospores. Cytometry *71A*, 143–153.

Cronin, U.P., and Wilkinson, M.G. (2008a). *Bacillus cereus* endospores exhibit a heterogeneous response to heat-treatment and low temperature storage. Food Microbiol. *25*, 235–243.

Cronin, U.P., and Wilkinson, M.G. (2008b). Monitoring changes in germination and permeability of *Bacillus cereus* endospores following chemical, heat and enzymatic treatments using flow cytometry. J. Rapid Methods Aut. Microbiol. *16*, 164–184.

Cronin, U.P., and Wilkinson, M.G. (2008c). Monitoring growth phase-related changes in phosphatidylcholine-specific phospholipase C production, adhesion properties and physiology of *Bacillus cereus* vegetative cells. J. Ind. Microbiol. Biotechnol. *35*, 1695–1703.

Cronin, U.P., and Wilkinson, M.G. (2008d). Physiological response of *Bacillus cereus* vegetative cells to simulated food processing treatments J. Food Protect. *71*, 2168–2176.

Cronin, U.P., and Wilkinson, M.G. (2009). The growth, physiology and toxigenic potential of *Bacillus cereus* in cooked rice during storage temperature abuse. Food Control *20*(9), 822–828.

Cronin, U.P., and Wilkinson, M.G. (2010). The potential of flow cytometry in the study of *Bacillus cereus*. J. Appl. Microbiol. *108*, 1–16.

Cutting, S.M., Hong, H.A., Baccigalupi, L., and Ricca, E. (2009). Oral vaccine delivery by recombinant spore probiotics. Int. Rev. Immunol. *28*(6), 487–505.

Diéguez-Hurtado, R., Martín, J., Martínez-Corral, I., Martínez, M.D., Megías, D., Olmeda, D., and Ortega, S. (2011). A Cre-reporter transgenic mouse expressing the far-red fluorescent protein Katushka. Genesis *49*(1), 36–45.

Dricks, A. (2002). Maximum shields: the assembly and function of the bacterial spore coat. Trends Microbiol. *10*, 251–254.

Errington, J. (2010). From spores to antibiotics via the cell cycle. Microbiology *156*, 1–13.

Gomes, R.C., Soares, R.M.A., Nakamura, C.V., Souto-Padrón, T., de Souza, R.F., de Azevedo, S.S., Luzia, T., Alviano, C.S., and Coelho, R.R. (2008). *Streptomyces lunalinharesii* spores contain chitin on the outer sheath. FEMS Microbiol. Lett. *286*, 118–123.

Gould, G.W. (1999). Bacterial endospores. In Encyclopedia of Food Microbiology, Robinson, R.K., Batt, C.A., and Patel, P.D., eds. Academic Press, New York, NY, USA. pp. 168–172.

Higgins, D., and Dworkin, J. (2012). Recent progress in *Bacillus subtilis* sporulation. FEMS Microbiol. Rev. *36*, 131–148.

Hornstra, L.M., ter Beek, A., Smelt, J.P., Wouter, W.K., and Brul, S. (2009). On the origin of heterogeneity in (preservation) resistance of *Bacillus* spores: Input for a 'systems' analysis approach of bacterial spore outgrowth. Int. J. Food Microbiol. *134*, 9–15.

de Jong, I.G., Veening, J.W., and Kuipers, O.P. (2010). Heterochronic phosphorelay gene expression as a source of heterogeneity in *Bacillus subtilis* spore formation. J. Bacteriol. *192*(8), 2053–2067.

Kennedy, D., Cronin, U.P., and Wilkinson, M.G. (2011). The response of *Escherichia coli*, *Listeria monocytogenes* and *Staphylococcus aureus* to simulated food processing treatments using fluorescence activated cell sorting and plate counting. Appl. Environ. Microbiol. *77*, 4657–4668.

Kim, J.-H., Roh, C., Lee, C.-W., Kyung, D., Choi, S.-K., Jung, H.-C., Pan, J.-G., and Kim, B.-G. (2007). Bacterial surface display of GPF_{uv} on *Bacillus subtilis* spores. J. Microbiol. Biotechnol. *17*, 677–680.

Kimura, Y., Kawasaki, S., Yoshimoto, H., and Takegawa, K. (2010). Glycine betaine biosynthesized from glycine provides an osmolyte for cell growth and spore germination during osmotic stress in *Myxococcus xanthus*. J. Bacteriol. *192*(5), 1467–1470.

Laflamme, C., Ho, J.Z., Veillette, M., de Latremoille, M.C., Verrault, D., Meriaux, A., and Duchaine, C. (2005). Flow cytometry analysis of germinating *Bacillus* spores, using membrane potential dye. Arch. Microbiol. *183*, 107–112.

Laflamme, C., Verrault, D., Ho, J.Z., and Duchaine, C. (2006). Flow cytometry sorting protocol of *Bacillus* spore using ultraviolet laser and utofluorescence as main sorting criterion. J. Fluoresc. *16*, 733–737.

Leser, T.D., Knarreborg, A., and Worm, J. (2008). Germination and outgrowth of *Bacillus subtilis* and *Bacillus licheniformis* spores in the gastrointestinal tract of pigs. J. Appl. Microbiol. *104*, 1025–1033.

Leuschner, R.G.K., and Lillford, P.J. (2000). Effects of hydration on molecular mobility in phase-bright *Bacillus subtilis* spores. Microbiology *149*, 49–55.

Lopes da Silva, T., Reis, A., Kent, C.A., Kosseva, M., Roseiro, J.C., and Hewitt, C.J. (2005). Stress-induced physiological responses to starvation periods as well as glucose and lactose pulses in *Bacillus licheniformis* CCMI 1034 continuous aerobic fermentation processes as measured by multi-parameter flow cytometry. Biochem. Eng. J. *24*, 31–41.

Lulko, A., Veening, J.-W., Buist, G., Smits, W.K., Blom, E.J., Beekman, A.C., Bron, S., and Kuipers, O.P. (2007). Production and secretion stress caused by overexpression of heterologous α-amylase leads to inhibition of sporulation and a prolonged motile phase in *Bacillus subtilis*. Appl. Environ. Microbiol. 73, 5354–5362.

Mathys, A., Chapman, B., Bull, M., Heinz, V., and Knorr, D. (2007). Flow cytometric assessment of *Bacillus* spore response to high pressure and heat. IFSET *8*, 519–527.

Paredes-Sabja, D., Setlow, P., and Sarker, M. 2011 Germination of spores of *Bacillales* and *Clostridiales* species: mechanisms and proteins involved. Trends Microbiol. *19*(2), 85–94.

Phillips, A.P., and Martin, K.L. (1983). Immunofluorescence analysis of *Bacillus* spores and vegetative cells by flow cytometry. Cytometry *4*, 123–131.

Rajwa, B., Dundar, M.M., Akoua, F., Bettasso, A., Patsekin, V., Hirleman, E.P., Bhunia, A.K., and Robinson, J.P. (2010). Discovering the unknown: detection of emerging pathogens using a label-free light-scattering system. Cytometry 77(12), 1103–1112.

Reis, A., Lopes da Silva, T., Kent, C.A., Kosseva, M., Roseiro, J.C., and Hewitt, C.J. (2005). Monitoring population dynamics of the thermophilic *Bacillus licheniformis* CCMI 1034 in batch and continuous cultures using multi-parameter flow cytometry. J. Biotechnol. *115*, 199–210.

Sakaue-Sawano A., Kurokawa, H., Morimura, T., Hanyu, A., Hama, H., Osawa, H., Kashiwagi, S., Fukami, K., Miyata, T., Miyoshi, H., *et al.* (2008). Visualizing spatiotemporal dynamics of multicellular cell-cycle progression. Cell. *132*, 487–498.

Shapiro, H.M., and Nebe-von-Caron, G. (2004). Multiparameter flow cytometry of bacteria. In Methods in Molecular Biology: Flow Cytometry Protocols, Hawley, T.S., and Hawley, R.G., eds. (Humana Press, New Jersey, USA) pp. 33–43.

Schumacher, W.C., Storozuk, C.A., Dutta, P.K., and Phipps, A.J. (2008). Identification and characterization of *Bacillus anthracis* spores by multiparameter flow cytometry. Appl. Environ. Microbiol. *74*, 5220–5223.

Setlow, P. (2006). Spores of *Bacillus subtilis*: their resistance to and killing by radiation, heat and chemicals. J. Appl. Microbiol. *101*, 514–525.

Smelt, J.P.P.M., Bos, A.P., and Kort, R. (2008). Modelling the effect of sub(lethal) heat treatment of *Bacillus subtilis* spores on germination rate and outgrowth to exponentially growing vegetative cells. Int. J. Food Microbiol. *128*, 34–40.

Smits, W.K., Eschevins, C.C., Susanna, K.A., Bron, S., Kuipers, O.P., and Hamoen, L.W. (2005). Stripping *Bacillus*: ComK auto-stimulation is responsible for the bistable response in competence development. Mol. Microbiol. *56*, 604–614.

Stojkovic, B., Torres, E.M., Prouty, A.M., Patel, H.K., Zhuang, L., Koehler, T.M., Ballard, J.D., and Blanke, S.R.(2008). High-throughput, single-cell analysis of macrophage interactions with fluorescently labelled *Bacillus anthracis* spores. Appl. Environ. Microbiol. *17*(16), 5201–5210.

Stopa, P.J. (2000). The flow cytometry of *Bacillus anthracis* spores revisited. Cytometry *41*, 237–244.

Stevenson, K.E., and Segner, W.P. (2001). Mesophilic aerobic spore formers. In Microbiological Examination of Foods, Downes, F.P., and Ito, K., eds. (American Public Health Association, Washington, DC, USA) pp. 223–227.

Straüber, H., and Müller, S. (2010). Viability states of bacteria – specific mechanisms of selected probes. Cytometry *77A*, 623–634.

Thompson, B.M., Waller, N.L., Fox, K.F., Fox, A., and Stewart, G.C. (2007). The BclB glycoprotein of *Bacillus anthracis* is involved in exosporium integrity. J. Bacteriol. *189*, 6704–6713.

Veening, J.-W., Hamoen, L.W., and Kuipers, O.P. (2005). Phosphatases modulate the bistable sporulation gene expression pattern in *Bacillus subtilis*. Mol. Microbiol. *56*, 1481–1494.

Veening, J.-W., Smits, W.K., Hamoen, L.W., and Kuipers, O.P. (2006). Single cell analysis of gene expression patterns of competence development and initiation of sporulation in *Bacillus subtilis* grown on chemically defined media. J. Appl. Microbiol. *101*, 531–541.

Vlamakis, H., Aguilar, C., Losick, R., and Kolter, R. (2008). Control of cell fate by the formation of an architecturally complex bacterial community. Genes Dev. 22, 945–953.

Weese, J.S. (2010). *Clostridium difficile* in food – innocent bystander or serious threat? Clin. Microbiol. Infect. *16*(1), 3–10.

Zwirglmaier, K. (2005). Fluorescence *in situ* hybridisation (FISH) – the next generation. FEMS Microbiol. Lett. *24*, 151–158.

Flow Cytometry of Yeasts and Other Fungi

7

Tim W. Overton

Abstract

Flow cytometry is used extensively to investigate many aspects of yeast physiology and cell biology. This chapter describes the most common applications of flow cytometry and fluorescence-activated cell sorting (FACS) to yeasts and other fungi, and reviews recent studies to capture current directions in the field. Pure and applied aspects are discussed, with a focus on the areas of fundamental cell biology, medical applications, monitoring during bioprocesses including brewing, and the use of yeast as a molecular biology tool.

Novel flow cytometric and microfluidic systems which utilize yeast as a model organism are also described and their relevance to previous research in the area and their possible application for future fundamental and applied research are discussed.

Introduction

Flow cytometry has been used to study aspects of fungal cells since the 1970s (for example, Hutter and Eipel, 1979). Yeast cells have always been recognized as a good subject for investigations using flow cytometry, due to their widespread use, both industrially and as a model organism in cell and molecular biology, their non-pathogenicity and their comparatively large size. These advantages continue to make yeast a choice for the development of new flow-based single-cell assays in the field of microfluidics, a topic which will be briefly discussed at the end of this chapter.

Today, flow cytometry and fluorescence-activated cell sorting are used to investigate many aspects of yeast and other fungi in a wide variety of experimental settings, from basic cell biology and molecular biology, to bioprocess monitoring and clinical microbiology. In each case, the advantages of flow cytometry (FCM) and fluorescence-activated cell sorting (FACS) are utilized to great effect: high speed data acquisition; non-reliance on growth; investigation and isolation of single cells; and high statistical significance of data from interrogation of large numbers of individual cells.

Scope of this chapter

This chapter discusses recent research using flow cytometry and cell sorting where yeast and fungal cells are the interrogated particles. Studies involving flow cytometric analysis of the mode of action of yeast extracts or metabolites on other cells are not included. The basics of flow cytometry are covered elsewhere in this volume and in other texts (for example, Shapiro, 2003) and will not be covered in detail here.

The chapter starts by describing the application of basic flow cytometry (FCM) and FACS methods to answer fundamental questions regarding fungal cell biology and molecular biology, then discusses the application of these methods in the fields of medical research (including antifungal compound development and basic research in the area of fungal diseases) and in the bioprocess industries. The use of yeast as a screening system for cell engineering and protein engineering, using techniques such as yeast surface display, will be covered, and finally novel single-cell investigative systems using yeast as model particles will be discussed.

Flow cytometry and FACS to investigate cell and molecular biology

The basic cell biology of yeasts and fungi can be studied at a number of levels, from gross cellular physiology to molecular biology. FCM and FACS are used to probe fungal cells at every physical level, from scatter measurements on a whole-cell scale to investigation of individual molecular components. This section describes methodologies using dyes that investigate cellular viability, vitality and physiology, DNA and RNA quantification methods and the use of autofluorescent proteins. An overview of such FCM methods in relation to yeast or fungal cells is given in Fig. 7.1.

Physiological studies using viability and other dyes

At its simplest, FCM scatter measurements can be used to monitor yeast cell shape, much in the way that light microscopy could be used; an obvious advantage is that FCM generates quantitative data for far larger numbers of cells than light microscopy could ever be used to study. Georgieva *et al.* (2008) investigated the impact of the actin-related protein Act3p/Arp4 on cell size and shape in *Saccharomyces cerevisiae.* Two temperature sensitive *act3/arp4* mutants were initially found to be morphologically distinct from the wild-type using microscopy. FCM was used to quantify differences in size and granularity, analysed using a quadrant on a forward scatter (FSC) versus side scatter (SSC) plot. This revealed that both mutations generated populations with a higher proportion of cells having higher FSC

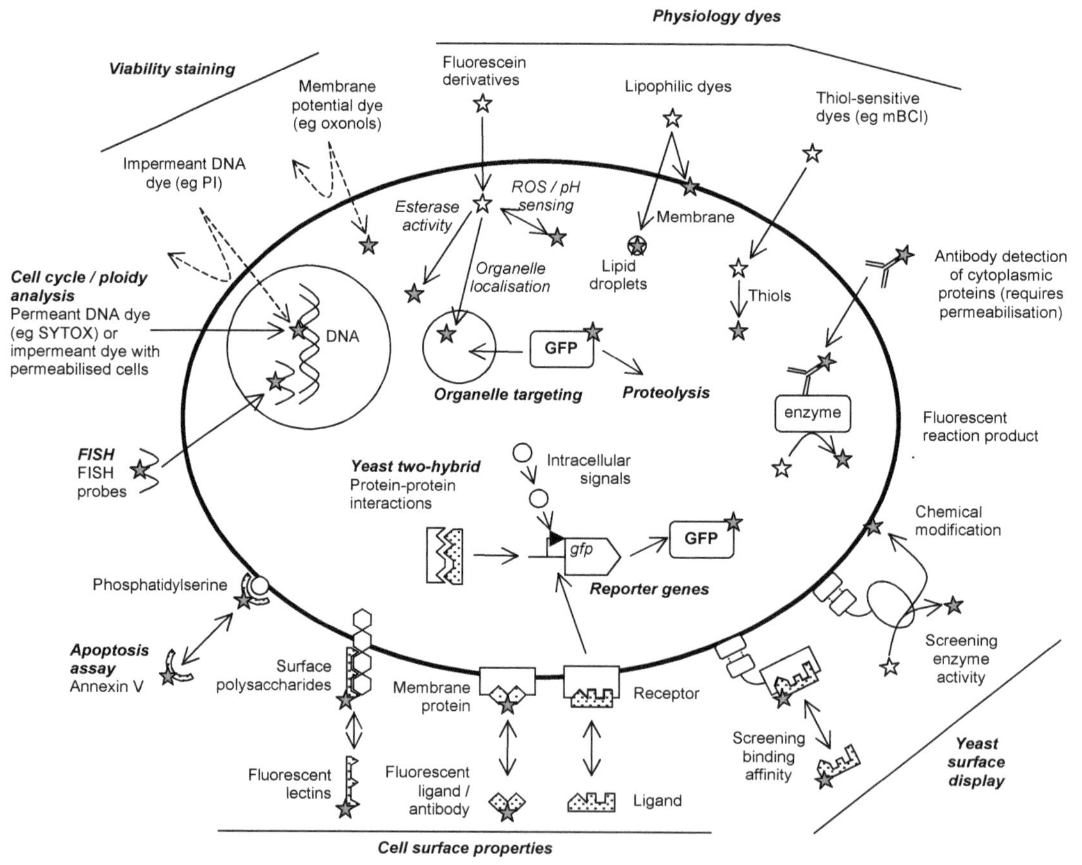

Figure 7.1 Overview of flow cytometric methods used in yeasts and fungi. Filled stars represent fluors, white stars non-fluorescent fluor precursors.

and SSC than the wild-type (thus were larger and more granular), although the distribution in each mutant was different, indicating a different morphological phenotype. Further experiments revealed that the mutants possessed larger nuclei and were more sensitive to nuclease activity, suggesting that Act3p/Arp4 has a role in controlling chromatin structure.

However, most FCM studies use some form of fluorescent measurement to investigate particular cell characteristics. Some commonly used fluorescent dyes are listed in Table 7.1. Dual-staining systems for cell viability and membrane integrity are the mainstay of microbial multiparameter FCM (Shapiro, 2003; Hewitt and Nebe-Von-Caron, 2004), and frequently function by monitoring aspects of the cell membrane. Impermeant DNA dyes such as propidium iodide (PI) allow determination of cell membrane integrity, staining occurring only in cells with a compromised membrane, while oxonol dyes such as DiBAC$_4$(3) determine plasma membrane potential (reviewed by Shapiro, 2000). PI has been combined with the permeant DNA stain SYTO 9 in commercially available viability stains (FungaLight™, Invitrogen). Many of these dye systems are well-documented and are used in more applied FCM research, as will be discussed later. It should be noted that, while many dye systems could be measured using fluorescence microscopy or bulk fluorimetry, FCM generates quantitative data for large numbers of single cells. Also, since the cells themselves are being interrogated, background fluorescence from unincorporated dyes in solution is reduced or eliminated.

Fluorescein derivatives are popular dual viability- and cell integrity-markers. 5-carboxy fluorescein diacetate acetoxylmethyl ester (CFDA-AM) is used to monitor intracellular esterase activity, a marker of viability. The non-fluorescent AM ester derivative of CFDA is readily taken up by cells, whereupon cellular esterases cleave the molecule, liberating the fluorescent carboxyfluorescein, which is charged and membrane impermeant; loss of fluorescence therefore indicates a compromised membrane. PI is a commonly used counterstain for CFDA-AM; this dye combination is also available commercially (FungaLight™, Invitrogen). Recent work has optimized dual-staining systems for the yeast *Pichia anomala* (used as a biocontrol agent in the form of a foliar spray on nut trees to eliminate *Aspergillus flavus*; Hernlem and Hua, 2010), evaluating different mitochondrial membrane potential-sensitive oxonol dyes. A combined intracellular esterase activity and membrane potential assay was developed utilizing CFDA-AM with a red-fluorescing oxonol, DiBAC$_4$(5), related to the widely used green oxonol DiBAC$_4$(3) (Shapiro, 2000).

Several fluorescein derivatives have been utilized to probe yeast organelle physiology. The peroxide-sensitive dichlorodihydrofluorescein diacetate (H_2DCFDA) was used to measure levels of intracellular reactive oxygen species (ROS) in response to a mutation in the *Hansenula polymorpha* peroxisome peroxiredoxin Pmp20 (Aksam *et al.*, 2008). In addition, FCM analysis was used to monitor cell death in response to exposure to methanol by apoptosis and necrosis. FITC-conjugated Annexin V was used to detect the presence of phosphatidylserine on the outer leaflet of the cell membrane, indicative of apoptosis, and PI was used to monitor membrane integrity, which is compromised in both apoptosis and necrosis. The Annexin/PI assay was compared to a traditional plate count for survival; proportions of cells stained with PI were broadly comparable to plate count results.

The very widely used pH-sensitive fluorescein derivative 2,7-bis(2-carboxyethyl)-5(6)-carboxyfluorescein (BCECF) has been used to monitor *S. cerevisiae* vacuole pH (Johnson *et al.*, 2010). Ester derivatives (BCECF-AM) were shown to localize at the vacuole and were hydrolysed by the high vacuolar esterase activity, trapping the charged fluor within the vacuole. This system was used to screen for inhibitors of vacuolar proton-translocating ATPases (V-ATPases), which regulate vacuolar pH. These are thought to be anticancer targets, as they have a role in the development of abnormal pH gradients that exist in cancer cells, and their inhibition has been shown to have potential in anticancer therapy. Carboxyfluorescein diacetate (CFDA) has also been used to monitor *S. cerevisiae* intracellular pH (pH_i or ICP) as a measure of vitality (Weigert *et al.*, 2009). pH-dependent changes to the emission spectrum of carboxyfluorescein can used to quantify pH_i by

Table 7.1 Fluorescent dyes discussed in this chapter

Dye	Measured parameter	Basis of function	Maximum excitation and emission wavelengths (nm)
Metabolic activity indicators			
CFDA-AM (carboxyfluorescein diacetate acetoxymethyl ester)	Esterase activity	Intracellular esterases hydrolyse to membrane-impermeant carboxyfluorescein	Carboxyfluorescein: 492/518
	Cellular integrity	Carboxyfluorescein leaks from compromised cells	
FUN-1	Viability	Membrane permeant green fluorescent stain Forms red vacuolar structures in metabolically active cells	488 nm excitation. Emits in range 510–560nm (dead cells); 560–610 nm (viable cells)
Nucleic acid dyes			
PI (propidium iodide)	DNA	Fluoresces upon intercalating DNA or RNA	530/625 in presence of DNA
	Membrane integrity	Membrane impermeant	
SYTOX Green	DNA	Fluoresces upon binding DNA or RNA	504/523 in presence of DNA
	Membrane integrity	Membrane impermeant	
EB (ethidium bromide)	DNA	Fluoresces upon intercalating DNA or RNA	518/605 in presence of DNA
	Membrane integrity	Membrane impermeant	
SYTO 9	DNA	Fluoresces upon binding DNA or RNA; membrane permeant	483/503 in presence of DNA
SYTO RNASelect	RNA	Selectively stains RNA; membrane permeant	490/530 in presence of RNA
Acridine orange	DNA and RNA	Fluoresces upon binding DNA/RNA	500/526 in presence of DNA
	pH	Concentrates in acidic organelles	460/650 in presence of RNA
DAPI (4′,6-Diamidino-2-phenylindole)	DNA and RNA	Fluoresces upon binding DNA/RNA	358/461 bound to DNA
Lipophilic dyes			
DiI (1,1′-dioctadecyl-3,3,3′,3′-tetramethylindocarbocyanine perchlorate)	Cell membranes	Increases in fluorescence when integrated into membranes	549/565 in MeOH
DiD-DS (1,1′-dioctadecyl-3,3,3′,3′-tetramethylindodicarbocyanine-5,5′-disulfonic acid)	Cell membranes	Increases in fluorescence when integrated into membranes	555/570 in MeOH
BODIPY-PC (2-(4,4-difluoro-5,7-dimethyl-4-bora-3a,4a-diaza-s-indacene-3-pentanoyl)-1-hexadecanoyl-sn-glycero-3-phosphocholine)	Cell membranes	Increases in fluorescence when integrated into membranes	502/511 in MeOH
Nile Red	Lipid droplets	Almost non-fluorescent in polar solvents but highly fluorescent in non-polar solvents	552/636 in MeOH

Table 7.1 Continued

Dye	Measured parameter	Basis of function	Maximum excitation and emission wavelengths (nm)
Membrane potential			
$DiBAC_4(3)$ (bis-(1,3-dibutylbarbituric acid) trimethine oxonol)	Plasma membrane potential	Only enter cells with depolarized membrane potential. Fluoresces upon binding to proteins and membranes	493/516 in MeOH
$DiBAC_4(5)$ (bis-(1,3-dibutylbarbituric acid) pentamethine oxonol)	Plasma membrane potential	Only enter cells with depolarized membrane potential. Fluoresces upon binding to proteins and membranes	590/616 in MeOH
Rh123 (Rhodamine 123)	Mitochondrial activity	Sequestered by active mitochondria. Fluoresces upon binding to proteins and membranes	507/529 in MeOH
	Multidrug efflux	Exported from the cell by drug efflux pumps	
JC-1 (5,5′,6,6′-tetrachloro-1,1′,3,3′-tetraethylbenzimidazol ylcarbocyanine iodide)	Mitochondrial membrane potential	Green fluorescent monomer at low membrane potentials, aggregates at high membrane potential and fluoresces red	514/529 (monomer) 585/590 (aggregate)
Thiol sensing dyes			
mBCl (monochlorobimane)	Reacts with thiol groups	Non-fluorescent until conjugated to thiols	394/490 when conjugated to glutathione
Reactive species sensors			
DHE (dihydroethidium)	Superoxide	Blue fluor until oxidized to 2-hydroxyethidium, when it intercalates DNA and emits red fluorescence	Reduced: 355/420 Oxidized: 355/605
H_2DCFDA (dichlorodihydrofluorescein diacetate)	Reactive oxygen radicals	Deacetylated inside cell by esterases and oxidized by ROS to form dichlorofluorescein	504/525 on exposure to ROS
pH-sensitive dyes			
CFDA (5-carboxyfluorescein diacetate)	pH	Fluorescence emission spectrum is pH-dependent. Ratio of emission at 525 nm/575 nm is measured	pH-dependent
BCECF, AM (2′,7′-bis-(2-carboxyethyl)-5-(and-6)-carboxyfluorescein, acetoxymethyl ester)	pH	Fluorescence emission is dependent upon pH. pKa is higher than that of CFDA	pH-dependent

measuring the ratio of emission at 525 nm (pH-dependent) to 575 nm (largely pH-independent).

Lipids can also be monitored using FCM. Nicola *et al.* (2009) studied extracellular vesicle formation by the opportunistic human pathogen *Cryptococcus neoformans*. Dialkylcarbocyanine dyes (DiI, DiD-DS and BODIPY-PC; Table 7.1) were used to detect lipid in vesicle membranes using FCM and fluorescence microscopy, and counterstaining with the RNA-specific STYO RNASelect dye determined that the vesicles contained RNA. The vesicles were observed to be of different sizes, up to around 200 nm; a useful reminder that FCM can be used to investigate very small particles. Whereas viability-determining dyes are useful for a wide variety of applications, these lipid-specific dyes currently have limited applications, although the emerging field of fungal biodiesel synthesis offers another possible application (Li Q *et al.*, 2008).

Determination of DNA content – cell cycle, ploidy and genome size

A long-standing use of FCM is the rapid analysis of the cell cycle using a DNA-specific dye to measure the DNA content of cells within a population (reviewed by Darzynkiewicz *et al.*, 2004, 2010). Fluorescence data corresponding to cellular DNA content are converted to the percentage of cells in the G_1, S and G_2/M phases using computer algorithms (Fig. 7.2A). Cells with sub-G_1 DNA content are often interpreted as apoptotic (Ben and Arcangioli, 2009; Darzynkiewicz *et al.*, 2010). Dyes such as DAPI (4′,6-diamidino-2-phenylindole) have been used to monitor cellular DNA levels, although recent studies have used PI and SYTOX stains to a greater extent. Use of impermeant dyes requires membrane permeabilization before use for cell cycle analysis.

Several recent papers have investigated molecular aspects of cell cycle regulation using FCM as an analytical tool. Two studies have used

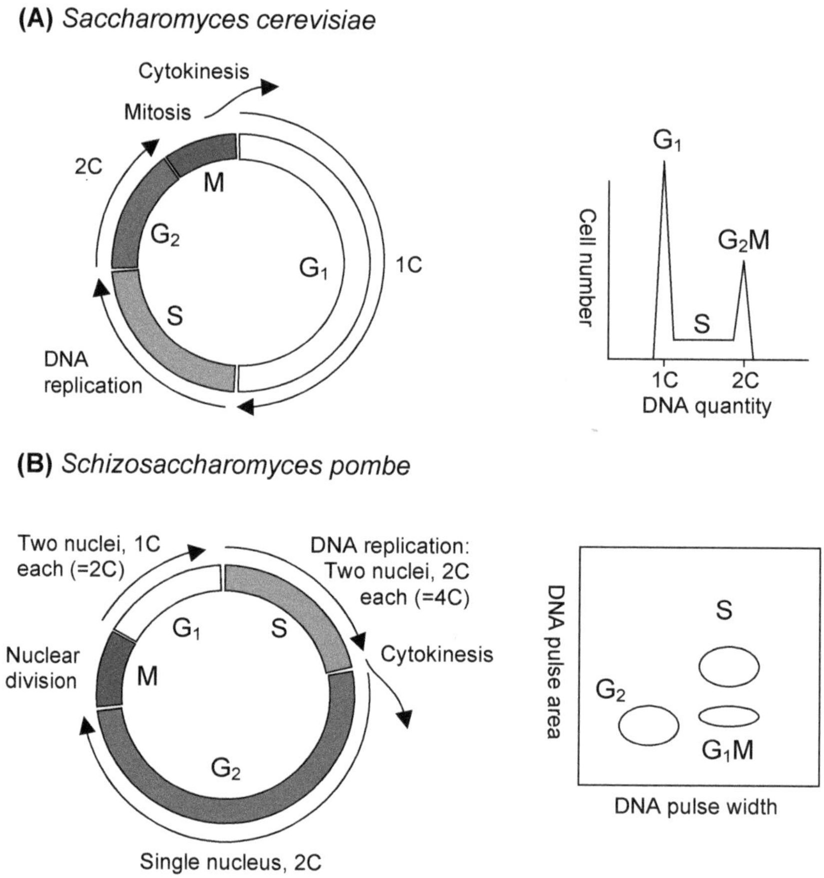

Figure 7.2 Cell cycle analysis in yeast. (A) In *S. cerevisiae*, analysis of the cell cycle depends upon quantifying DNA; genomic DNA is replicated in S phase, leading to G_2 and M phase cells having twice the quantity of DNA as G_1 phase cells. S phase cells have an intermediate quantity of DNA. Cells with a DNA content below G_1 are frequently apoptotic (Darzynkiewicz *et al.*, 2010). (B) In *S. pombe*, cytokinesis occurs after S phase, so G_2 and G_1 cells contain the same quantity of DNA although different numbers of nuclei (Knutsen *et al.*, 2011). Pulse width analysis was used to measure DNA quantity (DNA pulse area) and the number of nuclei (DNA pulse width).

high-throughput approaches to identify *S. cerevisiae* genes for involvement in cell cycle regulation. In each case, a library of mutants cultured within 96-well plates (with a single clonal population in each well) were interrogated by FCM using SYTOX Green to measure cellular DNA content (Fig. 7.3A). Percentages of cells in G_1 and G_2/M phases for each strain were compared to the wild-type, and strains found to accumulate in a particular growth phase were investigated further. White *et al.* (2009) employed a loss of function approach; FCM was used to screen a library of 268 strains, each deleted for a known or predicted non-essential transcription factor. 48 knockout strains were selected and confirmed by further analysis to have a cell cycle progression defect. Niu *et al.* (2008) used an overexpression approach; 5556 strains, each over-expressing a different ORF (amounting to 91% of the genome) were screened. They found that overexpression of 108 genes led to altered cell cycle progression, 82 of which had not been implicated in cell cycle regulation before. It should be noted that such screens depend upon the assumption that DNA content accurately reflects cell cycle progression, and that mutations leading to altered cellular physiology (for example, aggregation) would prevent FCM being used as a screen (White *et al.*, 2009). As well as high-throughput screening studies, DNA content analysis continues to be used as a tool to investigate individual components of the cell cycle regulatory machinery (Matmati *et al.*, 2009; Tao *et al.*, 2009).

Cell cycle analysis can be performed using a combination of FCM and fluorescent microscopy. Calvert *et al.* (2008) developed methods for cell cycle analysis using multispectral imaging flow cytometry (MIFC). The principle advantage of this technique for cell cycle analysis is the ability to record not only fluorescence intensity

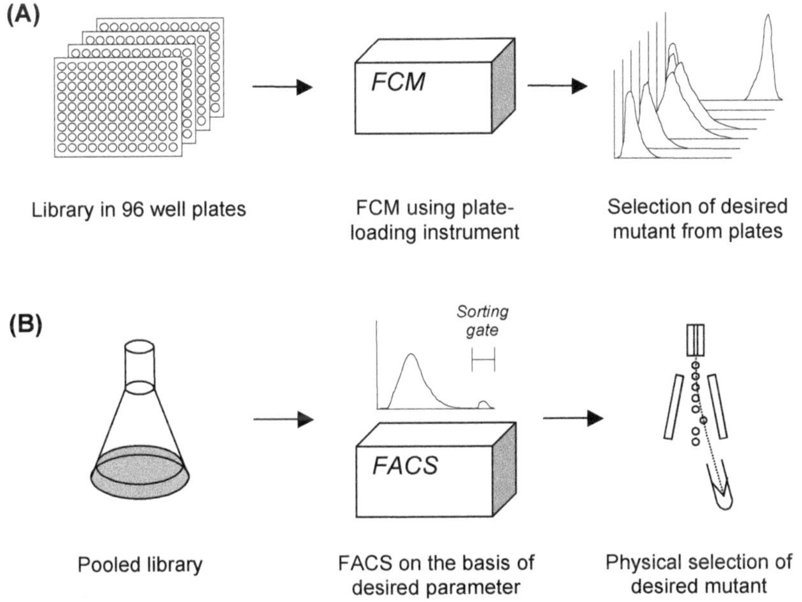

Figure 7.3 High throughput screening approaches using FCM and FACS. Broadly speaking, two approaches are used to screen libraries: screening multiple clones by FCM; or screening a mixture of clones using FACS. These methods have been used in many approaches discussed in the text, from analysis of proteins involved in cell cycle progression to cell engineering. (A) The FCM screening approach relies upon libraries of cells to be screened in 96-well plates, one clone per well. These are then analysed using FCM, ideally using an instrument with a plate-loader, and the desired clone is selected on the basis of the FCM data. Library construction is often a limiting factor, requiring robotic systems for high throughput. (B) The FACS approach uses a pooled library grown in one vessel which is interrogated using FCM and sorted by FACS on the basis of a desired characteristic. Sorted clones can then be further characterized. One possible drawback is that poorly growing clones would not be present in the sorted library.

data but also morphological data, both in terms of the morphology as determined in the visible spectrum (cell shape) and the morphology of the fluorescence signal (corresponding to the distribution of DNA within the cell). In this study, the authors studied *S. cerevisiae* bud morphology during the budding process. Comparing MIFC data with comparable FCM cell cycle data (with the percentage of G_1, S and G_2/M phase cells calculated using two different algorithms), it was noted that FCM methods tended to overestimate the percentage of cells in G_2/M phase and underestimate those in G_1 phase.

The allocation of cells to S phase remains a problem for DNA histogram analysis, since the S population overlaps the G_1 and G_2/M populations, so the dividing lines between populations must be estimated. The Watson Pragmatic model (Watson *et al.*, 1987) also identified the overestimation of the G_2/M population. Cell cycle analysis was initially devised for mammalian cell analysis, and use for yeast cells is complicated by their lower DNA content giving rise to higher coefficients of variance (CVs) and their propensity to remain attached following cell division. The authors discussed the limitations of using FCM for yeast cell cycle analysis, and highlighted the advantages of using an MIFC approach.

The fission yeast *Schizosaccharomyces pombe* presents some additional challenges for DNA content and cell cycle analysis using FCM. *S. pombe* undergoes nuclear division in M phase but cytokinesis after S phase. Therefore G_1 and G_2 phase cells have the same DNA content, although G_1 cells are binuclear while G_2 cells are mononuclear (Fig. 7.2B). Traditional DNA content analysis, as used in *S. cerevisiae,* has been limited to differentiation between 1N and 2N states (Chen and Runge, 2009) or monitoring apoptotic cells with sub-1N quantities of DNA in response to oxidative damage (Ben Hassine and Arcangioli, 2009). In order to differentiate between mononuclear and binuclear cells, a pulse width analysis method for FCM cell cycle analysis of *S. pombe* has recently been developed (Knutsen *et al.*, 2011). Pulse width analysis was used to determine both the quantity of DNA (the area under the SYTOX Green fluorescence peak) and number of nuclei (the width of the peak), thereby differentiating G_2, G_1/M and S phase cells, and allowing monitoring of the progression of cells through M phase. This simultaneous detection of DNA quantity and morphology without the need for scanning cytometry instruments further expands the applications of FCM within this area of research.

Not all DNA quantification is related to analysis of the cell cycle. Solieri *et al.* (2008) used FCM to measure ploidy in *Zygosaccharomyces* spp. yeasts and to estimate genome size. These workers compared fluorescence measurement of PI-stained DNA against equivalent fluorescence values for yeast species with known genome size and ploidy. Similar methods were used by Jacques *et al.* (2010) to estimate genome size and ploidy in *Debaryomyces hansenii,* a yeast used for food maturation e.g. soft cheese ripening and which is important in food spoilage. In the human pathogen *Cryptococcus neoformans,* ploidy was measured following mating between haploid and non-haploid strains (Skosireva *et al.*, 2010). PI was used to determine cellular DNA content in fusant cells compared to control haploid and diploid cells. Raithatha and Stuart (2009) investigated treatment of sporulating *S. cerevisiae* with KOH prior to DAPI, PI and SYTOX staining to eliminate non-nuclear binding, which resulted in broadening of DNA content histogram peaks and thus would affect data generated by cell cycle analysis. The biochemical basis of the observed non-nuclear binding remains to be revealed.

In an investigation of the effect of a Geminiviral Rep protein on cell cycle progression, Kittelmann *et al.* (2009) used FCM to monitor cellular morphology and the quantity of DNA in *S. pombe* expressing Rep variants. PI staining determined that Rep expression led to increased DNA content while scatter measurements indicated changes in morphology caused by probable blockage of cell division. Methods of measuring the DNA content of *S. pombe* have also been used to monitor rates of DNA synthesis following UV light-induced DNA damage (Callegari *et al.*, 2010). SYTOX Green stained DNA was measured by fluorimetry and the resultant data were analysed to correct for cell background fluorescence (dependent upon cell size) to determine absolute DNA content per cell.

As an alternative to simply measuring DNA content using SYTOX dyes, it is also possible to measure incorporation of nucleotide analogues into replicating DNA. Hua and Kearsey (2011) assessed the suitability of the thymidine analogue 5-ethynyl-2′-deoxyuridine (EdU) to monitor DNA replication in *S. pombe*. EdU is non-fluorescent but can be derivatized using click chemistry with azide-modified dyes, in this case Alexa Fluor 488 azide. Although EdU is toxic and causes cell cycle arrest, it was readily incorporated into DNA and easily detectable using both FCM and fluorescence microscopy. The authors concluded that EdU is a useful tool for analysis of DNA replication during a single cell cycle, for example DNA repair and recombination events.

Cell cycle analysis is a very useful tool for modelling the growth of budding yeast (reviewed by Porro *et al*., 2009). Combined analysis of DNA content, cellular protein content and bud scar number (using fluorescently labelled wheat germ agglutinin, a lectin that binds to chitin present at high levels in bud scars; Chen and Contreras, 2004, 2007) gives rise to datasets that can be analysed and used to model the kinetics of *S. cerevisiae* growth.

Probing cell surface components

FCM has been used to measure the binding properties of cell surface receptors; for example, the interaction between the peptide pheromone α-factor and its cognate G-protein coupled receptor (Tantry *et al*., 2010). Different quantities of fluorescently labelled ligand (in this case, NBD-labelled) were incubated with yeast cells and K_d values calculated from resultant binding curves. This method is similar in principle to yeast surface display, covered later in this chapter, but investigates native yeast proteins.

As well as analysis of surface proteins, FCM can be used to analyse other surface components. Wanchoo *et al*. (2009) used fluorescently labelled lectins to probe carbohydrates on the surface of the insect pathogen *Beauveria bassiana* at different stages of its life cycle. Use of different lectins allowed identification of different sugar residues, enabling quantitative measurements of the carbohydrate components of the cell surface, while combination with fluorescent microscopy allowed determination of the localization of specific areas of carbohydrate. This method revealed that some surface carbohydrates are present uniformly in fungal cells (for example, *N*-acetylglucosamine and *N*-acetylneuraminic acid detected by WGA-Alexa Fluor 488 in aerial condida) whereas other combinations had a great deal of heterogeneity (such as Sialic acids detected by Fluorescein-labelled *Sambucus nigra* bark lectin in Blastospores). Since surface carbohydrates are important in the pathogenesis of *Beauveria bassiana*, understanding their distributions at different stages of the fungal life cycle is essential for developing models of pathogenicity. These methods would also be broadly applicable for research into human pathogens whose pathogenicity depends upon surface carbohydrates. Cross-reactivity of antiserum to surface glycoproteins was probed in the saprophyte *Cladosporium resinae*, a contaminant in oil storage tanks (Calixto *et al*., 2010). Anti-*C. resinae* sera binding to fungal conidia was assessed by use of a FITC-labelled secondary antibody; cross-reactivity was determined by measuring decreases in binding following incubation of sera with glycoproteins from other fungi, demonstrating that certain immunogenic epitopes were shared between different organisms. Other immunological studies on yeast and fungal surface components are discussed in 'Medical applications of FCM', below.

Use of autofluorescent proteins

Originally cloned from the jellyfish *Aequorea victoria*, green fluorescent protein (GFP) and other functionally related autofluorescent proteins are widely used tools in many areas of biology, and since their discovery have been utilized in diverse experimental systems. GFP is ideal for use as a FCM reporter due to its fluorescence in live, intact cells, its largely monomeric nature, high fluorescence and lack of cofactors required for activity. The fluorochrome comprises of three cyclised amino acids located inside the β-barrel structure of the protein. The development of GFP variants with different spectral and chemical properties has been widely reviewed (for example, Zhang *et al*., 2002; Müller-Taubenberger and Anderson, 2007; Wang H *et al*., 2009), so the focus here is on recent FCM studies using GFP in yeast and other fungi.

GFP has been used as a reporter gene in yeast; promoters are transcriptionally fused to the *gfp* gene carried on a plasmid or integrated into the genome, and promoter activity is measured by monitoring GFP fluorescence (Cormack *et al.*, 1997). While this can be measured using bulk fluorimetry, FCM offers the ability to monitor heterogeneities in gene expression within members of a population. Coleman *et al.* (2008) used FCM and GFP fusions to monitor stress-induced promoter activities (and cell cycle progression using PI staining) in response to changes in buoyancy brought about by Ficoll, a density gradient centrifugation medium. This sensing strategy allows measurement of physiology at a molecular level and permits monitoring of many intracellular or extracellular events by correct choice of promoter or generation of a mutant promoter with the desired sensing properties.

The high-throughput screening capabilities arising from combining GFP with FCM are extremely powerful. Neklesa and Davis (2009) used GFP as a reporter of the TOR (target of rapamycin) response, a signal processing complex that responds to multiple stimuli in order to regulate cell growth. A library of ≈5100 *S. cerevisiae* single-gene deletion mutants was screened under three growth conditions for over- or under-expression of GFP, leading to the isolation of novel genes involved in the amino acid starvation response. This technique holds out great promise for identifying members of signal transduction pathways, and could potentially be applied to a wide variety of organisms.

Ishii *et al.* (2008) used GFP as a reporter gene to investigate G-protein coupled receptor (GPCR) signalling. In a development of previous studies (Ishii *et al.*, 2006), a chromosomal GFP fusion, activated by a MAP kinase signalling cascade originating at an α-factor-sensitive GPCR, was used to monitor signalling. FCM allowed measurement of not only the magnitude of the response (via determination of GFP fluorescence) but also the number of cells within a population responding to α-factor. One end point of the α-factor activated MAPK signalling cascade is cell cycle arrest; surprisingly, the authors did not use FCM to analyse cell cycle progression. In addition, FACS was used to sort activated and non-activated cells in experiments using plasmid-encoded GPCRs, revealing that an observed loss in activation over time was due to plasmid loss. The usefulness of the GFP/FACS method in this respect was highlighted, since other reporter genes (e.g. *lacZ* or *HIS3*) would not have as readily allowed single cell analysis or cell sorting. The authors noted the possibility of using this system to screen heterologous GPCRs, which are potential drug targets, in yeast. This would represent a novel type of yeast surface display technology (see later in this chapter), where measurement of function would be via an endogenous signalling pathway rather than affinity or extracellular enzyme activity, and display would be mediated by the nature of the GPCR rather than fusion to a surface protein.

Recently, a dicistronic reporter system was developed (Edwards and Wandless, 2010), utilizing GFP and mCherry, an engineered monomeric red fluorescent protein (Shaner *et al.*, 2004). This system was used to investigate the expression of dicistronic genes, the effect of gene order, and to develop methods of modulating expression of the second gene, specifically additional promoters or internal ribosome entry site sequences. In addition, a yeast codon-optimized mCherry variant was generated. A potential disadvantage of mCherry in FCM studies is that it requires a yellow laser source for efficient excitation (optimal wavelength 587 nm), which is not a feature on many older instruments. Some newer multicolour flow cytometers offer greater flexibility in laser choice, and a yellow laser is often the fourth laser option offered on instruments (after the more common blue, red and UV/violet). Yellow lasers are also useful for excitation of other dyes such as Texas Red (Kapoor *et al.*, 2008).

A variation on cellular DNA content analysis was presented by Ishii *et al.* (2009), who used detection of plasmid-encoded GFP as a functional marker of plasmid stability in *S. cerevisiae*. This simple technique was shown to be robust and useful both as a reporter of expression level during recombinant protein production and as an indicator of the number of cells within a population expressing GFP. The drawback of this method is that cells that have recently lost their plasmid may still be GFP-positive, although it is not clear what temporal resolution is required for plasmid maintenance assays.

GFP is an especially useful fusion partner in cell biology, as it allows measurement of both protein quantity (through fluorescence) and also location, using fluorescence or confocal fluorescence microscopy. FCM provides opportunities for quantification and high-throughput approaches. Marie *et al.* (2008) used GFP fused to two *S. cerevisiae* sterol-regulating transcription factors, Upc2p and Ecm22p. This allowed measurement of the concentration of each transcription factor (using FCM) as well as determination of cellular localization (using microscopy) upon inhibition of sterol synthesis, confirming their role in ergosterol homeostasis and revealing that they localize to the nucleus upon sterol depletion. Imaging flow cytometry would be an ideal technique to further such studies, allowing concurrent measurement of GFP-tagged protein quantity and location within the cell.

A high throughput study of peroxisome biogenesis in *S. cerevisiae* utilized GFP fused to a protein targeted to the peroxisome matrix, Pot1p (Saleem *et al.*, 2010). A library of 4000 yeast deletion mutants was screened for the inability to generate green fluorescence as a functional reporter of defects in peroxisome biogenesis. Again, fluorescence microscopy was used to provide spatial information, while FCM was used as a high-throughput screen to generate quantitative data (Fig. 7.3A).

GFP can also be used to monitor protein stability and pathways of degradation. A high-throughput approach was used to identify *S. cerevisiae* genes involved in the degradation of a mutant allele of thiopurine *S*-methyltransferase (TPMT*3A), an enzyme involved in the breakdown of some thiopurine immunosuppressant drugs (Li F *et al.*, 2008). TPMT*3A fused to GFP was expressed in a pooled library of 4667 *S. cerevisiae* single-gene deletion mutant strains. Three successive rounds of FACS selected GFP^+ cells as degradation-defective mutants, resulting in 24 isolated knockout mutants (Fig. 7.3B). These mutants had deletions in several pathways including ubiquitin-mediated proteolysis and vesicle trafficking. This approach represents a considerable time saving over alternative methodologies: the first step alone, transforming a pool of 4667 strains with an expression plasmid, would take a fraction of the time of individually transforming 4667 strains. FACS screening allows high-throughput methods to be used without the requirement for liquid handling robotics systems, which are costly.

Medical applications of FCM

Since FCM and FACS are used routinely in medical applications (for example, studies in the field of immunology), it is natural that there are many studies that have utilized these methods for studying microbial infections, including those caused by fungi. The applications of FCM in clinical microbiology have been comprehensively reviewed by Álvarez-Barrientos *et al.* (2000). Of current particular interest is *Candida albicans*, a member of the common oral microflora and an opportunistic pathogen, causing candidiasis (commonly oral or vaginal) and potentially more systemic and invasive diseases including septicaemia, especially in immunocompromised individuals including those undergoing anti-cancer chemotherapy or following organ transplant (reviewed by Warnock, 2007). It is currently the most common human fungal pathogen, and a frequent hospital-acquired infection. Several other *Candida* spp. are emerging fungal pathogens (Miceli *et al.*, 2011). As such, much research focuses on *Candida* in the areas of antifungal therapeutics, clinical diagnostics and modes of pathogenicity. In addition, *Aspergillus* spp., the causative agent of invasive aspergillosis, threaten immunocompromised patients, especially leukaemia sufferers and recipients of stem cell transplantations, and are therefore the subject of current research.

Killing fungal cells: antifungal and drug testing using FCM

The ability to monitor viability rapidly and without the requirement for growth in selective culture media makes FCM an ideal diagnostic for inactivation of yeasts and other fungi (reviewed by Vale-Silva and Buchta, 2006). The development of new antifungals is a priority for the treatment of fungal diseases due to undesirable side effects of current therapies and rapidly developing resistance in clinical settings. Several groups have published papers on the use of FCM to determine the efficacy and mode of action of potential antifungals from various natural sources. As with basic cell biology research, many different approaches have been used. Park *et al.* (2010a,b) used scatter measurements to determine morphological changes in *C. albicans* protoplasts upon exposure to lignan glycosides extracted from tree bark, and DiBAC$_4$(3) to monitor plasma membrane potential. The latter dye was also used to monitor membrane potential on exposure of *C. albicans* to ginseng triterpene glycosides (Sung and Lee, 2008), a biflavanoid plant extract (Lee *et al.*, 2009) and viral-, fish- and insect-derived peptides (Lee and Lee, 2009; Sung *et al.*, 2008a,b; Hwang *et al.*, 2010).

Pinto *et al.* (2009) used PI staining to determine membrane permeability in *C. albicans* treated with clove essential oil and its most common component, eugenol. Standard culture-based minimum inhibitory concentration (MIC) assays indicated an MIC value of 0.64 μl/ml; PI staining and FCM revealed around 20% of cells were PI$^+$ at this concentration, increasing to around 100% of cells at a concentration of 2.5 μl/ml. Owing to the rapidity of permeabilization to PI, they concluded that the oils killed the cells via membrane damage. This antifungal activity was also assessed in *Aspergillus fumigatus*, which was stained using the metabolic vitality dye FUN1 (Millard *et al.*, 1997) since PI staining did not function in this organism. Again, metabolic activity data generated using FUN1 and FCM were broadly comparable to culture-based MIC data (around 60% of cells being metabolically inactive following six hours incubation with the MIC of eugenol). The susceptibility of *C. albicans* to other antifungal therapies has also been tested using FCM. Kim *et al.* (2009) investigated susceptibility to silver nanoparticles using DiBAC$_4$(3) and PI to monitor changes in membrane potential and cell cycle progression, respectively. Other *Candida* spp., frequently emerging pathogens, are also the subject of antifungal agent testing. The response of *C. krusei* to baicalein, a plant-derived flavone, was assessed using the mitochondrial membrane potential-sensitive dye JC-1 (Kang *et al.*, 2010).

To expand the range of therapeutic options for treating fungal diseases, alternatives to conventional antifungal drugs are being developed. Photodynamic therapy relies upon incorporation of a photosensitizer into target cells, followed by irradiation with light which induces a photochemical reaction that damages the cell (Donnelly *et al.*, 2008). A common strategy is increasing the quantity of intracellular porphyrin by administering extracellular δ-aminolaevulinic acid (ALA) or its methyl ester derivative. This approach was employed to study photoinactivation of *C. albicans*, with FCM being used to measure intracellular porphyrin fluorescence (Oriel and Nitzan, 2010).

Antimicrobial peptides (AMPs), some of which are involved in innate immunity, offer a potential mechanism for treatment of fungal disease. AMPs function through a variety of mechanisms including pore formation in the cell membrane (Peters *et al.*, 2010). Since microbes are able to become resistant to most antimicrobial therapies, AMPs included, information as to the mode of action of these peptides can be gained from a study of resistance mechanisms. Following chemical random mutagenesis and selection using growth-based methods, Lis and Bobek (2008) isolated a *C. albicans* mutant resistant to an AMP derived from the human salivary mucin MUC7. FCM was then used to determine AMP killing activity via PI staining for membrane integrity; data were comparable to growth-based killing assays. In addition, by labelling the AMP with FITC, simultaneous FCM determination of cell viability (using PI) and AMP–cell interaction could be assessed. Combining these data with fluorescence microscopy allowed differentiation between cells with surface-attached AMP, which were PI$^-$ with lower FITC fluorescence, and cells with internalized AMP (PI$^+$ and high FITC fluorescence). A similar method was also used to measure interactions between the AMP PAF26 and *S. cerevisiae* (López-García *et al.*, 2010).

Antisense oligonucleotides (AONs) are another developing alternative antifungal therapy. Some catalytic self-splicing RNA molecules (so-called group 1 introns) are essential for cell viability; inhibiting this self-splicing activity by AON binding represents a potential drug treatment target. The mode of action of an AON directed against the 26S rRNA of *C. albicans* was investigated using FCM (Zhang *et al.*, 2009). FITC-labelled AON and PI were used to show that AON uptake and cell death were correlated, whereas in an intron-negative mutant, AON uptake did not lead to cell death.

Antifungal resistance and susceptibility

As well as being an important part of the development of new antifungal therapeutics, the determination of susceptibility of clinical isolates to currently used antifungals can be an important step in selection of therapies for effective treatment of infected patients. FCM is ideally suited as a method to determine this, being more rapid than current growth-based techniques (Vale-Silva and Buchta, 2006). A previously developed rapid FCM method to determine fluconazole susceptibility of *Candida* spp. using Acridine Orange as a viability marker (Rudensky *et al.*, 2005) was adapted for direct use on blood samples without the requirement for a culturing step (Rudensky *et al.*, 2008). Data from the FCM method were shown to be in excellent agreement with the commonly used culture-based macrodilution method, thereby allowing more rapid testing (the same day as blood samples were taken, compared to 48–72 hours later) within a clinical environment to determine the antifungals suitable for individual patient treatments.

The elucidation of mechanisms of drug resistance is an important area of research in this field. A common antifungal resistance mechanism in *Candida* spp. is the enhanced efflux of antifungal agents (Niimi *et al.*, 2010). Efflux in *C. albicans* was measured with FCM and fluorimetry using rhodamine 123 as the marker dye (Sun *et al.*, 2010). The intracellular Rh123 accumulation was enhanced upon exposure to a lichen-derived antifungal, but was decreased in an azole-resistant clinical isolate that exhibited increased efflux. The *CDR1* efflux pump is a major factor in efflux-derived drug resistance. *CDR1* expression can be induced by exposure to many different factors, including increased temperature, steroid hormones and azole antifungal drugs. Schmidt *et al.* (2008) utilized a *CDR1* promoter–*gfp* fusion and FCM to measure expression of the efflux pump upon exposure to different agricultural herbicides to determine whether they would pre-adapt *C. albicans* in the soil to antifungal resistance. They found that, unlike certain steroid hormones, which induce efflux pump expression without inhibiting growth, azole antifungals and herbicides only induce *CDR1* expression under conditions when growth was inhibited.

Other aspects of fungal biology have been investigated in relation to resistance. Secreted aspartyl peptidases (Saps) are a known virulence factor in *Candida* spp.; expression of one of these proteins, Sap2, was measured using Alexa Fluor 488-conjugated antibodies and FCM in *C. albicans* clinical isolates displaying a 'trailing endpoint' phenotype (where antifungal susceptibility is observed at 24 hours incubation, but resistance is seen at 48 hours; Braga-Silva *et al.*, 2009). No significant correlation was observed between Sap expression and resistance, although the method can be used in future to investigate Sap expression in relation to pathogenicity. Finally, Yang *et al.* (2010) used the Annexin/PI dual staining technique to detect early apoptotic versus necrotic cells in two clinical isolates of *C. albicans* following treatment with amphotericin B.

Diagnostics and identification

FCM offers advantages in the field of medical diagnostics in terms of rapidity, specificity, statistical significance of data generated and non-reliance on growth. Fluorescence *in situ* hybridization (FISH) is a technique used to identify DNA sequences (frequently ribosomal RNA genes) contained within cells using labelled sequence-specific nucleotide probes, and is routinely used in many areas of study including medical diagnostics (Peters *et al.*, 2006; Shepard *et al.*, 2008), environmental biotechnology (Czechowska *et al.*, 2008) and food safety (Bottari *et al.*, 2006). Probes can be designed to detect organisms with varying specificity, from single species to whole phyla. In common with FCM, FISH allows investigation of heterologous mixtures of different organisms without the requirement for growth in the laboratory, thus allowing investigation of unculturable organisms and multi-species communities that cannot be grown in pure culture. The measurable output of FISH is fluorescence within a cell; this is usually measured using fluorescence microscopy. Coupling FISH with FCM detection offers many advantages over fluorescent microscopy in terms of speed and the lack of observer bias (Kempf *et al.*, 2005).

Optimization of FISH, whether using FCM as a detection method or not, is required for each experimental system. Probe design, cell fixation and permeabilization methods and hybridization conditions all need careful consideration and experimentation to achieve acceptable detection sensitivity and specificity (Fig. 7.4). Bisha *et al.* (2011) optimized FISH-FCM methods for detection of *Candida* spp. in blood samples using two different fluorescently labelled oligonucleotide probes. The paper described the increase in fluorescence when unlabelled helper probes, which assist in the FISH probes binding to target DNA sequences, were used. Next, the fixation protocol (using neutral buffered formalin, either alone or with ethanol, or a commercial fixative HistoChoice MN) and NaCl concentration of the hybridization buffer (0.7 M vs. 0.9 M) were optimized. FCM was used to measure the ratio of the geometric mean fluorescence to the robust coefficient of variance (CV); higher ratios indicating increased hybridization quality. The presence of red blood cells in samples led to aggregation, so a detergent lysis step using 0.1% Triton X-100 was incorporated before fixation and hybridization; this was not found to interfere with FISH detection of fungal cells. It should be noted that methods need to be optimized for investigation of different clinical samples such as respiratory washes or tissue biopsies (Fig. 7.4). Hybridization time was also optimized using FCM, with an aim to reduce assay time while allowing differentiation between *C. albicans* and non-*Candida* cells (*Rhodotorula glutinis*). Finally, differentiation between *C. albicans* and other *Candida* spp. was investigated, a critical step in diagnosis of pathogens in a clinical setting.

FCM has recently been used in combination with a commercial PNA-FISH kit to rapidly and specifically identify *C. albicans* (Trnovsky *et al.*, 2008). In this method, the probe consists of a peptide nucleic acid (PNA); nucleic acid bases are arranged on a peptide, rather than sugar phosphate, backbone. PNAs are uncharged, thus allowing stronger binding to their DNA target due to the lack of electrostatic repulsion and, in many cases, greater binding specificity (Ray and Norden, 2000; Bottari *et al.*, 2006). In this study, PNA-FISH-FCM offers a very rapid (one hour) diagnostic method that does not rely upon growth. The PNAs used were labelled with a green fluor, which was measured using FCM; a simple threshold value was set to differentiate between *C. albicans* (high green fluorescence) and non-*C. albicans* (low fluorescence). The method correctly identified nine reference strains of *C. albicans* and 20 other non-*C. albicans* yeast species. When tested with 150 clinical isolates, it correctly identified all *C. albicans* strains and 97% of the non-*C. albicans* strains; three strains generated

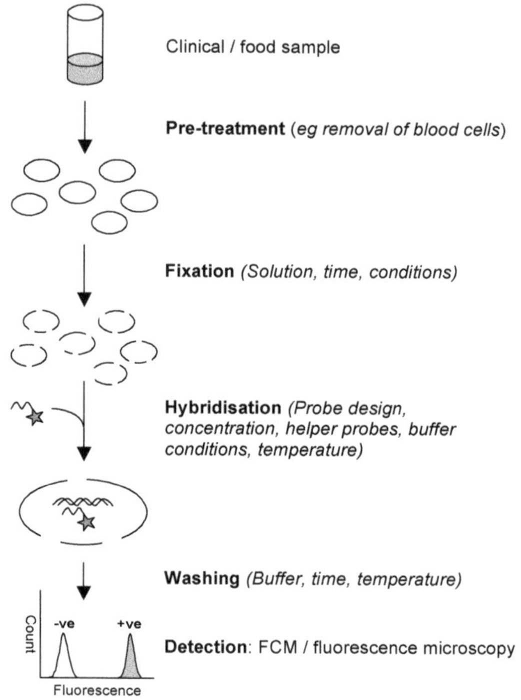

Figure 7.4 Optimization of FISH-FCM methods. Each step of the FISH procedure needs to be optimized in turn; variables are listed in italics. Bisha *et al.* (2011) offers an insight into optimization for FISH-FCM for the detection of *Candida* spp. in whole blood samples. Critical outcomes of FISH-FCM methods are specific detection of the organism of interest, lack of cross-reactivity with closely related organisms, rapidity of the assay and compatibility with clinical or other samples.

inconclusive FCM data, which were all grown on chromogenic media and were found to contain aggregates. It could be suggested that PNA-FISH-FCM could be used as an initial assessment, and isolates giving inconclusive data (i.e. not falling clearly either side of the threshold fluorescence value) could be further analysed by fluorescence microscopy.

Immunological detection is an alternative diagnostic route for medically important microbes in clinical samples, typically in the form of immunofluorescence staining with detection using epifluorescence microscopy; drawbacks of this detection method are similar to those for FISH. Immunofluorescence labelling was used in combination with FCM to detect the opportunistic fungal pathogen *Pneumocystis jirovecii*, which causes pneumonia in immunocompromised individuals and cannot be cultured *in vitro*, in clinical respiratory samples (Barbosa *et al.*, 2010). FCM was found to detect *P. jirovecii* in 10% more of the clinical samples than the fluorescence microscopy detection method; clinical diagnosis suggested that these were true-positives, and that FCM represented a more sensitive detection method.

Fungal pathogen–host interactions

Surface components of fungal pathogens are frequently a target of pathogenicity research. As adherence to host cells is frequently essential for pathogenicity, adhesins are a potential drug target. The ability of four isolates of *Sporothrix schenckii*, the pathogen that causes the mycosis sporotrichosis, to bind FITC-labelled fibronectin was measured using FCM (Teixeira *et al.*, 2009); the strain with lowest virulence in a mouse model also showed lowest fibronectin binding.

It has been noted that opportunistic *C. albicans* infections frequently occur during antibacterial treatments; this phenomenon has been attributed to decreases in natural bacterial microflora at infection sites and interference with the antifungal immune responses. Grüger *et al.* (2008) investigated the influence of fluoroquinolone antibiotics on phagocytosis of *C. albicans* by polymorphonuclear neutrophil (PMN) granulocytes. BCECF was used to label live *C. albicans* cells, which were mixed with blood; FCM was used to identify PMNs (using forward- and side-scatter) and measure their green fluorescence as an indicator of fungal phagocytosis. In addition, the ROS-sensitive dye dihydrorhodamine was used to measure the neutrophil oxidative burst on exposure to *C. albicans*. Although the FCM technique could not differentiate between phagocytosed and surface-attached *C. albicans* cells, it was noted that fluorescence microscopy had previously shown that surface-associated yeast cells quickly became phagocytosed. Similar methods were used to study the phagocytosis of *Aspergillus fumigatus* condida in different strains of mice, both *in vivo* in lung homogenates and *in vitro* using PMN phagocytosis assays (Svirshchevskaya *et al.*, 2009). This study also utilized scanning cytometry to visualize phagocytosed conidia.

Phagocytosis studies have also been used to investigate innate immunity. Brouwer *et al.* (2008) used FCM to determine phagocytosis of FITC-labelled *C. albicans* (as well as three bacterial species) by PMNs as an indicator of opsonisation via different complement activation pathways. FCM and specific antibodies were also used to measure the deposition of mannose-binding lectin, an initiator of the lectin pathway of complement activation, and complement factor C3 on *C. albicans* cells. It was concluded that the lectin pathway of complement activation plays a significant role in the innate immune response to *C. albicans* infection. Flow cytometric C3 binding assays were also used to study antibody-dependent complement activation in response to *C. albicans* (Boxx *et* al., 2009).

Fungal pathogen defences against complement-mediated killing have also been studied by FCM. Several pathogens have been shown to subvert the regulation of the complement cascade by binding host-derived complement inhibitors, thereby inactivating complement and increasing pathogen survival (reviewed by Blom *et al.*, 2009). This phenomenon has been studied in the opportunistic pathogens *Aspergillus fumigatus*, *A. terreus* and *C. albicans* with binding to the complement inhibitors factor H, factor H-like protein and C4b-binding protein being measured *in vitro* using FCM (Vogl *et al.*, 2008). Characterizing the pathogen surface ligands which bind to host complement inhibitors is a further potential route to drug development.

These studies demonstrate the tremendous power of FCM as a technique for monitoring immune cell–pathogen interactions, since many methods have already been developed, using CD and other cell surface receptor staining, to characterize cells of the immune system.

FCM for bioprocess monitoring and development

Many facets of FCM have been utilized by the process industries for monitoring. The common factor of the application of FCM in bioprocessing is measurement of single-cell events that are particularly relevant, in terms of product quality, quantity and/or cell biology. Frequently, these processing aspects are absolutely interdependent. As with other applications of FCM, bioprocess monitoring and design frequently benefit from the speed of analysis and the lack of a requirement for growth. The field of single-cell analysis of microbial industrial processes has been reviewed previously; interested readers are directed to Mattanovich and Borth (2006) and Díaz *et al.* (2010).

Fermented foodstuffs: brewing and oenology

Amongst the earliest fungal bioprocesses, beer brewing and winemaking are both fields in which FCM is being used to refine processes and understand the underpinning biology. The advantages of FCM stated earlier in the chapter allow rapid measurement of aspects of yeast physiology that are essential for brewers for both process design and in-progress monitoring. FCM was first used in brewing research in the 1970s, and some of the developments over time were reviewed by Stewart (2009). Physiological dyes are the most commonly used method, as they give information comparable to the standard brewers' diagnostic test, methylene blue staining, and additional information about physiology beyond dead/live screening (reviewed by Díaz *et al.*, 2010). Common research questions include measurement of yeast viability and physiology during fermentation, development of more intensive fermentation processes, and re-pitching (the re-use of spent yeast from one fermentation process for subsequent fermentations). As with most bioprocesses, yeast physiology affects product quality, both in terms of overall fermentation efficiency and secondary metabolite production responsible for aroma and flavour compounds in the beer.

In the field of brewing process intensification, Chlup *et al.* (2008) used FCM to evaluate changes in *S. cerevisiae* physiology following disc-stack centrifugation. This technique can be used for cell separation at several stages in brewing, including following primary fermentation, although the high shear forces exerted during centrifugation have the capacity to damage cells, leading to loss of viability and cell breakage, in turn affecting beer quality and stability. Yeast viability, cell surface mannan residues, pH_i and intracellular glycogen and trehalose concentrations were monitored by FCM following high speed or low speed centrifugation, allowing optimization of the separation process and determination of suitability at different stages of the brewing process.

Flocculation is a critical separation step at the end of fermentation in the brewing process, whereby yeast cells adhere to form flocs which sediment or rise to the surface in lager or ale yeasts, respectively (reviewed by Verstrepen *et al.*, 2003). Failure to flocculate means yeast remain in suspension and die, decreasing beer quality, and requiring additional filtering or clarification steps, adding cost and time to the process. After several repitches, some commercial yeast cease flocculating, so identification of the point at which yeast should no longer be repitched is critical in process optimization. Cell surface proteins (flocculins) and mannans are thought to be critical for flocculation. Heine *et al.* (2009) used fluorescently labelled lectins concanavalin A and *Pisum sativum* agglutinate to monitor mannan and glucan residues on the cell surface in order to differentiate flocculent or powdery (non-flocculating) yeast strains. Unfortunately, FCM alone did not sufficiently differentiate flocculant and powdery yeasts in industrial conditions, but the application of alternative lectins or other cell surface detection methods might allow this in future.

The bud scar counting method of yeast cell age assessment (Chen and Contreras, 2004) has been used to investigate brewing (Kuřec *et al.*, 2009). FCM was used to measure WGA-Alexa

Fluor 488 binding to bud scars; mean fluorescence data were found to correlate with bud scar counts determined by fluorescence microscopy. The FCM technique was used to monitor yeast age at different depths of the sedimentation zone in model fermenters at different pitching rates (inoculation concentrations). Bud scar number was also assessed in long-term immobilized yeast brewing systems (Kuřec *et al.*, 2009). The method was suggested for routine assessment of yeast age to determine when yeast should be repitched or replaced with fresh yeast in order to maintain product quality. Other FCM analyses applied to beer brewing include cell cycle analysis (Novak *et al.*, 2007) and use of physiological dyes such as PI and oxonols to monitor physiology during fermentation and the effects of mixing (reviewed by Nienow *et al.*, 2010). As well as beer brewing, FCM has been used to measure ploidy in sake brewing yeasts (Katou *et al.*, 2008).

FCM has also been successfully applied to analysis of fermentation in winemaking. Most wine fermentations utilize active dry yeast, and viability determination using PI was found to be useful for optimizing rehydration (Rodríguez-Porrata *et al.*, 2008). Many wine fermentations are microbiologically heterogeneous, with wild yeast strains and bacteria (frequently from grape skins) present (Walker, 2000). Rodriguez and Thornton (2008) used FCM to identify and measure viability of *S. cerevisiae* (using polyclonal antibodies and PI) and identify *Oenococcus oeni* (using antibodies) in Chardonnay fermentations. *O. oeni* is a desirable lactic acid bacterium responsible for the malolactic fermentation that converts malic acid to lactic acid, improving mouthfeel and overall wine quality (reviewed by Torriani *et al.*, 2011). Since *O. oeni* enumeration by viable count methods takes a week or longer, FCM offers a very useful rapid monitoring technique to ensure wine quality during the fermentation.

Further molecular winemaking studies have also been carried out using FCM. Intracellular ROS are a major factor in wine yeast viability, and increase as fermentation progresses and therefore ethanol concentration increases. Intracellular ROS were measured using dihydroethidium (Landolfo *et al.*, 2008), membrane integrity was measured using PI, and data were compared to bulk analysis methods. ROS and PI permeability profiles during model wine fermentations were found to differ according to yeast strain, and so offer the ability to screen for optimal strains. In a further study, the effect of lipid supplementation on intracellular ROS and viability was measured (Landolfo *et al.*, 2010).

Food: elimination of spoilage

In common with studies of fungicides and fungal inactivation methods for medical applications, FCM has also been used to monitor fungicidal treatments in food processing. Viability staining has been employed to investigate fungicidal treatments; for example, the use of PI and FDA to monitor viability of *S. cerevisiae* following ultraviolet irradiation (Schenk *et al.*, 2011), or use of the proprietary BactQuick (Sysmex, Japan) stain for investigation of *S. cerevisiae* responses to ultrasound and γ-radiation (Oyane *et al.*, 2009). In addition to relatively simple viability measurements, FCM has also been used to study mechanisms of fungal inactivation. Stratford *et al.* (2009) used carboxyfluorescein diacetate succinimidyl ester (CFDA-SE) to monitor changes in the cytoplasmic pH of *Aspergillus niger* condida in response to organic acid treatment, a popular food preservation technique. CFDA-SE was chosen over CDFA (Weigert *et al.*, 2009) due to its uniform localization within the *A. niger* condida; CFDA preferentially localized to the vacuoles, which would not have resulted in accurate cytoplasmic pH determination.

Fluorescence *in situ* hybridization (FISH) is becoming a more commonly used technique in food and environmental microbiology (reviewed by Bottari *et al.*, 2006; Amann and Fuchs, 2008). Fermented beverages are subject to spoilage from microorganisms, and FISH-FCM has been used as a rapid diagnostic test. The spoilage yeast *Brettanomyces bruxellensis*, which is naturally present on grape skins, generates phenols, acetic acid and other undesirable flavour and aroma molecules (Malfeito-Ferreira, 2011). A FISH-FCM method for detecting low concentrations (10^2 cells/ml) of *Brettanomyces* in wine has recently been developed (Serpaggi *et al.*, 2010); this low detection concentration means that the *Brettanomyces* can be detected before spoilage occurs. The authors highlight the usefulness of early detection of this spoilage yeast to winemakers, and suggest that the technique could be extended to detect other spoilage organisms. Paradoxically, *Brettanomyces* is key to production of some beers generated using spontaneous fermentation processes such as lambics, so these FISH-FCM techniques could also be used for process monitoring. In addition, *Candida* spp. present in foodstuffs can be detected using FISH (Bisha *et al.*, 2011); rather than spoilage, the emphasis here is on minimizing the effects of this microorganism on human health, especially that of immunocompromised individuals.

Bioprocessing fermentation monitoring

Other fungal fermentations are also monitored using FCM. Microbial oils are a potential feedstock for biodiesel production (Li Q *et al.*, 2008). In microbial oil fermentations, the ability to monitor viability and lipid production makes FCM a very useful process analytical tool. Nile red is a commonly used lipophilic dye to detect intracellular microbial oils. This technique was used to monitor microbial oil production in the oleaginous yeasts *Waltomyces lipofer* (Raschke and Knorr, 2009) and *Rhodotorula glutinis* (da Silva *et al.*, 2010). Both studies combined lipid staining with viability assessment (CFDA/PI and PI, respectively). FCM was also used to monitor cell cycle progression in continuous *S. cerevisiae* ethanol fermentations (Shen *et al.*, 2009).

Kottmeier *et al.* (2009) used FCM to validate biomass data collected by a high-throughput microplate-based process development platform, BioLector, which measures back-scatter optical data as an estimation of the biomass contained within each well. It was noted that, in *Hansenula polymorpha* fermentations, online back-scatter data did not correlate well with oxygen uptake data. FCM was used to determine that forward- and side-scatter became more heterogeneous over the fermentation and that overall cell scatter decreased. This paper highlights the importance of measuring multiple aspects of cell physiology, and that FCM is a complementary technique to be used along side other optical methods.

Recombinant protein production in yeast and filamentous fungi

Yeasts and filamentous fungi are popular microbial hosts for recombinant protein production due to their relatively simple growth requirements and ability to post-translationally modify proteins (Porro *et al.*, 2011). *Pichia pastoris* is a methylotrophic yeast originally developed for single-cell protein use, but is also an attractive host for recombinant protein production due to its high secretion capacity and glycosylation capability. Hyka *et al.* (2010) utilized FCM to assess the viability and physiology of *P. pastoris* in high-cell-density cultivation, using a combination of dyes (PI, EB, DiBAC$_4$(3) and CFDA) to assess the effects of the nature of the recombinant protein produced, the promoter diving expression, culture pH and methanol feeding profile. PI was determined to be the optimal assay of viability, and the value of FCM as an analysis technique for bioprocess monitoring (in particular, the lack of reliance on growth, the speed of analysis allowing in-process decisions to be made and the quantification of heterogeneity within the culture, which is of particular interest to bioprocess engineers) was reinforced. PI was also used to assess *P. pastoris* viability in response to antifoam in shake flask cultures (Routledge *et al.*, 2011).

Immunofluorescence labelling can be used in conjunction with FCM as a convenient method of determining recombinant protein production levels. Resina *et al.* (2009) used antibodies to detect expression of recombinant lipase in *P. pastoris*. In addition, the heat shock protein HSP70 was detected using antibody labelling to monitor the cellular response to misfolded protein in the endoplasmic reticulum (the unfolded protein response or UPR). Using this technique, productivity and cellular physiology can be measured, leading to improvements to strains and processes. A disadvantage of antibody detection of intracellular proteins is that it requires cell permeabilization whereas proteins present on the surface of the cell can be detected in whole, live cells. Immunofluorescence labelling can also be used to measure the activity of recombinant antibody fragments generated in yeast fermentations. Wan *et al.* (2008) used FCM to measure binding of an anti-CD25 scFv fragment generated in *P.*

pastoris. FITC-labelled scFv was detected binding to human-derived cell lines either expressing or not expressing CD25, a potential cancer marker and drug target, and activated and unstimulated human monocytes.

Cell engineering and screening using FACS

The high-throughput nature of FCM and FACS allow them to be used as efficient screening systems. Broadly speaking, two main forms of screening exist: screening for improved strains for bioprocesses; and screening for optimized proteins and enzymes, as a form of protein engineering. In each case, the objective is the same: to isolate the cell containing the genetic material giving rise to a desired phenotype. In order for this to happen, a physical link must exist between the cell and the screening characteristic. Frequently, organisms used in industrial bioprocesses are optimized by some sort of strain engineering or selection. Traditionally, this took the form of classical strain development, a systematic process of successive rounds of mutagenesis and selection of the most productive organisms for a particular process. FACS allows this procedure to be carried out quantitatively at very high throughputs. Whereas classical strain development methods frequently rely upon measurement of a product concentration or an enzyme activity (for example, SDS-PAGE analysis for recombinant protein production or citric acid production by *Aspergillus niger*), FACS selection methods require that a fluorescence signal is ascribed to the desired phenotype (Fig. 7.3B). These methods are discussed below in the context of not only engineering cellular phenotype, but also engineering molecules using surface display techniques. The relationship between classical strain development methods and more modern genomic and molecular techniques, including the roles of FCM and FACS, is reviewed by Patnaik (2008).

Screening for bioprocess producer strains

In many cases, bioprocess strains are isolated from environmental samples. FISH is useful for discovery of potentially useful microbes (for example, cellulose degradation (O'Sullivan *et al.*, 2007) or crude oil bioremediation (Harayama *et al.*, 2004)) and is ideally suited to analysis of community structures in microbial consortia used in bioprocesses. The yeast and bacterial components of an activated sludge system were monitored in response to commonly encountered industrial waste stream characteristics such as low pH and high organic loading (Zheng *et al.*, 2010). It is anticipated that FISH-FCM will become increasingly popular for the analysis of mixed microbial populations within the areas of environmental microbiology and environmentally derived biotechnology, complementing high-throughput sequencing approaches. A disadvantage of FISH is that cells need to be permeabilized (and therefore killed) in order for the oligonucleotide probes to access their genomic DNA, so physical sorting via FACS of viable cells is not possible. It would still be possible to recover DNA from these sorted cells for further analysis, for example PCR isolation of genes encoding novel enzymes or pathways.

Overall cell characteristics have been engineered using a cytostat, a continuous culture bioreactor similar in principle to a turbidostat but having the feed pump controlled by an automated flow cytometer, allowing maintenance of a constant cell density (Kacmar *et al.*, 2006). This apparatus allows maintenance of low biomass concentrations and more rapid mutant selection than with chemostat methods. Gilbert *et al.* (2009) used the cytostat to select an acetate-resistant *S. cerevisiae* strain for lignocellulosic ethanol production; acetate is a common growth inhibitor found in feedstocks for this bioprocess. Use of FCM as an at-line technique allowed monitoring of cell number and also cell scatter throughout the cell engineering process; acetate-resistant cells were observed to be smaller than the wild-type. FCM was subsequently used to investigate cell cycle progression in the wild-type strain and generated mutants.

The generation of recombinant human antibodies by *P. pastoris* is a potentially commercially important bioprocess, and selection of optimally producing cells is a major step in process development. Antibody-secreting cells transiently display a small percentage of the secreted antibody on their surface, so selection of cells with large amounts of surface-displayed antibody will allow isolation of highly secreting clones (Lin *et al.*, 2010). Immunogenic detection (using Alexa Fluor 488-labelled antibodies) and FACS was used to screen for *P. pastoris* cells with large amounts of recombinant IgG on their surface (Fig. 7.3B); this method was shown to increase antibody secretion titres by up to 300%. The use of GFP fusions allows selection of optimally producing clones in recombinant protein production systems (for example, CHO cell selection; Pilbrough *et al.*, 2009). This strategy was used to screen optimal production of a kinase insert domain receptor–GFP fusion in *P. pastoris* (Wang Y *et al.*, 2009); the GFP fluorescence of 50 individual clones was analysed by FCM and optimal clones selected on this basis (Fig. 7.3A). This study also utilized peroxisome targeting of the recombinant protein, a strategy that is especially useful when the recombinant protein being generated is toxic to the host cell. The added advantage of a GFP tag is that localization can be measured using fluorescence microscopy, in this case by co-expression of peroxisomally targeted blue fluorescent protein.

Although *P. pastoris* has many advantages as a recombinant protein host, there is a lack of promoters from which RP synthesis can be directed. Stadlmayr *et al.* (2010b) fused a library of promoters, chosen either from a microarray study of *Pichia* gene expression or using a literature comparison with known promoters in other yeast species, to GFP as a reporter. FCM was used to measure GFP expression as an indicator of promoter activity under different growth conditions, and promoters with desired characteristics were used for further studies including fed-batch expression of human serum albumin.

Multiple strategies for screening cells for small-molecule production have been developed. More simple techniques are used to screen for small molecule metabolites that reliably remain inside the cell rather than those that are secreted or diffuse from the cell. Ukibe *et al.* (2008) used FACS to screen for enhanced carotenoid production in the yeast *Xanthophyllomyces dendrorhous*. The carotenoid was astaxanthin, desired for pharmaceutical and food industries as an antioxidant and as a feed supplement in Salmon aquaculture, which is usually generated at low concentrations in *X. dendrorhous*. As astaxanthin is fluorescent, FACS lends itself to optimization of production strains; however, other carotenoids also exhibit fluorescence (although with different emission spectra) so characterization of optimal emission detection wavelengths was required using comparison of known strains and HPLC analysis of carotenoid production. Random chemical mutagenesis was followed by a conventional plate-based screen and a FACS screen; FACS screening was more efficient and isolated more overproducing mutants than the conventional screen.

To screen for production of non-fluorescent metabolites, alternative methods are used. Glutathione (GSH; γ-glutamyl-L-cysteinyl-glycine) is a tripeptide thiol with antioxidant properties that protects cells against a range of stresses. It is currently made industrially using fermentation for the food, drug and cosmetic industries (Li *et al.*, 2004). To monitor intracellular GSH concentrations, the thiol-specific monochlorobimane can be used, which forms an adduct with GSH and fluoresces at 490 nm. This reaction was used in an FCM assay to screen for *S. cerevisiae* mutants with increased intracellular GSH concentrations following random chemical mutagenesis in order to optimize production (Wang *et al.*, 2010).

Screening for extracellular species

For metabolites that do not reliably remain inside the cell, or are actively secreted, methods have been developed that physically associate the reaction products with the producing cell, such that isolation of the cell still allows isolation of the genetic material giving rise to the productive phenotype. Kwan *et al.* (2011) took a synthetic chemistry approach to screening enzyme activity by generating synthetic substrates for β-glucosidase enzymes expressed in *P. pastoris*. Upon cleavage, these glycoside substrates generate a quinone methide, which reacts with nucleophiles close to the site of enzymatic catalysis, thus fluorescently labelling the cell (Fig. 7.5A). This allowed clear identification by FCM of *P. pastoris* cells expressing recombinant Thai rosewood β-glucosidase, and FACS sorting of expressing and non-expressing cells from mixtures. The highly reactive nature of the quinone methide allows for rapid reaction with the yeast cell, although it presented some

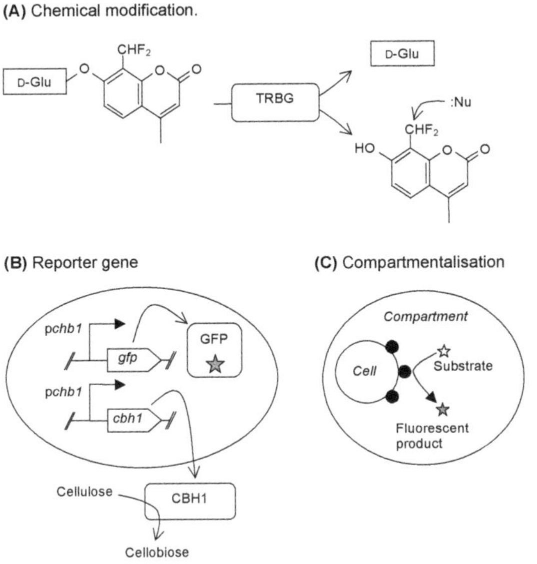

Figure 7.5 Screening for soluble or extracellular products. (A) Chemical modification of the cell by a reaction product. Thai rosewood β-glucosidase (TRBG) expression in *Pichia pastoris* was assayed using a synthetic glycoside which upon cleavage forms a quinone methide. The CHF_2 group indiscriminately reacts with nucleophiles (:Nu) within the cell, immobilizing the fluorescent products close to the site of catalysis (Kwan *et al.*, 2011). (B) Use of a reporter gene to monitor synthesis of a *Trichoderma reesei* secreted cellobiohydrolase, CBH1. The *cbh1* promoter was used to regulate expression of GFP; selection of mutants with increased GFP expression correlated with increased cellobiohydrolase activity (Throndset *et al.*, 2010a). (C) *In vitro* compartmentalization for detection of cellular activity giving rise to diffusible molecules. Single cells are compartmentalized, either in alginate (Zhou *et al.*, 2009) or in a water-oil emulsion (reviewed by Schaerli and Hollfelder, 2009). Compartmentalized cells are selected on the basis of generation of a fluorescent product which is detected by FCM; compartmentalized cells with the desired enzyme activity are sorted using FACS.

adverse cytotoxic effects in a similar experiment in *E. coli*. This method could be extended in order to screen for other enzyme activities, and possibly surface-displayed enzymes, although the secondary effects caused by toxicity must always be considered.

FACS has been used to screen for optimal production of native secreted enzymes. The filamentous fungus *Trichoderma reesei* is utilized industrially in the generation of endoglucanases and exoglucanases, both of which are used in the breakdown of cellulose in the textile and paper industries and increasingly in the production of lignocellulosic biofuels (Schuster and Schmoll, 2010). Throndset *et al.* (2010a) used FACS to select *T. reesei* mutant spores that overexpressed cellobiohydrolase 1 (*cbh1*), an exoglucanase. Since this enzyme is secreted, FCM could not be used to measure either the presence of enzyme (using a translational GFP fusion) or enzyme activity (using a fluorescent substrate analogue), so an alternative method was developed (Fig. 7.5B). The *cbh1* promoter was fused upstream of *gfp* and integrated into the genome; thus, GFP fluorescence was used as a reporter for cellobiohydrolase 1 expression. A mutant strain selected on the basis of high GFP fluorescence displayed increased conversion of lignocellulosic biomass to sugars.

In a second study, the *cbh1* screening method was used to select for strains with enhanced secretion of glucoamylase, the enzyme that catalyses the hydrolysis of starch (Throndset *et al.*, 2010b). A strain containing dual fusions was employed: p*cbh1*::*gla1*, where the glucoamylase gene was under the transcriptional control of the cellobiohydrolase 1 promoter; and p*cbh1*::*gfp*, whereby promoter activity could be measured by GFP fluorescence. Initial experiments demonstrated that glucoamylase activity correlated with GFP fluorescence. Spores were mutagenized using UV light and the resultant germlings were sorted by FACS into ten 96-well plates; the germlings with the top 0.01% of GFP fluorescence were selected (Fig. 7.3A). The GFP fluorescence and glucoamylase activity of these clones was measured, resulting in the isolation of a clone with 50% greater glucoamylase activity. The screen was repeated, giving rise to a clone with 70% greater glucoamylase activity than the starting strain. Since ultraviolet irradiation was used as the mutagenic method in both studies, random mutations would be expected at multiple genomic loci, thus demanding further characterization. Indeed, comparative genomic hybridization revealed widespread genomic rearrangements in the enhanced glucoamylase strain.

Alternative methods of screening for the production of soluble molecules by surface-displayed proteins are described below.

Molecular screening methods using yeast and FACS

Yeast cells have long been used as molecular screening tools. Yeast surface display is probably the most widely used microbiological application of flow cytometry, which is frequently used to aid molecular biology, drug discovery and environmental biotechnology (Feldhaus and Siegel, 2004; Pepper *et al.*, 2008; Shibasaki *et al.*, 2009).

Yeast surface display – screening for molecular engineering

At its most fundamental, surface display is a method of targeting heterologous proteins to the surface of a biological particle (for example a phage, bacterium or cell). The displayed protein can then be assayed according to binding affinity or enzymatic activity. In each case, the displayed protein is physically linked to the gene sequence giving rise to the protein, so that selection of phenotype allows selection of genotype. Many surface display technologies have been developed over time with different biological particles, including phage display (reviewed by Pande *et al.*, 2010), bacterial display technologies (reviewed by Daugherty, 2007) such as anchored periplasmic expression (ApEX; Harvey *et al.*, 2004) and autodisplay (Jose and Meyer, 2007), and mammalian display (Zhou *et al.*, 2010). The advantages of yeast surface display over other display technologies for the selection of eukaryotic proteins primarily originates from the ability of yeast to secrete functional, correctly folded and post-translationally modified eukaryotic proteins in a way that bacteria cannot. Recent studies have compared display technologies; for example, Bowley *et al.* (2007) applied yeast surface display and phage display to the selection of HIV-1 gp120-specific single chain antibody fragments with YSD being shown to be far superior.

The commonly used Boder and Wittrup yeast surface display method (Boder and Wittrup, 1997) utilizes the α-agglutinin anchor system (Fig. 7.6A). Proteins to be displayed on the cell surface are translationally fused to Aga2p, a protein which is secreted by *S. cerevisiae* and binds (via two disulphide linkages) to the GPI-anchored protein Aga1p. A fusion tag (frequently myc or Xpress) allows detection and quantification of display using a specific fluorescently labelled antibody. This also allows comparison of the quantity of displayed protein with the observed activity or affinity, thus allowing differentiation

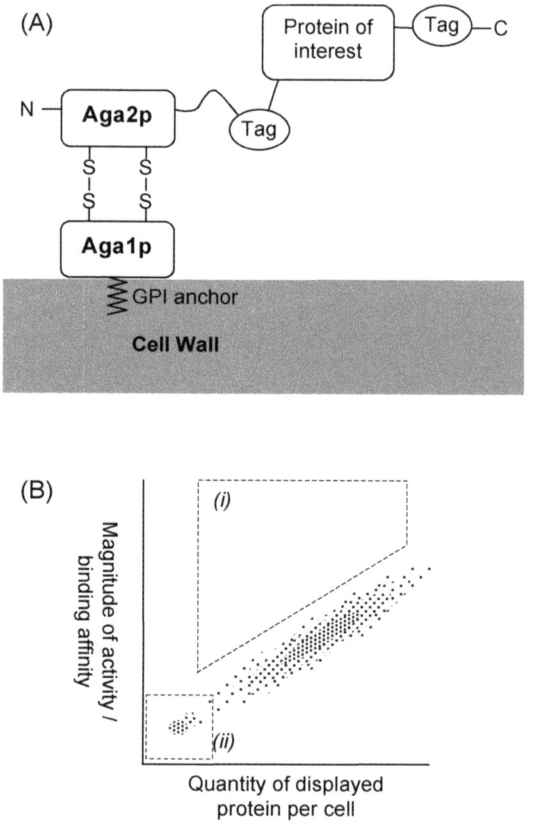

Figure 7.6 Yeast surface display. (A) The Boder and Wittrup (1997) *S. cerevisiae* agglutinin yeast surface display system; the protein of interest is fused to Aga2p via a flexible linker. The displayed protein may be fused to Aga2p at the N- or C-terminal (shown here), depending upon protein conformation and function. Aga2p is disulphide-bonded to the Aga1p protein, which is attached to the cell wall via a GPI anchor. Alternative anchoring systems have been developed and are discussed in the text. An affinity tag may be fused to the protein of interest (either N- or C-terminal) to quantify surface displayed protein. (B) FCM data collected during a typical YSD experiment, showing fluorescence data collected to measure both the quantity of displayed protein per cell (detected using anti-fusion tag antibodies) and the affinity or activity of the displayed protein (as discussed in the text). Region (i) in the plot denotes surface-displayed proteins with enhanced affinity or activity, which would be selected by FACS. Region (ii) comprises cells failing to display protein, or displaying protein with inaccessible fusion tags.

of cells displaying a large quantity of protein from cells displaying protein with high specific activity or affinity (Fig. 7.6B). Alternative surface-display fusions are being developed to address problems occasionally encountered with the α-agglutinin anchor; these are reviewed elsewhere (Pepper *et al.*, 2008). As well as *S. cerevisiae*, yeast surface display systems have been used in other species, including *P. pastoris* (Wang *et al.*, 2007; Jacobs *et al.*, 2008) and *Yarrowia lipolytica* (Yue *et al.*, 2008). YSD in *P. pastoris* is seen by some as especially useful, since the organism is widely used for recombinant protein production in the biopharmaceutical industry.

This 'display' of proteins is frequently applied to libraries of randomly mutagenized proteins, which can then be selected on the basis of affinity or enzymatic activity using FACS or magnetic-activated cell sorting (MACS; Miltenyi *et al.*, 1990). MACS relies upon nanometre-scale superparamagnetic beads conjugated to antibodies (Fig. 7.7A). The antibodies can be directed against specific cell surface epitopes or displayed proteins, allowing direct capture of desired cells. Alternatively, indirect capture can be used; cells are labelled with primary antibodies, which are bound by bead-attached antibodies. High-gradient magnetic fields are used to separate cells bound to the beads from unbound cells (Fig. 7.7B); removal of the magnetic field allows elution of the bound cells. Multiple rounds of selection are used, often with increasing stringency. MACS and FACS approaches are often combined, and diversification of mutant libraries by additional rounds of random or directed mutagenesis can be used between rounds of selection (Fig. 7.7C).

As the field of YSD is highly developed and has been thoroughly reviewed (Feldhaus and Siegel, 2004; Pepper *et al.*, 2008; Shibasaki *et al.*, 2009), discussion here will be limited to selected recent advances in the field as related to FCM and FACS. The fundamentals of recombinant protein production, secretion and trafficking have been investigated using surface display and FCM. The unfolded protein response (UPR) is a eukaryotic mechanism by which unfolded protein in the endoplasmic reticulum is sensed and either refolded or degraded. Guerfal *et al.* (2010) investigated the possibility of increasing expression of the *HAC1* gene, which induces the UPR, in order to enhance folding of four recombinant proteins in *P. pastoris*: mouse interferon-γ; human interferon-β; human thrombomodulin; and human erythropoietin. FCM was used to measure the amount of surface-displayed protein using anti-V5 antibodies to detect a fused V5 tag as a measurement of secretion efficiency in the presence and absence of overexpressed *HAC1*. In cells constitutively overexpressing *HAC1*, there was little effect on the quantity of surface-displayed recombinant protein, whereas inducible *HAC1* overexpression decreased the surface display of all proteins except human thrombomodulin. These findings correlate with other recombinant protein research; overexpression of folding and chaperone systems frequently have no effect on, or actually decrease, yields of correctly folded recombinant protein, and such systems need to be optimized on a case-by-case basis (Martínez-Alonso *et al.*, 2010).

Maturation of antibody fragments with YSD

A major application of YSD has been selection of antibody fragments for diagnostic and potential clinical use, a process referred to as affinity maturation. Owing to the complexity of human antibody structure, it is often easier to express fragments of antibodies rather than whole antibodies; the posttranslational modification pathways of yeast allow correct folding and some degree of glycosylation. Fig. 7.8A summarizes some of the antibody fragments that have been screened using YSD. Displayed antibody fragments are selected for affinity, then the DNA sequences of the V_H and V_L domains can be cloned into full-length IgGs for production in mammalian cells. FCM experiments to screen for affinity usually use a fluor-labelled antigen, or a secondary detection method (such as fluorescent antibodies or streptavidin-conjugated fluor and biotinylated antigen) to detect binding. In addition, a method to detect the quantity of displayed antibody fragment is required, for example a c-myc tag attached to the fragment detected by fluorescently labelled anti-myc antibodies (Fig. 7.8B) to allow quantification of both surface display and affinity (Fig. 7.6B).

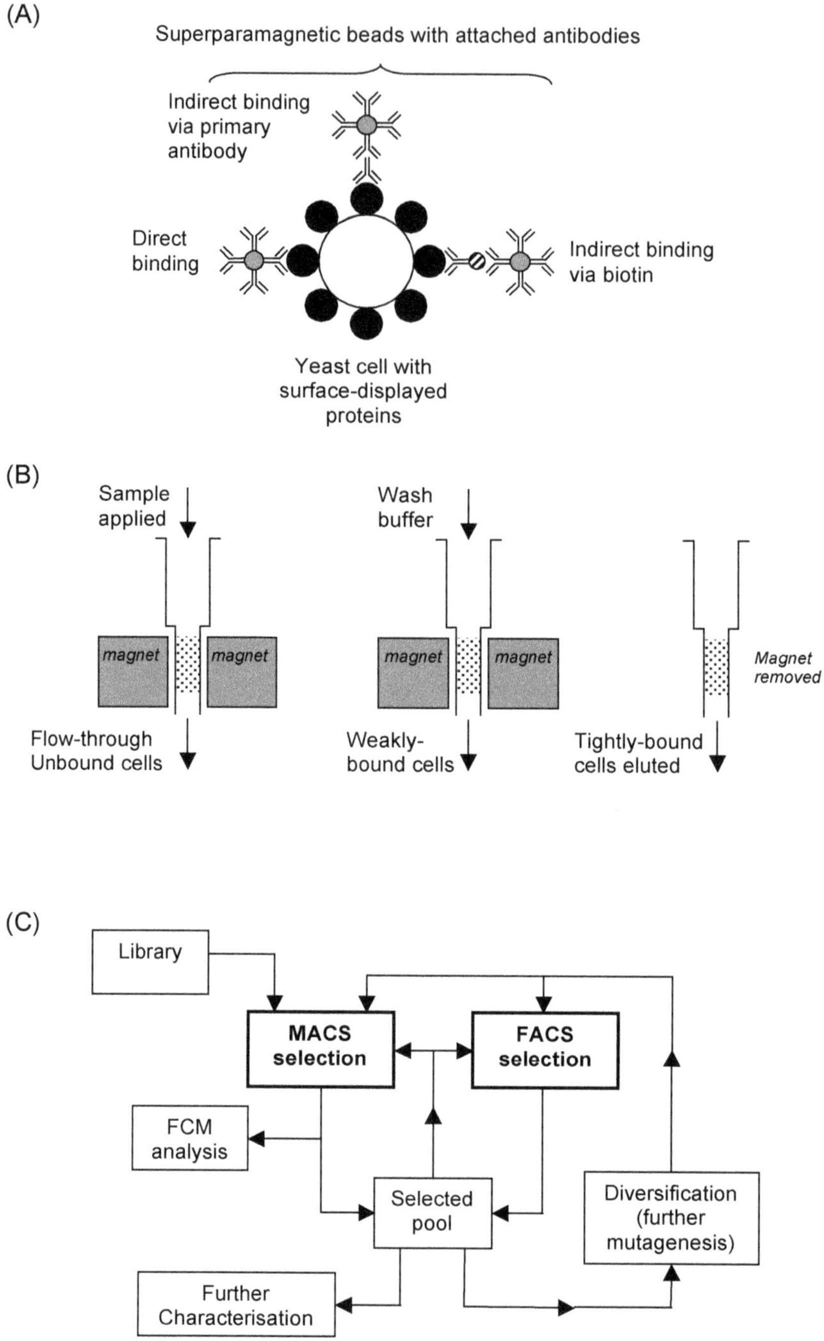

Figure 7.7 Magnetic cell sorting. (A) Cells displaying desired surface proteins are bound by antibodies conjugated to superparamagnetic beads. Binding is either direct (antibodies directed against the desired surface protein are used) or indirect (binding is via a primary antibody or a biotinylated primary antibody). (B) The sample is applied to a column containing a ferromagnetic matrix within a magnetic field. The matrix concentrates the magnetic field, allowing extremely tight binding of the beads. The column is then washed to remove weakly bound cells; altering the wash buffer can increase stringency. Desired cells are eluted following removal of the magnetic field. (C) Possible MACS/FACS workflow for library screening. Each selection method has advantages and disadvantages, as outlined in the text. Library diversification (further rounds of random mutagenesis between selections) can be used to increase the number of variants present for selection.

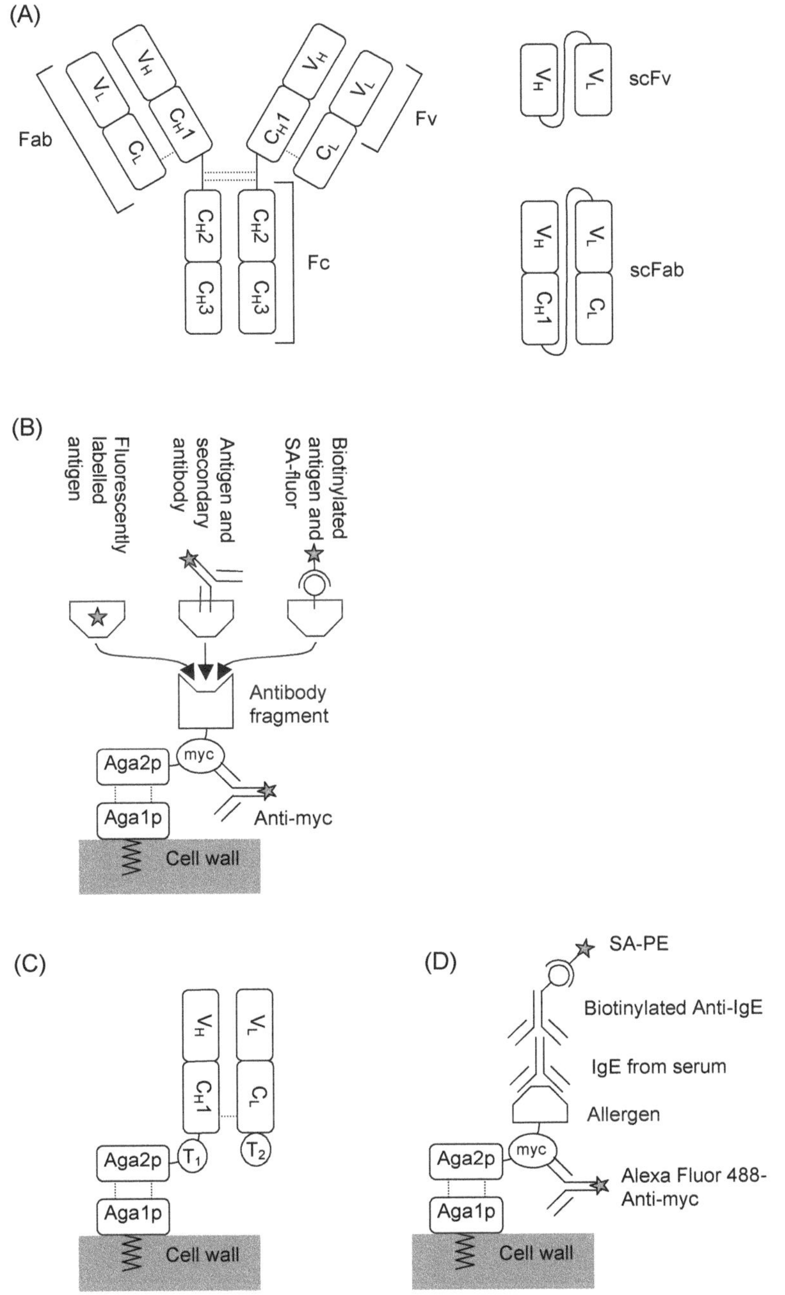

Figure 7.8 Antibody fragments used during yeast surface display-mediated affinity maturation. (A) The structure of a mature human Immunoglobulin G (IgG), with disulphide bonds in dotted lines and constituent fragments shown, and single-chain variants of the variable domains (scFvs; Feldhaus *et al.*, 2003) and Fab fragment (scFab; Walker *et al.*, 2009). (B) Possible strategy for yeast surface display-mediated antibody maturation. The myc tag is used to detect the quantity of displayed antibody fragment. The antigen is either fluorescently labelled or detected by affinity (labelled antibody or biotinylated and detected by a streptavidin (SA)-tagged fluor). (C) A common methodology for surface display of Fab fragments, with the heavy chain surface-displayed as a fusion protein and disulphide-bonded light chain (Weaver-Feldhaus *et al.*, 2004). Both fragments may be affinity tagged (T_1 and T_2) for detection of surface display and association of light chain. (D) Allergen surface display with IgE selection (Borodina *et al.*, 2010). Surface-displayed wasp allergens were probed with serum-derived IgEs, which were detected using biotinylated anti-IgEs and streptavidin (SA)-PE. Display was quantified using a myc tag and Alexa Fluor 488 anti-myc antibodies.

There is controversy surrounding the best approach to screening using antibody fragments; scFvs (Fig. 7.8A) are relatively simple and widely used (for example, Feldhaus *et al.*, 2003), but the conversion of selected scFvs to Fabs or IgGs can be problematical. Fab fragments have two polypeptide chains and so are slightly more complex to work with, usually requiring surface display of the heavy chain and disulphide bonding of the secreted light chain (Fig. 7.8C; Weaver-Feldhaus *et al.*, 2004). Lou *et al.* (2010) generated combinations of shuffled light and constant heavy chains via yeast mating (Weaver-Feldhaus *et al.*, 2004) in order to increase affinity to botulism neurotoxin. A haploid yeast strain expressing surface displayed V_H-C_H1 was mated with a strain expressing a library of secreted shuffled light chain variants, generating a diploid library surface-displaying Fabs. FACS was used to select high affinity Fabs from this library, with an average 37-fold increase in affinity. The high-affinity V_H and V_L domains were transferred to recombinant human IgGs and expressed in CHO cells, resulting in IgGs with picomolar affinity.

Walker *et al.* (2009) combined the benefits of scFvs and Fabs to generate the first YSD single-chain Fab (scFab) library, which was screened against HIV-1 gp120 glycoprotein (Fig. 7.8A). The scFabs were found to have higher affinity than corresponding scFvs with identical V_L and V_H domains, and the scFab method was found to select far more antigen-specific clones (as compared to background binders) than the scFv approach. The single-vector scFab system was noted to be simpler than the mating-based Fab approach, eliminating heavy-chain only clones that are generated when the Fab light chains fail to be secreted or do not bind to the surface-displayed heavy chain, although multimerisation of scFabs remained a potential problem. It is clear that each variant of single-chain antibody fragment has its advantages and disadvantages, and each will probably continue to be used for affinity maturation.

Wozniak-Knopp *et al.* (2010) bypassed the scFv versus Fab discussion altogether in their development of an antigen-binding Fc fragment. A library of Fc fragments, randomly mutated in two surface-exposed loops in the C_H3 domain, was surface displayed and selected for affinity to the receptor HER2/neu, as well as CD64 and Protein A to select for retention of overall Fc structure. Biological tests demonstrated that Fc function was unaffected by the gain in antigen binding. This resulted in an Fcab (Fc antigen-binding) scaffold as an alternative to Fab fragments. YSD has also been used to select llama single-domain V_HH antibody fragments (also known as nanobodies) against botulism neurotoxin (Dong *et al.*, 2010) and GFP in a glycosylation-engineered *P. pastoris* (Ryckaert *et al.*, 2010). The ultimate development of surface-displayed antibody selection is surface display of monoclonal antibodies (MAbs). Although MAbs have not been expressed on the yeast surface, a recent paper has described a technique for MAb surface display on mammalian cells (Zhou *et al.*, 2010). This would potentially bypass the disadvantages of yeast display for affinity maturation, but relies upon mammalian cell culture and so is technically more complex. As with phage, bacterial and yeast display, methods need to be selected on the basis of individual requirements.

Although YSD can be used to screen antibody fragments efficiently, transferring the selected fragments from yeast-displayed to soluble form can be problematical, since the initial screen does not test soluble activity. Gray *et al.* (2010) developed two FCM methods to evaluate soluble scFv activity: first, a competitive inhibition assay whereby yeast surface-displayed scFvs competed against soluble scFvs to bind PE-labelled antigen; and an 'epitope binning' assay to detect when two scFvs (one yeast-displayed, one soluble and FITC-labelled) bind to non-overlapping sites on a single antigen. The latter assay could potentially be utilized to select scFvs that bind to non-overlapping sites on an antigen by FACS. In addition, a competitive assay was developed to measure unlabelled antigen concentrations by competing with PE-labelled antigen binding to surface displayed scFvs. These assays were developed to replace conventional ELISA assays, which typically require IgG antibodies generated by the immunization of animals, thus enabling a fully *in vitro* workflow for the isolation, screening and production of scFv fragments in yeast. A central problem with screening soluble antibody fragments is the lack of physical connection between the phenotype of the fragment and its genotype.

Compartmentalization methods (detailed below; Fig. 7.5C) might address this in the future.

Antigen–antibody interactions can be studied from the opposite viewpoint. The interaction of surface-displayed wasp venom allergens and human IgEs from patient antiserum was measured using FCM (Borodina *et al.*, 2010). Three venom antigens were translationally fused to myc tags and surface-displayed, allowing detection using Alexa Fluor 488-conjugated anti-myc antibodies. Each antigen-displaying yeast strain was incubated with patient serum and IgE binding was quantified using biotinylated anti-IgE and streptavidin-PE (Fig. 7.8D). A second study utilized a similar technique to identify antigens in the saliva of a tick, *Ixodes scapularis*, that gives rise to immunity in its rabbit host (Schuijt *et al.*, 2011). A cDNA library from the salivary gland of *I. scapularis* was expressed using YSD coupled to the Xpress epitope. The resultant yeast library was incubated with IgGs purified from the sera of immune rabbits conjugated to Alexa Fluor 488. MACS was used to sort yeast cells that bound the rabbit IgGs (utilizing anti-rabbit beads), and FCM was used to quantify antigen–antibody interactions by measuring Alexa Fluor 488 fluorescence as compared to Xpress-PE fluorescence (denoting surface display). Five unique clones were identified corresponding to five salivary proteins, which were then further investigated as potential vaccine candidates. Other facets of immune recognition have been studied using YSD. T cell receptors have been engineered using a YSD approach (Kieke *et al.*, 1999), and recently T-cell epitopes have been identified from a peptide library derived from viral cDNA (Wen *et al.*, 2008).

Problems with YSD screening of naïve libraries with FACS

Despite the advantages of YSD methods summarized above, there remain some drawbacks of the method when screening naïve libraries for *de novo* interactions (Ackerman *et al.*, 2009). The detection of interactions between the displayed protein and its ligand is dependent upon a fluorescent label, and in some cases direct interactions between the fluor and the surface-displayed protein generates a large number of false positives. Weak protein–ligand interactions (in the micromolar range) present a problem for detection; typically, nanomolar binding affinities are required for FACS detection, with large amounts of ligand being required for selection of weaker binders (cited as hundreds of micrograms). In addition, screening of large naïve libraries is problematical due to the throughput limitation of FACS (typically 10^8 cells per hour); given that a 10-fold coverage of the library is desirable for detection of all clones, a rate of 10^8 cells per day is thought reasonable. However, many naïve libraries are larger than this, and some estimate that up to 10^{10} cells are needed to obtain high-affinity binders. With these drawbacks in mind, Ackermann *et al.* (2009) developed magnetic bead-based isolation methods for naïve library selection, utilizing the multivalent nature of both YSD and affinity bead methods to give very high sensitivity (μM) with small quantities of ligand immobilized onto the beads. FACS was still used to eliminate truncated displayed proteins (via a C-terminal c-myc tag). The authors also described a workflow combining magnetic bead methods for isolation of initial candidates from libraries and selection of candidates with nM binding affinity, followed by FACS selection.

Screening other protein interaction partners with YSD

As well as investigating and selecting protein–protein interactions, YSD has been used recently to study interactions between surface-displayed proteins and other molecules. FACS was used to select phosphatidylinositol-binding proteins from a surface-displayed human cDNA library (Bidlingmaier and Liu, 2007; Bidlingmaier *et al.*, 2011). The YSD library was incubated with biotinylated phosphatidylinositols (the 4,5 bisphosphate PtdIns(4,5)P_2 or the 3,4,5 trisphosphate PtdIns(3,4,5)P_3) followed by PE or Alexa Fluor 647 conjugated to streptavidin and highly fluorescent cells were isolated using three successive rounds of FACS. The use of two different dyes eliminated the possibility of isolating clones displaying proteins that bind to one of the dyes. FCM was then used to characterize the binding of nine isolated clones to PtdIns(4,5)P_2 and PtdIns(3,4,5)P_3. As phosphatidylinositols are very important eukaryotic signalling molecules, responsible for many aspects of cellular regulation, understanding their mode of action and the proteins to which they bind will potentially aid development of novel therapeutics.

On a wider scale, the study of interactions between proteins and non-protein ligands is central to drug discovery and development, so generic high-throughput methods such as these are critical to effective screening programmes. Ryckaert *et al.* (2008) developed a system for screening lectins, a family of carbohydrate binding proteins, using YSD in *S. cerevisiae* and glycoengineered *P. pastoris*. Initially, known lectins were surface-displayed on *S. cerevisiae* and carbohydrate binding was quantified by incubation with FITC-labelled polyacrylamide (PAA) glycoconjugates followed by FCM. Successful display of lectins was monitored using a C-terminal V5 tag. A domain of sialoadhesin did not bind to its cognate sugar, despite successful surface display. This problem was remedied by displaying the domain in *P. pastoris* using an alternative display mechanism (α-agglutinin), reinforcing the requirement for correct folding and orientation of the displayed protein for activity.

Three surface-displayed cDNA libraries were then screened by two rounds of FACS for binding to *N*-acetyl-lactosamine-PAA-FITC, resulting in between 10 and 20 isolated clones. False positives were eliminated by screening individual isolated clones for binding to PAA-FITC; each true positive clone corresponded to a known β-galactoside binding protein. The paper highlighted the ability to select naturally occurring lectins from cDNA libraries and develop lectins with novel and engineered carbohydrate-binding specificities (through screening using a desired sugar), potentially an important step in the development of glycosylation-dependent therapeutics. This methodology is similar in principle, but in the reverse orientation, to Wanchoo *et al.* (2009), who used fluorescent lectins to probe surface carbohydrates.

Jarjour *et al.* (2009) used YSD to probe the sequence-specific protein–DNA interactions of homing nucleases, a class of DNA-cleaving enzymes that recognize between 16 and 22 base pairs. The paper describes the development of two experimental systems; a binding assay and a DNA cleavage assay. First, the endonuclease I-AniI was surface-displayed using an Aga2p fusion together with N-terminal HA and C-terminal myc tags. Cells were incubated with biotinylated double stranded oligonucleotide substrates, and analysed by FCM using FITC conjugated anti-myc antibodies to detect surface display and Streptavidin-PE to detect oligonucleotide binding. For the cleavage assay, the two ends of the substrate oligonucleotide were labelled with streptavidin-PE and Alexa Fluor 647, and the substrate molecule tethered at the streptavidin end to I-AniI (via the HA or myc epitope). Cleavage of the fragment liberated an Alexa Fluor 647-labelled oligonucleotide, and thus the Alexa Fluor 647:streptavidin–PE ratio decreased. Using these techniques, the authors were able to measure the K_d of binding and the catalytic activity for oligonucleotides mutagenized at specific locations. FACS was also used to select enzymes with increased DNA binding affinity by selecting cells with elevated PE fluorescence. These techniques offer the ability to generate DNA-binding and cleaving enzymes with engineered sequence specificity.

In addition to proteins, other molecules have been surface-displayed using YSD. By fusion of a biotin-binding peptide (BAP) to the Flo428 cell wall anchor protein, and coexpressing the *E. coli* biotin ligase BirA, *S. cerevisiae* cells displaying biotin were generated (Tanaka *et al.*, 2010). This allowed surface association of streptavidin-conjugated FITC or streptavidin followed by biotinylated FITC, both of which were measured using FCM. The biotin–streptavidin interaction is extremely strong ($K_d = 10^{-15}$ M) and is already used for many applications in biology (Laitinen *et al.*, 2007). The ability to surface-display non-protein and other synthetic molecules on yeast cells allows greater possibilities for screening by FACS and biocatalysis, for example immobilization of inorganic catalysts to cell surfaces.

Stadlmayr *et al.* (2010a) used the YSD approach to enhance the ability of *P. pastoris* to secrete Fab fragments. A library of *P. pastoris* cDNA was integrated into the chromosome under the regulation of the GAP promoter, giving rise to a library of 10^6 clones. This clone library was used to express a surface-displayed Human anti-HIV-1 Fab fragment; clones with enhanced Fab display were selected by FACS using a fluorescently labelled antibody. The cloned cDNA giving rise to enhanced surface display was sequenced for the clones selected after five rounds of selection, and also analysed by a microarray technique after each round of selection to determine pool variability throughout the procedure.

Screening enzyme activity by measuring reaction products

FACS has recently been used to screen for surface-displayed enzyme activity by measuring reaction products. As stated above, measurement of metabolites is limited by diffusion of small molecules away from the site of generation. There are several methods that have been developed to remedy this problem in the context of surface-displayed proteins.

Diffusible molecules generated by yeast cells could be contained using experimental systems similar in principle to the *in vitro* compartmentalization (IVC) approach (reviewed by Schaerli and Hollfelder, 2009, Yang and Withers, 2009). Droplet and IVC-based approaches have been heralded as important for the development of microfluidic experimental systems, and FCM has a potential role in this development process. Briefly, single cells are encapsulated in compartments (Fig. 7.5C) and interrogated by FCM to measure a substrate or product of the cell present in the compartment; this allows measurement of molecules that do not remain entrapped within the cell, but retains a physical link between the selected compartment and the genetic basis of the desired phenotype. It should be noted that most IVC studies are entirely cell-free, in that the compartments contain enzymes or genes together with transcriptional apparatus for the generation of enzymes, but cell encapsulation has been demonstrated in a number of studies (discussed in detail by Schaerli and Hollfelder, 2009).

Many IVC approaches use water-in-oil-in-water (w-o-w) double emulsions as compartments, which are relatively simple to generate and amenable to FCM and FACS using standard equipment [for example, Bernath *et al.* (2004) and Mastrobattista *et al.* (2005) used FACS on cell-free w-o-w emulsions]. Although yeast applications of this technique are so far rare, two bacterial studies used w-o-w emulsions to improve the ability of the mammalian serum paraoxonase enzyme to cleave thiobutyrolactone (Aharoni *et al.*, 2005) and to generate proteases with enhanced resistance to inhibitors (Tu *et al.*, 2011). In both cases, randomly mutated libraries were screened by the use of non-fluorescent precursors which became fluorescent following enzyme activity: CPM, a thiol-sensitive dye whose fluorescence was activated by the product of thiobutyrolactone cleavage; and a rhodamine 110-containing peptide that fluoresces upon proteolysis. In both cases, it was critical that the fluorescent product remained entrapped within the compartment, and did not diffuse into the oil phase. It is anticipated that future development in this field will include yeast applications for w-o-w compartmentalization.

Zhou *et al.* (2009) used alginate microbeads (mean diameter of 8.5 μm) to encapsulate *S. cerevisiae* cells expressing surface-displayed *Aspergillus aculeatus* β-glucosidase, an enzyme with industrial significance in lignocellulosic bioethanol manufacture. The fluorescent marker was TokyoGreen®-βGlu. The resultant experimental system was able to differentiate between microbeads containing β-glucosidase displaying yeast and non-displaying yeast, and FACS was used to enrich enzyme-positive microbeads up to 82-fold. However, the system is still a work in progress and was not used to screen any libraries in the study. The authors note that it is currently limited by fluorescent product diffusion out of the microbeads, and indicate that bead coatings will be tested to reduce diffusion. This promising technique offers screening of surface-displayed enzyme libraries by activity rather than interaction.

An alternative method of screening a surface-displayed enzyme for activity rather than structure is modification of the cell surface through enzyme activity, exemplified by engineering of enantioselective horseradish peroxidase (HRP) derivatives (Lipovšek *et* al., 2007; Antipov *et al.*, 2008) and is similar in principle to the method described in Fig. 7.5A. In the first paper (Lipovšek *et al.*, 2007), the authors displayed HRP on the yeast surface using an Aga2p fusion. HRP activity was measured using FCM, utilizing an Alexa 488 fluor-labelled chiral phenol substrate, tyrosinol. The link between the reaction product and the cell (and thus the gene sequence) was that oxidation of tyrosinol by HRP yields radicals which bind to tyrosine residues of native yeast surface proteins, labelling the yeast cell surface with the fluor. Randomly mutagenized and active site-mutagenized HRP libraries were generated and surface displayed. Successive FACS selection rounds were employed, using L- or D-tyrosinol labelled with Alexa 488; both L- and D- preferential enzymes were selected on the basis of high or low fluorescence in the presence of the relevant substrates. The active-site mutagenized library yielded HRP derivatives with up to 7.5-fold increased enantioselectivity over the wild-type enzyme, and both L- and D-selective enzyme variants. The randomly mutagenized library did not yield enzymes with increased enantioselectivity.

In the second paper, a method was developed to allow simultaneous detection of incorporation of both enantiomeric substrates (Antipov *et al.*, 2008). L- and D-Tyrosinol were linked to two different Alexa fluors, 488 and 647, and FACS was used to select for cells preferentially labelled with one fluor or the other. Using this technique, enzyme derivatives with up to a 64-fold improvement in enantioselectivity were selected. The advantage of this technique over IVC methods is that it does not require optimization of an encapsulation method; however, it requires that a product of the enzymatic reaction of interest is capable of direct and irreversible interaction with the cell surface.

Flow cytometric yeast two-hybrid – intracellular protein interaction screening

Yeast two-hybrid methods are used to measure and screen protein–protein interactions inside yeast cells (reviewed by Koegl and Uetz, 2007), and have potential for accelerating drug discovery and development (reviewed by Lentze and Auerbach, 2008). On a molecular biology level, yeast two-hybrid methods have been used to map the interactome, revealing vast numbers of protein–protein interactions both between yeast proteins and heterologously expressed proteins (reviewed by Parrish *et al.*, 2006): for example, Ito *et al.* (2001) mapped over 4500 interactions between 3278 *S. cerevisiae* proteins using this approach.

The principle of yeast two-hybrid is shown in Fig. 7.9A. The DNA-binding (BD) and transcriptional activator (AD) domains of a yeast transcriptional activator, GAL4, are translationally fused to two potentially interacting proteins, the 'Bait' and 'Prey'. Only when the bait and prey proteins interact is the activation domain recruited to the promoter, and the reporter gene is synthesized. Using plasmid-encoded bait and prey partners, many thousands or millions of potential prey proteins from cDNA libraries can be screened for interactions with a single bait protein. The traditional screening approach is either using LacZ as reporter, allowing detection of Lac^+ colonies on indicator agar plates, or a metabolic enzyme allowing growth on defined agar medium. However, these plate-based approaches are time- and space-consuming and difficult to automate and implement at very high throughput levels. Chen *et al.* (2008) devised an FACS-based method for yeast two-hybrid using GFP as a reporter.

As with yeast surface display, FACS screening of large cDNA libraries (typically 10^9 clones) would take many days. To address this problem, the FCM-yeast two-hybrid method was subsequently modified (Chen *et al.*, 2009) such that interacting bait–prey pairs directed expression of both GFP and a surface-displayed haemagglutinin (sdHA) epitope. The sdHA epitope allows immunomagnetic separation (such as MACS) to decrease the size of the library to be screened by FACS, thus increasing throughput (Fig. 7.9B). Large enrichment ratios were observed by two rounds of MACS (9100-fold), allowing isolation of rare events from large libraries. The false-positive rate of the novel method was also noted to be much lower than conventional yeast two-hybrid screening methods. A disadvantage of this system is that culture of yeast two-hybrid pools can introduce growth-dependent bias; the authors suggest analysis of larger numbers of positive cells than by conventional methods to overcome this, and cite high-throughput sequencing methods as a route to achieve this. Paradoxically, little advantage in terms of screening specificity was observed by following the MACS sorting with FACS, and the authors noted that FCM would be better suited as a quantification and secondary confirmation method, replacing conventional yeast two-hybrid reporter assays.

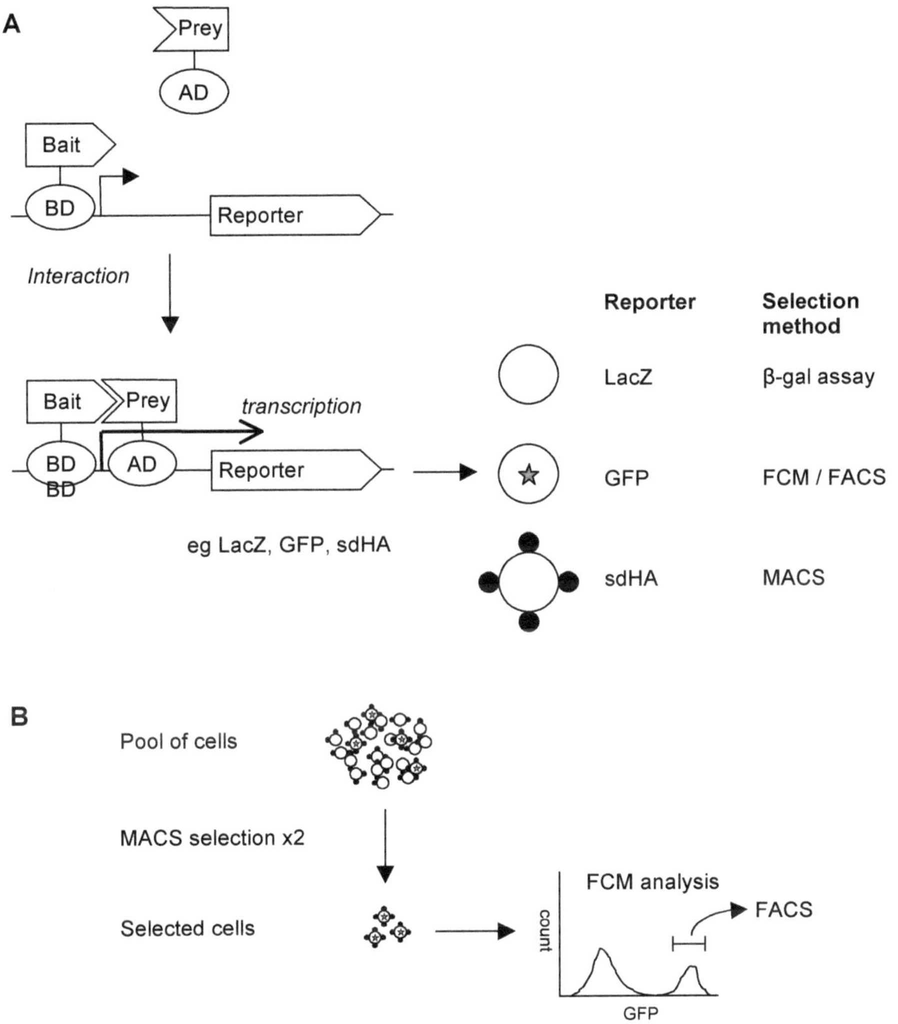

Figure 7.9 Yeast-two hybrid system for detecting intracellular protein–protein interactions. (A) Bait and prey proteins (potential interaction partners) are fused to the DNA binding domain (BD) and activation domain (AD) of the GAL4 transcription activator respectively. Interaction of the bait and prey proteins allows recruitment of the AD to the promoter and transcription activation, expressing the reporter. Potential reporters are discussed in the text: LacZ (measured using β-galactosidase assay and sorted using agar plate-based methods); GFP (assayed using FCM and sorted using FACS, Chen *et al*., 2008); and surface-displayed haemagglutinin (sdHA, detected using anti-HA antibodies and sorted using MACS; Chen *et al*., 2009). (B) MACS and FCM strategy applied by Chen *et al*. (2009). Utilizing concurrent expression of reporters GFP and sdHA, cells were selected by MACS on the basis of sdHA affinity to anti-HA antibodies using secondary antibodies on microbeads. FCM was used to measure GFP fluorescence. FACS selection following MACS did not significantly improve specificity.

New methods and technology related to, and developed from, FCM and FACS

The utility of yeast as a model biological particle has meant that many groups developing novel cytometric methods use it as a test organism. Many of these methods are based on flowing particles in a sheath fluid, but come under the heading of microfluidics rather than flow cytometry *per se*. Microfluidic optical flow cytometry devices are currently being developed by several groups. Kostner and Vellekoop (2008a,b) developed two such instruments: a near-field chip-based cytometer, with a photodiode detector, measuring 15 mm × 20 mm; and a far-field device utilizing a DVD pickup head for optical excitation and detection, measuring 9 mm × 13 mm. Both devices were used to measure scatter of *S. cerevisiae* cells, along with beads and larger cells. Sitton and Srienc (2009) developed a Teflon core flow cell instrument with fibre optic-based collection optics which was used to measure scatter and cell cycle progression (via Sytox green-labelled DNA) in *S. cerevisiae*. While not able to generate the variety of data obtainable using a modern multi-colour flow cytometer, these microfluidic devices are cheap, small in size and simple, allowing specific flow cytometric questions to be asked in an accessible and portable format.

As well as optical measurement techniques, microfluidic flow cytometers are being developed that utilize impedance spectroscopy (Schade-Kampmann *et al.*, 2008) and microwave detection of cell capacitance (Nikolic-Jaric *et al.*, 2009) to offer label-free interrogation of biological particles. Label-free sorting based on intrinsic cellular properties has been achieved using dielectrophoresis (DEP; reviewed by Meighan *et al.*, 2009). On entering a non-uniform electric field, particles experience a physical force depending upon their dielectric properties; the dielectric properties of live and dead yeast are different, making them an ideal model system (Vahey and Voldman, 2008; Braschler *et al.*, 2008; Valero *et al.*, 2010). Recent alternative approaches to microfluidic yeast sorting include optical tweezers (Lin *et al.*, 2008) lateral displacement (Inglis *et al.*, 2010) and sheath flow (Lin *et al.*, 2009). In each of these studies, the microfluidic devices are designed to sort particles on the basis of an inherent characteristic, rather than interrogate multiple parameters as in FCM. As such, they represent simplified cell sorters for specific applications.

Solid phase cytometry utilizes a laser scanning cytometer to detect fluorescently labelled particles immobilized on a solid substrate, for example a filter. Rather than interrogate particles in a flow stream as in FCM, the particles are first passed through the filter, then interrogated. It is useful for detection of microbes present at dilute concentrations, such as airborne microbes (Vanhee *et* al., 2008) and microbes in water samples (Lisle *et al.*, 2004). Nishimura *et al.* (2008) describe the use of such a system to detect *C. albicans* using specific FISH probes and DAPI.

Conclusion

As can be seen, FCM and FACS impact on many aspects of fungal biology, both in terms of fundamental research and highly applied fields. The methods offer the ability to investigate and treat fungal disease, optimize and develop bioprocesses and screen large libraries for the isolation of novel proteins and cellular factories. Fundamental biological and medical research continues to utilize FCM and FACS to answer questions about fungal cells in a multitude of ways. As can be seen, utility of these techniques is constantly expanding. In particular, diagnosis of fungal disease and determination of antifungal susceptibility are extremely promising advances that have the potential to greatly aid treatment of mycoses.

Library screening presently remains a major use of FACS with yeast, and will probably continue to be due to its many advantages over other display technologies. Developments in the screening of extracellular activities will allow much more use of these techniques to develop novel organisms for industrial biotechnology and biofuel applications. It is anticipated that the industrial repertoire of biologically catalysed reactions will greatly increase in the future, due in part to the requirement to replace processes that currently utilize fossil fuel feedstocks. Screening in yeast followed by either yeast- or bacterial-based bioprocesses, also optimized using FCM, is a potential route to novel process development. The development of new miniaturized flow cytometric devices will also bring FACS within the reach of more researchers than ever before.

Acknowledgement

Many thanks to Colin Thomas for assistance in identifying relevant papers and critical comments on the manuscript.

References

Ackerman, M., Levary, D., Tobon, G., Hackel, B., Orcutt, K.D., and Wittrup, K.D. (2009). Highly avid magnetic bead capture: an efficient selection method for *de novo* protein engineering utilizing yeast surface display. Biotechnol. Progr. *25*, 774–783.

Aharoni, A., Amitai, G., Bernath, K., Magdassi, S., and Tawfik, D.S. (2005). High-throughput screening of enzyme libraries: thiolactonases evolved by fluorescence-activated sorting of single cells in emulsion compartments. Chem. Biol. *12(12)*, 1281–1289.

Aksam, E.B., Jungwirth, H., Kohlwein, S.D., Ring, J., Madeo, F., Veenhuis, M., and van der Klei, I.J. (2008). Absence of the peroxiredoxin Pmp20 causes peroxisomal protein leakage and necrotic cell death. Free Radical Bio. Med. *45*, 1115–1124.

Alvarez-Barrientos, A., Arroyo, J., Canton, R., Nombela, C., and Sanchez-Perez, M. (2000). Applications of flow cytometry to clinical microbiology. Clin. Microbiol. Rev. *13*, 167–195.

Amann, R., and Fuchs, B.M. (2008). Single-cell identification in microbial communities by improved fluorescence *in situ* hybridization techniques. Nat. Rev. Microbiol. *6*, 339–348.

Antipov, E., Cho, A.E., Wittrup, K.D., and Klibanov, A.M. (2008). Highly L and D enantioselective variants of horseradish peroxidase discovered by an ultrahigh-throughput selection method. Proc. Natl. Acad. Sci. U.S.A. *105*, 17694–17699.

Barbosa, J., Bragada, C., Costa-de-Oliveira, S., Ricardo, E., Rodrigues, A.G., and Pina-Vaz, C. (2010). A new method for the detection of *Pneumocystis jirovecii* using flow cytometry. Eur. J. Clin. Microbiol. *29*, 1147–1152.

Ben Hassine, S., and Arcangioli, B. (2009). Tdp1 protects against oxidative DNA damage in non-dividing fission yeast. EMBO J. *28*, 632–640.

Bernath, K., Hai, M., Mastrobattista, E., Griffiths, A.D., Magdassi, S., and Tawfik, D.S. (2004). *In vitro* compartmentalization by double emulsions: sorting and gene enrichment by fluorescence activated cell sorting. Anal. Biochem. *325*(1), 151–157.

Bidlingmaier, S., and Liu, B. (2007). Interrogating yeast surface-displayed human proteome to identify small molecule-binding proteins. Mol. Cell Proteomics. *6*, 2012–2020.

Bidlingmaier, S., Wang, Y., Liu, Y., Zhang, N., and Liu, B. (2011). Comprehensive analysis of yeast surface displayed cDNA library selection outputs by exon microarray to identify novel protein–ligand interactions. Mol. Cell Proteomics. *10*, M110.

Bisha, B., Kim, H.J., and Brehm-Stecher, B.F. (2011). Improved DNA-FISH for cytometric detection of *Candida* spp. J. Appl. Microbiol. *110*, 881–892.

Blom, A.M., Hallstrom, T., and Riesbeck, K. (2009). Complement evasion strategies of pathogens-acquisition of inhibitors and beyond. Mol. Immunol. *46*, 2808–2817.

Boder, E.T., and Wittrup, K.D. (1997). Yeast surface display for screening combinatorial polypeptide libraries. Nat. Biotechnol. *15*, 553–557.

Borodina, I., Jensen, B.M., Sondergaard, I., and Poulsen, L.K. (2010). Display of wasp venom allergens on the cell surface of *Saccharomyces cerevisiae*. Microb. Cell Fact. *9*, 74.

Bottari, B., Ercolini, D., Gatti, M., and Neviani, E. (2006). Application of FISH technology for microbiological analysis: current state and prospects. Appl. Microbiol. Biotechnol. 73, 485–494.

Bowley, D.R., Labrijn, A.F., Zwick, M.B., and Burton, D.R. (2007). Antigen selection from an HIV-1 immune antibody library displayed on yeast yields many novel antibodies compared to selection from the same library displayed on phage. Protein Eng. Des. Sel. *20*, 81–90.

Boxx, G.M., Nishiya, C.T., Kozel, T.R., and Zhang, M.X. (2009). Characteristics of Fc-independent human antimannan antibody-mediated alternative pathway initiation of C3 deposition to *Candida albicans*. Mol. Immunol. *46*, 473–480.

Braga-Silva, L.A., Mesquita, D.G.A., Ribeiro, M.D., Carvalho, S.M.F., Fracalanzza, S.E.L., and Santos, A.L.S. (2009). Trailing end-point phenotype antibiotic-sensitive strains of *Candida albicans* produce different amounts of aspartyl peptidases. Braz. J. Med. Biol. Res. *42*, 765–770.

Braschler, T., Demierre, N., Nascimento, E., Silva, T., Oliva, A.G., and Renaud, P. (2008). Continuous separation of cells by balanced dielectrophoretic forces at multiple frequencies. Lab Chip *8*, 280–286.

Brouwer, N., Dolman, K.M., van Houdt, M., Sta, M., Roos, D., and Kuijpers, T.W. (2008). Mannose-binding lectin (MBL) facilitates opsonophagocytosis of yeasts but not of bacteria despite MBL binding. J. Immunol. *180*, 4124–4132.

Calixto, R., Mattos, B., Bittencourt, V., Lopes, L., Souza, L., Sassaki, G., Cipriani, T., Silva, M., and Barreto-Bergter, E. (2010). beta-Galactofuranose-containing structures present in the cell wall of the saprophytic fungus *Cladosporium* (*Hormoconis*) *resinae*. Res. Microbiol. *161*, 720–728.

Callegari, A.J., Clark, E., Pneuman, A., and Kelly, T.J. (2010). Postreplication gaps at UV lesions are signals for checkpoint activation. Proc. Natl. Acad. Sci. U.S.A. *107*, 8219–8224.

Calvert, M.E.K., Lannigan, J.A., and Pemberton, L.F. (2008). Optimization of yeast cell cycle analysis and morphological characterization by multispectral imaging flow cytometry. Cytometry A *73A*, 825–833.

Chen, B.R., and Runge, K.W. (2009). A new *Schizosaccharomyces pombe* chronological lifespan assay reveals that caloric restriction promotes efficient cell cycle exit and extends longevity. Exp. Gerontol. *44*, 493–502.

Chen, C., and Contreras, R. (2004). The bud scar-based screening system for hunting human genes extending life span. Ann. N. Y. Acad. Sci. *1019*, 355–359.

Chen, C., and Contreras, R. (2007). Identifying genes that extend life span using a high-throughput screening system. Methods Mol. Biol. *371*, 237–248.

Chen, J., Zhou, J., Bae, W., Sanders, C.K., Nolan, J.P., and Cai, H. (2008). A yEGFP-based reporter system for nigh-throughput yeast two-hybrid assay by flow cytometry. Cytometry A *73A*, 312–320.

Chen, J., Zhou, J.H., Sanders, C.K., Nolan, J.P., and Cai, H. (2009). A surface display yeast two-hybrid screening system for high-throughput protein interactome mapping. Anal. Biochem. *390*, 29–37.

Chlup, P.H., Bernard, D., and Stewart, G.G. (2008). Disc stack centrifuge operating parameters and their impact on yeast physiology. J. Inst. Brewing *114*, 45–61.

Coleman, C.B., Allen, P.L., Valles, J.M., and Hammond, T.G. (2008). Transcriptional regulation of changes in growth, cell cycle, and gene expression of *Saccharomyces cerevisiae* due to changes in buoyancy. Biotechnol. Bioeng. *100*, 334–343.

Cormack, B.P., Bertram, G., Egerton, M., Gow, N.A., Falkow, S., and Brown, A.J. (1997). Yeast-enhanced green fluorescent protein (yEGFP)a reporter of gene expression in *Candida albicans*. Microbiology *143*, 303–311.

Czechowska, K., Johnson, D.R., and van der Meer, M. Jr. (2008). Use of flow cytometric methods for single-cell analysis in environmental microbiology. Curr. Opin. Microbiol. *11*, 205–212.

Darzynkiewicz, Z., and Huang, X. (2004). Analysis of cellular DNA content by flow cytometry. Curr. Protoc. Immunol. *60* (Unit 5.7), 5.7.1–5.7.18.

Darzynkiewicz, Z., Halicka, H.D., and Zhao, H. (2010). Analysis of cellular DNA content by flow and laser scanning cytometry. Adv. Exp. Med. Biol. *676*, 137–147.

Daugherty, P.S. (2007). Protein engineering with bacterial display. Curr. Opin. Struct. Biol. *17*, 474–480.

Diaz, M., Herrero, M., Garcia, L.A., and Quiros, C. (2010). Application of flow cytometry to industrial microbial bioprocesses. Biochem. Eng. J. *48*, 385–407.

Dong, J.B., Thompson, A.A., Fan, Y.F., Lou, J.L., Conrad, F., Ho, M.F., Pires-Alves, M., Wilson, B.A., Stevens, R.C., and Marks, J.D. (2010). a single-domain llama antibody potently inhibits the enzymatic activity of botulinum neurotoxin by binding to the non-catalytic alpha-exosite binding region. J. Mol. Biol. *397*, 1106–1118.

Donnelly, R.F., McCarron, P.A., and Tunney, M.M. (2008). Antifungal photodynamic therapy. Microbiol. Res. *163*, 1–12.

Edwards, S.R., and Wandless, T.J. (2010). Dicistronic regulation of fluorescent proteins in the budding yeast *Saccharomyces cerevisiae*. Yeast 27, 229–236.

Feldhaus, M., and Siegel, R. (2004). Flow cytometric screening of yeast surface display libraries. Methods Mol. Biol. *263*, 311–332.

Feldhaus, M.J., Siegel, R.W., Opresko, L.K., Coleman, J.R., Feldhaus, J.M., Yeung, Y.A., Cochran, J.R., Heinzelman, P., Colby, D., Swers, J., *et al.* (2003). Flow-cytometric isolation of human antibodies from a nonimmune *Saccharomyces cerevisiae* surface display library. Nat. Biotechnol. *21*, 163–170.

Georgieva, M., Harata, M., and Miloshev, G. (2008). The nuclear actin-related protein act3p/arp4 influences yeast cell shape and bulk chromatin organization. J. Cell. Biochem. *104*, 59–67.

Gilbert, A., Sangurdekar, D.P., and Srienc, F. (2009). Rapid strain improvement through optimized evolution in the cytostat. Biotechnol. Bioeng. *103*, 500–512.

Gray, S.A., Weigel, K.M., Miller, K.D., Ndung'u, J., Buscher, P., Tran, T., Baird, C., and Cangelosi, G.A. (2010). Flow cytometry-based methods for assessing soluble scFv activities and detecting antigens in solution. Biotechnol. Bioeng. *105*, 973–981.

Gruger, T., Morler, C., Schnitzler, N., Brandenburg, K., Nidermajer, S., Horre, R., and Zundorf, J. (2008). Influence of fluoroquinolones on phagocytosis and killing of *Candida albicans* by human polymorphonuclear neutrophils. Med. Mycol. *46*, 675–684.

Guerfal, M., Ryckaert, S., Jacobs, P.P., Ameloot, P., Van, C.K., Derycke, R., and Callewaert, N. (2010). The HAC1 gene from *Pichia pastoris*: characterization and effect of its overexpression on the production of secreted, surface displayed and membrane proteins. Microb. Cell Fact. *9*, 49.

Harayama, S., Kasai, Y., and Hara, A. (2004). Microbial communities in oil-contaminated seawater. Curr. Opin. Biotechnol. *15*, 205–214.

Harvey, B.R., Georgiou, G., Hayhurst, A., Jeong, K.J., Iverson, B.L., and Rogers, G.K. (2004). Anchored periplasmic expression, a versatile technology for the isolation of high-affinity antibodies from *Escherichia coli*-expressed libraries. Proc. Natl. Acad. Sci. U.S.A. *101*, 9193–9198.

Heine, F., Stahl, F., Strauber, H., Wiacek, C., Benndorf, D., Repenning, C., Schmidt, F., Scheper, T., von Bergen, M., Harms, H., and Muller, S. (2009). Prediction of flocculation ability of brewing yeast inoculates by flow cytometry, proteome analysis, and mRNA profiling. Cytometry A *75*, 140–147.

Hernlem, B., and Hua, S.S. (2010). Dual fluorochrome flow cytometric assessment of yeast viability. Curr. Microbiol. *61*, 57–63.

Hewitt, C.J., and Nebe-von-Caron, G. (2004). The application of multi-parameter flow cytometry to monitor individual microbial cell physiological state. Adv. Biochem. Eng. Biotechnol. *89*, 197–223.

Hua, H., and Kearsey, S.E. (2011). Monitoring DNA replication in fission yeast by incorporation of 5-ethynyl-2′-deoxyuridine. Nucleic Acids Res. *39*, e60.

Hutter, K.J., and Eipel, H.E. (1979). Microbial determinations by flow cytometry. J. Gen. Microbiol. *113*, 369–375.

Hwang, B., Hwang, J.S., Lee, J., and Lee, D.G. (2010). Antifungal properties and mode of action of psacotheasin, a novel knottin-type peptide derived from *Psacothea hilaris*. Biochem. Biophys. Res. Comm. *400*, 352–357.

Hyka, P., Zullig, T., Ruth, C., Looser, V., Meier, C., Klein, J., Melzoch, K., Meyer, H.P., Glieder, A., and Kovar, K. (2010). Combined use of fluorescent dyes and flow cytometry to quantify the physiological state of *Pichia pastoris* during the production of heterologous proteins in high-cell-density fed-batch cultures. Appl. Env. Microbiol. *76*, 4486–4496.

Inglis, D.W., Herman, N., and Vesey, G. (2010). Highly accurate deterministic lateral displacement device and its application to purification of fungal spores. Biomicrofluidics *4*, 024109.

Invitrogen (2010). Molecular Probes Handbook, A Guide to Fluorescent Probes and Labeling Technologies. Invitrogen. Waltham, MA, USA.

Ishii, J., Matsumura, S., Kimura, S., Tatematsu, K., Kuroda, S., Fukuda, H., and Kondo, A. (2006). Quantitative and dynamic analyses of G protein-coupled receptor signaling in yeast using Fus1, enhanced green fluorescence protein (EGFP), and His3 fusion protein. Biotechnol. Prog. *22*, 954–960.

Ishii, J., Tanaka, T., Matsumura, S., Tatematsu, K., Kuroda, S., Ogino, C., Fukuda, H., and Kondo, A. (2008). Yeast-based fluorescence reporter assay of G protein-coupled receptor signalling for flow cytometric screening: FAR1-disruption recovers loss of episomal plasmid caused by signalling in yeast. J. Biochem. *143*, 667–674.

Ishii, J., Izawa, K., Matsumura, S., Wakamura, K., Tanino, T., Tanaka, T., Ogino, C., Fukuda, H., and Kondo, A. (2009). A simple and immediate method for simultaneously evaluating expression level and plasmid maintenance in yeast. J. Biochem. *145*, 701–708.

Ito, T., Chiba, T., Ozawa, R., Yoshida, M., Hattori, M., and Sakaki, Y. (2001). A comprehensive two-hybrid analysis to explore the yeast protein interactome. Proc. Natl. Acad. Sci. U.S.A. *98*, 4569–4574.

Jacobs, P.P., Ryckaert, S., Geysens, S., De Vusser, K., Callewaert, N., and Contreras, R. (2008). Pichia surface display: display of proteins on the surface of glycoengineered *Pichia pastoris* strains. Biotechnol. Lett. *30*, 2173–2181.

Jacques, N., Sacerdot, C., Derkaoui, M., Dujon, B., Ozier-Kalogeropoulos, O., and Casaregola, S. (2010). Population polymorphism of nuclear mitochondrial DNA insertions reveals widespread diploidy associated with loss of heterozygosity in *Debaryomyces hansenii*. Eukaryot. Cell *9*, 449–459.

Jarjour, J., West-Foyle, H., Certo, M.T., Hubert, C.G., Doyle, L., Getz, M.M., Stoddard, B.L., and Scharenberg, A.M. (2009). High-resolution profiling of homing endonuclease binding and catalytic specificity using yeast surface display. Nucleic Acids Res. *37*, 6871–6880.

Johnson, R.M., Allen, C., Melman, S.D., Waller, A., Young, S.M., Sklar, L.A., and Parra, K.J. (2010). Identification of inhibitors of vacuolar proton-translocating ATPase pumps in yeast by high-throughput screening flow cytometry. Anal. Biochem. *398*, 203–211.

Jose, J., and Meyer, T.F. (2007). The autodisplay story, from discovery to biotechnical and biomedical applications. Microbiol. Mol. Biol. Rev. *71*, 600–619.

Kacmar, J., Gilbert, A., Cockrell, J., and Srienc, F. (2006). The cytostat: A new way to study cell physiology in a precisely defined environment. J. Biotechnol. *126*, 163–172.

Kang, K., Fong, W.P., and Tsang, P.W.K. (2010). antifungal activity of baicalein against *Candida krusei* does not involve apoptosis. Mycopathologia *170*, 391–396.

Kapoor, V., Karpov, V., Linton, C., Subach, F.V., Verkhusha, V.V., and Telford, W.G. (2008). Solid state yellow and orange lasers for flow cytometry. Cytometry A *73*, 570–577.

Katou, T., Kitagaki, H., Akao, T., and Shimoi, H. (2008). Brewing characteristics of haploid strains isolated from sake yeast Kyokai No. 7. Yeast *25*, 799–807.

Kempf, V.A., Mandle, T., Schumacher, U., Schafer, A., and Autenrieth, I.B. (2005). Rapid detection and identification of pathogens in blood cultures by fluorescence *in situ* hybridization and flow cytometry. Int. J. Med. Microbiol. *295*, 47–55.

Kieke, M.C., Shusta, E.V., Boder, E.T., Teyton, L., Wittrup, K.D., and Kranz, D.M. (1999). Selection of functional T cell receptor mutants from a yeast surface-display library. Proc. Natl. Acad. Sci. U.S.A. *96*, 5651–5656.

Kim, K.J., Sung, W.S., Suh, B.K., Moon, S.K., Choi, J.S., Kim, J., and Lee, D.G. (2009). Antifungal activity and mode of action of silver nano-particles on *Candida albicans*. Biometals *22*, 235–242.

Kittelmann, K., Rau, P., Gronenborn, B., and Jeske, H. (2009). Plant geminivirus rep protein induces rereplication in fission yeast. J. Virol. *83*, 6769–6778.

Knutsen, J.H., Rein, I.D., Rothe, C., Stokke, T., Grallert, B., and Boye, E. (2011). Cell-cycle analysis of fission yeast cells by flow cytometry. PLOS ONE *6*, e17175.

Koegl, M., and Uetz, P. (2007). Improving yeast two-hybrid screening systems. Brief. Funct. Genomic. Proteomic. *6*, 302–312.

Kostner, S., and Vellekoop, M. (2008a). Cell analysis in a microfluidic cytometer applying a DVD pickup head. Sens. Actuators B Chem. *132*, 512–517.

Kostner, S., and Vellekoop, M.J. (2008b). Microsystems for optical cell detection: near versus far field. Part. Part. Syst. Char. *25*, 92–98.

Kottmeier, K., Weber, J., Muller, C., Bley, T., and Buchs, J. (2009). Asymmetric division of *Hansenula polymorpha* reflected by a drop of light scatter intensity measured in batch microtiter plate cultivations at phosphate limitation. Biotechnol. Bioeng. *104*, 554–561.

Kurec, M., Baszczynski, M., Lehnert, R., Mota, A., Teixeira, J.A., and Branyik, T. (2009). Flow cytometry for age assessment of a yeast population and its application in beer fermentations. J. Inst. Brewing *115*, 253–258.

Kwan, D.H., Chen, H.M., Ratananikom, K., Hancock, S.M., Watanabe, Y., Kongsaeree, P.T., Samuels, A.L., and Withers, S.G. (2011). Self-immobilizing fluorogenic imaging agents of enzyme activity. Angew. Chem. Int. Ed. Engl. *50*, 300–303.

Laitinen, O.H., Nordlund, H.R., Hytonen, V.P., and Kulomaa, M.S. (2007). Brave new (strept)avidins in biotechnology. Trends Biotechnol. *25*, 269–277.

Landolfo, S., Politi, H., Angeozzi, D., and Mannazzu, I. (2008). ROS accumulation and oxidative damage to cell structures in *Saccharomyces cerevisiae* wine strains during fermentation of high-sugar-containing medium. Biochim. Biophys. Acta. *1780*, 892–898.

Landolfo, S., Zara, G., Zara, S., Budroni, M., Ciani, M., and Mannazzu, I. (2010). Oleic acid and ergosterol supplementation mitigates oxidative stress in wine strains of *Saccharomyces cerevisiae*. Int. J. Food Microbiol. *141*, 229–235.

Lee, J., and Lee, D.G. (2009). Antifungal properties of a peptide derived from the signal peptide of the HIV-1 regulatory protein, Rev. FEBS Lett. *583*, 1544–1547.

Lee, J., Choi, Y., Woo, E.R., and Lee, D.G. (2009). Isocryptomerin, a novel membrane-active antifungal compound from *Selaginella tamariscina*. Biochem. Biophys. Res. Comm. *379*, 676–680.

Lentze, N., and Auerbach, D. (2008). The yeast two-hybrid system and its role in drug discovery. Expert. Opin. Ther. Targets. *12*, 505–515.

Li, F., Wang, L.W., Burgess, R.J., and Weinshilboum, R.M. (2008). Thiopurine S-methyltransferase pharmacogenetics: autophagy as a mechanism for variant allozyme degradation. Pharmacogenet. Genom. *18*, 1083–1094.

Li, Q., Du, W., and Liu, D. (2008). Perspectives of microbial oils for biodiesel production. Appl. Microbiol. Biotechnol. *80*, 749–756.

Li, Y., Wei, G., and Chen, J. (2004). Glutathione: a review on biotechnological production. Appl. Microbiol. Biotechnol. *66*, 233–242.

Lin, C.C., Chen, A., and Lin, C.H. (2008). Microfluidic cell counter/sorter utilizing multiple particle tracing technique and optically switching approach. Biomed. Microdev. *10*, 55–63.

Lin, C.H., Lee, C.Y., Tsai, C.H., and Fu, L.M. (2009). Novel continuous particle sorting in microfluidic chip utilizing cascaded squeeze effect. Microfluid. Nanofluid. *7*, 499–508.

Lin, S., Shen, Z., Zha, D.X., Sharkey, N., Prinz, B., Hamilton, S., Pavoor, T.V., Bobrowicz, B., Shaikh, S.S., Rittenhour, A.M., *et al.* (2010). Selection of *Pichia pastoris* strains expressing recombinant immunoglobulin G by cell surface labeling. J. Immunol. Meth. *358*, 66–74.

Lipovsek, D., Antipov, E., Armstrong, K.A., Olsen, M.J., Klibanov, A.M., Tidor, B., and Wittrup, K.D. (2007). Selection of horseradish peroxidase variants with enhanced enantioselectivity by yeast surface display. Chem. Biol. *14*, 1176–1185.

Lis, M., and Bobek, L.A. (2008). Proteomic and metabolic characterization of a *Candida albicans* mutant resistant to the antimicrobial peptide MUC7 12-mer. FEMS Immunol. Med. Microbiol. *54*, 80–91.

Lisle, J.T., Hamilton, M.A., Willse, A.R., and McFeters, G.A. (2004). Comparison of fluorescence microscopy and solid-phase cytometry methods for counting bacteria in water. Appl. Environ. Microbiol. *70*, 5343–5348.

Lopez-Garcia, B., Gandia, M., Munoz, A., Carmona, L., and Marcos, J.F. (2010). A genomic approach highlights common and diverse effects and determinants of susceptibility on the yeast *Saccharomyces cerevisiae* exposed to distinct antimicrobial peptides. BMC Microbiol. *10*, 209.

Lou, J., Geren, I., Garcia-Rodriguez, C., Forsyth, C.M., Wen, W., Knopp, K., Brown, J., Smith, T., Smith, L.A., and Marks, J.D. (2010). Affinity maturation of human botulinum neurotoxin antibodies by light chain shuffling via yeast mating. Prot. Eng. Des. Select. *23*, 311–319.

Malfeito-Ferreira, M. (2011). Yeasts and wine off-flavours: a technological perspective. Annals Microbiol. *61*, 95–102.

Marie, C., Leyde, S., and White, T.C. (2008). Cytoplasmic localization of sterol transcription factors Upc2p and Ecm22p in *S. cerevisiae*. Fungal Genet. Biol. *45*, 1430–1438.

Martinez-Alonso, M., Garcia-Fruitos, E., Ferrer-Miralles, N., Rinas, U., and Villaverde, A. (2010). Side effects of chaperone gene co-expression in recombinant protein production. Microb. Cell Fact. *9*, 64.

Mastrobattista, E., Taly, V., Chanudet, E., Treacy, P., Kelly, B.T., and Griffiths, A.D. (2005). High-throughput screening of enzyme libraries: *in vitro* evolution of a beta-galactosidase by fluorescence-activated sorting of double emulsions. Chem. Biol. *12*(12), 1291–1300.

Matmati, N., Kitagaki, H., Montefusco, D., Mohanty, B.K., and Hannun, Y.A. (2009). Hydroxyurea sensitivity reveals a role for ISC1 in the regulation of G(2)/M. J. Biol. Chem. *284*, 8241–8246.

Mattanovich, D., and Borth, N. (2006). Applications of cell sorting in biotechnology. Microb. Cell Fact. *5*, 12.

Meighan, M.M., Staton, S.J., and Hayes, M.A. (2009). Bioanalytical separations using electric field gradient techniques. Electrophoresis *30*, 852–865.

Miceli, M.H., Diaz, J.A., and Lee, S.A. (2011). Emerging opportunistic yeast infections. Lancet Infect. Dis. *11*, 142–151.

Millard, P.J., Roth, B.L., Thi, H.P., Yue, S.T., and Haugland, R.P. (1997). Development of the FUN-1 family of fluorescent probes for vacuole labeling and viability testing of yeasts. Appl. Environ. Microbiol. *63*, 2897–2905.

Miltenyi, S., Muller, W., Weichel, W., and Radbruch, A. (1990). High gradient magnetic cell separation with MACS. Cytometry *11*(2), 231–238.

Muller-Taubenberger, A., and Anderson, K.I. (2007). Recent advances using green and red fluorescent protein variants. Appl. Microbiol. Biotechnol. *77*, 1–12.

Neklesa, T.K., and Davis, R.W. (2009). A genome-wide screen for regulators of TORC1 in response to amino acid starvation reveals a conserved Npr2/3 complex. PLOS Genet. *5*, e1000515.

Nicola, A.M., Frases, S., and Casadevall, A. (2009). Lipophilic dye staining of *Cryptococcus neoformans* extracellular vesicles and capsule. Eukaryot. Cell *8*, 1373–1380.

Nienow, A.W., McLeod, G., and Hewitt, C.J. (2010). Studies supporting the use of mechanical mixing in large scale beer fermentations. Biotechnol. Lett. *32*, 623–633.

Niimi, M., Firth, N.A., and Cannon, R.D. (2010). Antifungal drug resistance of oral fungi. Odontology *98*, 15–25.

Nikolic-Jaric, M., Romanuik, S.F., Ferrier, G.A., Bridges, G.E., Butler, M., Sunley, K., Thomson, D.J., and Freeman, M.R. (2009). Microwave frequency sensor for detection of biological cells in microfluidic channels. Biomicrofluidics *3*, 34103.

Nishimura, M., Shimakita, T., Matsuzaki, T., Tashiro, Y., and Kogure, K. (2008). Automatic counting of FISH-labeled microbes by an LED illuminated detecting apparatus. Fisheries Sci. *74*, 405–410.

Niu, W., Li, Z., Zhan, W.J., Iyer, V.R., and Marcotte, E.M. (2008). Mechanisms of cell cycle control revealed by a systematic and quantitative overexpression screen in *S. cerevisiae*. PLOS Genet. *4*, e1000120.

Novak, J., Basarova, G., Teixeira, J.A., and Vicente, A.A. (2007). Monitoring of brewing yeast propagation under aerobic and anaerobic conditions employing flow cytometry. J. Inst. Brewing *113*, 249–255.

O'Sullivan, C., Burrell, P.C., Clarke, W.P., and Blackall, L.L. (2007). A survey of the relative abundance of specific groups of cellulose degrading bacteria in anaerobic environments using fluorescence *in situ* hybridization. J. Appl. Microbiol. *103*, 1332–1343.

Oriel, S., and Nitzan, Y. (2010). Photoinactivation of *Candida albicans* by its own endogenous porphyrins. Curr. Microbiol. *60*, 117–123.

Oyane, I., Takeda, T., Oda, Y., Sakata, T., Furuta, M., Okitsu, K., Maeda, Y., and Nishimura, R. (2009). Comparison between the effects of ultrasound and gamma-rays on the inactivation of *Saccharomyces cerevisiae*: Analyses of cell membrane permeability and DNA or RNA synthesis by flow cytometry. Ultrason. Sonochem. *16*, 532–536.

Pande, J., Szewczyk, M.M., and Grover, A.K. (2010). Phage display: concept, innovations, applications and future. Biotechnol. Adv. *28*, 849–858.

Park, C., Cho, J., Hwang, B., Hwang, I.S., Kim, M.R., Woo, E.R., and Lee, D.G. (2010a). Styraxjaponoside A and B, antifungal lignan glycosides isolated from *Styrax japonica* S. et Z. Biomol. Therapeutics *18*, 420–425.

Park, C., Woo, E.R., and Lee, D.G. (2010b). Anti-*Candida* property of a lignan glycoside derived from *Styrax japonica* S. et Z. via membrane-active mechanisms. Mol. Cells *29*, 581–584.

Parrish, J.R., Gulyas, K.D., and Finley, R.L. Jr. (2006). Yeast two-hybrid contributions to interactome mapping. Curr. Opin. Biotechnol. *17*, 387–393.

Patnaik, R. (2008). Engineering complex phenotypes in industrial strains. Biotechnol. Prog. *24*, 38–47.

Pepper, L.R., Cho, Y.K., Boder, E.T., and Shusta, E.V. (2008). A decade of yeast surface display technology: where are we now? Comb. Chem. High Throughput. Screen. *11*, 127–134.

Peters, B.M., Shirtliff, M.E., and Jabra-Rizk, M.A. (2010). Antimicrobial peptides: primeval molecules or future drugs? PLOS. Pathog. *6*, e1001067.

Peters, R.P., Savelkoul, P.H., Simoons-Smit, A.M., Danner, S.A., Vandenbroucke-Grauls, C.M., and van Agtmael, M.A. (2006). Faster identification of pathogens in positive blood cultures by fluorescence *in situ* hybridization in routine practice. J. Clin. Microbiol. *44*, 119–123.

Pilbrough, W., Munro, T.P., and Gray, P. (2009). Intraclonal protein expression heterogeneity in recombinant CHO cells. PLOS One *4*, e8432.

Pinto, E., Vale-Silva, L., Cavaleiro, C., and Salgueiro, L. (2009). Antifungal activity of the clove essential oil from Syzygium aromaticum on *Candida, Aspergillus* and dermatophyte species. J. Med. Microbiol. *58*, 1454–1462.

Porro, D., Vai, M., Vanoni, M., Alberghina, L., and Hatzis, C. (2009). Analysis and modeling of growing budding yeast populations at the single cell level. Cytometry A *75*, 114–120.

Porro, D., Gasser, B., Fossati, T., Maurer, M., Branduardi, P., Sauer, M., and Mattanovich, D. (2011). Production of recombinant proteins and metabolites in yeasts: when are these systems better than bacterial production systems? Appl. Microbiol. Biotechnol. *89*, 939–948.

Raithatha, S.A., and Stuart, D.T. (2009). A comparison of fluorescent DNA binding dyes for flow cytometric analysis of sporulating *Saccharomyces cerevisiae*. J. Microbiol. Meth. *78*, 357–359.

Raschke, D., and Knorr, D. (2009). Rapid monitoring of cell size, vitality and lipid droplet development in the oleaginous yeast *Waltomyces lipofer*. J. Microbiol. Meth. *79*, 178–183.

Ray, A., and Norden, B. (2000). Peptide nucleic acid (PNA): its medical and biotechnical applications and promise for the future. FASEB J. *14*, 1041–1060.

Resina, D., Maurer, M., Cos, O., Arnau, C., Carnicer, M., Marx, H., Gasser, B., Valero, F., Mattanovich, D., and Ferrer, P. (2009). Engineering of bottlenecks in *Rhizopus oryzae* lipase production in *Pichia pastoris* using the nitrogen source-regulated FLD1 promoter. N. Biotechnol. *25*, 396–403.

Rodriguez, S.B., and Thornton, R.J. (2008). Use of flow cytometry with fluorescent antibodies in real-time monitoring of simultaneously inoculated alcoholic-malolactic fermentation of Chardonnay. Lett. Appl. Microbiol. *46*, 38–42.

Rodriguez-Porrata, B., Novo, M., Guillamon, J., Rozes, N., Mas, A., and Otero, R.C. (2008). Vitality enhancement of the rehydrated active dry wine yeast. Int. J. Food Microbiol. *126*, 116–122.

Routledge, S.J., Hewitt, C.J., Bora, N., and Bill, R.M. (2011). Antifoam addition to shake flask cultures of recombinant *Pichia pastoris* increases yield. Microb. Cell Fact. *10*, 17.

Rudensky, B., Broidie, E., Yinnon, A.M., Weitzman, T., Paz, E., Keller, N., and Raveh, D. (2005). Rapid flow-cytometric susceptibility testing of *Candida species*. J. Antimicrob. Chemother. *55*, 106–109.

Rudensky, B., Broide, E., Berko, N., Wiener-Well, Y., Yinnon, A.M., and Raveh, D. (2008). Direct fluconazole susceptibility testing of positive *Candida* blood cultures by flow cytometry. Mycoses *51*, 200–204.

Ryckaert, S., Callewaert, N., Jacobs, P.P., Dewaele, S., Dewerte, I., and Contreras, R. (2008). Fishing for lectins from diverse sequence libraries by yeast surface display – an exploratory study. Glycobiology *18*, 137–144.

Ryckaert, S., Pardon, E., Steyaert, J., and Callewaert, N. (2010). Isolation of antigen-binding camelid heavy chain antibody fragments (nanobodies) from an immune library displayed on the surface of *Pichia pastoris*. J. Biotechnol. *145*, 93–98.

Saleem, R.A., Long-O'Donnell, R., Dilworth, D.J., Armstrong, A.M., Jamakhandi, A.P., Wan, Y., Knijnenburg, T.A., Niemisto, A., Boyle, J., Rachubinski, R.A., *et al.* (2010). Genome-wide analysis of effectors of peroxisome biogenesis. PLOS ONE *5*, e11953.

Schade-Kampmann, G., Huwiler, A., Hebeisen, M., Hessler, T., and Di Berardino, M. (2008). On-chip non-invasive and label-free cell discrimination by impedance spectroscopy. Cell Prolif. *41*, 830–840.

Schaerli, Y., and Hollfelder, F. (2009). The potential of microfluidic water-in-oil droplets in experimental biology. Mol. Biosyst. *5*, 1392–1404.

Schenk, M., Raffellini, S., Guerrero, S., Blanco, G.A., and Alzamora, S.M. (2011). Inactivation of *Escherichia coli*, *Listeria innocua* and *Saccharomyces cerevisiae* by UV-C light: Study of cell injury by flow cytometry. Lwt – Food Sci. Technol. *44*, 191–198.

Schmidt, M., Barker, S., Essmann, M., and Larsen, B. (2008). Effect of commonly used herbicides on the virulence factor Cdr1 in *Candida albicans*. Env. Toxicol. Chem. *27*, 2346–2351.

Schuijt, T.J., Narasimhan, S., Daffre, S., DePonte, K., Hovius, J.W., Van't, V.C., van der, P.T., Bakhtiari, K., Meijers, J.C., Boder, E.T., *et al.* (2011). Identification and characterization of *Ixodes scapularis* antigens that elicit tick immunity using yeast surface display. PLOS ONE *6*, e15926.

Schuster, A., and Schmoll, M. (2010). Biology and biotechnology of *Trichoderma*. Appl. Microbiol. Biotechnol. *87*, 787–799.

Serpaggi, V., Remize, F., Grand, A.S., and Alexandre, H. (2010). Specific identification and quantification of the spoilage microorganism *Brettanomyces* in wine by flow cytometry: a useful tool for winemakers. Cytometry A 77, 497–499.

Shaner, N.C., Campbell, R.E., Steinbach, P.A., Giepmans, B.N., Palmer, A.E., and Tsien, R.Y. (2004). Improved monomeric red, orange and yellow fluorescent proteins derived from *Discosoma* sp. red fluorescent protein. Nat. Biotechnol. *22*, 1567–1572.

Shapiro, H.M. (2000). Membrane potential estimation by flow cytometry. Methods. *21*, 271–279.

Shapiro, H.M. (2003). Practical Flow Cytometry. (John Wiley and Sons, Hoboken, NJ, USA).

Shen, Y., Zhao, X.Q., Ge, X.M., and Bai, F.W. (2009). Metabolic flux and cell cycle analysis indicating new mechanism underlying process oscillation in continuous ethanol fermentation with *Saccharomyces cerevisiae* under VHG conditions. Biotechnol. Adv. *27*, 1118–1123.

Shepard, J.R., Addison, R.M., Alexander, B.D., la-Latta, P., Gherna, M., Haase, G., Hall, G., Johnson, J.K., Merz, W.G., Peltroche-Llacsahuanga, H., *et al.* (2008). Multicenter evaluation of the *Candida albicans*/*Candida glabrata* peptide nucleic acid fluorescent *in situ* hybridization method for simultaneous dual-color identification of *C. albicans* and *C. glabrata* directly from blood culture bottles. J. Clin. Microbiol. *46*, 50–55.

Shibasaki, S., Maeda, H., and Ueda, M. (2009). Molecular display technology using yeast—arming technology. Anal. Sci. *25*, 41–49.

da Silva, T.L., Feijao, D., and Reis, A. (2010). Using multi-parameter flow cytometry to monitor the yeast *Rhodotorula glutinis* CCMI 145 batch growth and oil production towards biodiesel. Appl. Biochem. Biotechnol. *160*, 2166–2176.

Sitton, G., and Srienc, F. (2009). Flow cytometry without alignment of collection optics. Cytometry A *75A*, 990–998.

Skosireva, I., James, T.Y., Sun, S., and Xu, J.P. (2010). Mitochondrial inheritance in haploid x non-haploid crosses in *Cryptococcus neoformans*. Curr. Genet. *56*, 163–176.

Solieri, L., Cassanelli, S., Croce, M.A., and Giudici, P. (2008). Genome size and ploidy level: New insights for elucidating relationships in *Zygosaccharomyces* species. Fung. Genet. Biol. *45*, 1582–1590.

Stadlmayr, G., Benakovitsch, K., Gasser, B., Mattanovich, D., and Sauer, M. (2010a). Genome-scale analysis of library sorting (GALibSo): isolation of secretion enhancing factors for recombinant protein production in *Pichia pastoris*. Biotechnol. Bioeng. *105*, 543–555.

Stadlmayr, G., Mecklenbräuker, A., Rothmüller, M., Maurer, M., Sauer, M., Mattanovich, D., and Gasser, B. (2010b). Identification and characterisation of novel *Pichia pastoris* promoters for heterologous protein production. J. Biotechnol. *150*, 519–529.

Stewart, G.G. (2009). The Horace Brown medal lecture: forty years of brewing research. J. Inst. Brewing *115*, 3–29.

Stratford, M., Plumridge, A., Nebe-von-Caron, G., and Archer, D.B. (2009). Inhibition of spoilage mould conidia by acetic acid and sorbic acid involves different modes of action, requiring modification of the classical weak-acid theory. Int. J. Food Microbiol. *136*, 37–43.

Sun, L.M., Cheng, A.X., Wu, X.Z., Zhang, H.J., and Lou, H.X. (2010). Synergistic mechanisms of retigeric acid B and azoles against *Candida albicans*. J. Appl. Microbiol. *108*, 341–348.

Sung, W.S., and Lee, D.G. (2008). *In vitro* candidacidal action of Korean red ginseng saponins against *Candida albicans*. Biol. Pharm. Bull. *31*, 139–142.

Sung, W.S., Lee, J., and Lee, D.G. (2008a). Fungicidal effect of piscidin on *Candida albicans*: pore formation in lipid vesicles and activity in fungal membranes. Biol. Pharm. Bull. *31*, 1906–1910.

Sung, W.S., Lee, J., and Lee, D.G. (2008b). Fungicidal effect and the mode of action of piscidin 2 derived from hybrid striped bass. Biochem. Biophys. Res. Comm. *371*, 551–555.

Svirshchevskaya, E.V., Shevchenko, M.A., Huet, D., Femenia, F., Latge, J.P., Boireau, P., and Berkova, N.P. (2009). Susceptibility of mice to invasive aspergillosis correlates with delayed cell influx into the lungs. Int. J. Immunogenet. *36*, 289–299.

Tanaka, T., Masunari, S., Ishii, J., Wakamura, K., Segawa, M., Fukuda, H., and Kondo, A. (2010). Displaying non-natural, functional molecules on yeast surfaces via biotin–streptavidin interaction. J. Biotechnol. *145*, 79–83.

Tantry, S., Ding, F.X., Dumont, M., Becker, J.M., and Naider, F. (2010). Binding of fluorinated phenylalanine alpha-factor analogues to Ste2p: evidence for a cation-pi binding interaction between a peptide ligand and its cognate G protein-coupled receptor. Biochem. *49*, 5007–5015.

Tao, Z.H., Gao, P., and Liu, H.W. (2009). Studies of the expression of human poly(ADP-ribose) polymerase-1 in *Saccharomyces cerevisiae* and identification of PARP-1 substrates by yeast proteome microarray screening. Biochem. *48*, 11745–11754.

Teixeira, P.A.C., De Castro, R.A., Nascimento, R.C., Tronchin, G., Torres, A.P., Lazera, M., de Almeida, S.R., Bouchara, J.P., Penha, C.V.L.Y., and Lopes-Bezerra, L.M. (2009). Cell surface expression of adhesins for fibronectin correlates with virulence in *Sporothrix schenckii*. Microbiology *155*, 3730–3738.

Throndset, W., Kim, S., Bower, B., Lantz, S., Kelemen, B., Pepsin, M., Chow, N., Mitchinson, C., and Ward, M. (2010a). Flow cytometric sorting of the filamentous fungus *Trichoderma reesei* for improved strains. Enz. Microb. Technol. *47*, 335–341.

Throndset, W., Bower, B., Caguiat, R., Baldwin, T., and Ward, M. (2010b). Isolation of a strain of *Trichoderma reesei* with improved glucoamylase secretion by flow cytometric sorting. Enz. Microb. Technol. *47*, 342–347.

Torriani, S., Felis, G., and Fracchetti, F. (2011). Selection criteria and tools for malolactic starters development: an update. Annals Microbiol. *61*, 33–39.

Trnovsky, J., Merz, W., Della-Latta, P., Wu, F., Arendrup, M.C., and Stender, H. (2008). Rapid and accurate identification of *Candida albicans* isolates by use of PNA FISHFlow. J. Clin. Microbiol. *46*, 1537–1540.

Tu, R., Martinez, R., Prodanovic, R., Klein, M., and Schwaneberg, U. (2011). A flow cytometry-based screening system for directed evolution of proteases. J. Biomol. Screen. *16*(3), 285–294.

Ukibe, K., Katsuragi, T., Tani, Y., and Takagi, H. (2008). Efficient screening for astaxanthin-overproducing mutants of the yeast *Xanthophyllomyces dendrorhous* by flow cytometry. FEMS Microbiol. Lett. *286*, 241–248.

Vahey, M.D., and Voldman, J. (2008). An equilibrium method for continuous-flow cell sorting using dielectrophoresis. Anal. Chem. *80*, 3135–3143.

Valero, A., Braschler, T., Demierre, N., and Renaud, P. (2010). A miniaturized continuous dielectrophoretic cell sorter and its applications. Biomicrofluidics *4*, 022807.

Vale-Silva, L.A., and Buchta, V. (2006). Antifungal susceptibility testing by flow cytometry: is it the future? Mycoses *49*, 261–273.

Vanhee, L.M.E., Nelis, H.J., and Coenye, T. (2008). Enumeration of airborne bacteria and fungi using solid phase cytometry. J. Microbiol. Meth. *72*, 12–19.

Verstrepen, K.J., Derdelinckx, G., Verachtert, H., and Delvaux, F.R. (2003). Yeast flocculation: what brewers should know. Appl. Microbiol. Biotechnol. *61*, 197–205.

Vogl, G., Lesiak, I., Jensen, D.B., Perkhofer, S., Eck, R., Speth, C., Lass-Florl, C., Zipfel, P.F., Blom, A.M., Dierich, M.P., and Wurzner, R. (2008). Immune evasion by acquisition of complement inhibitors: The mould *Aspergillus* binds both factor H and C4b binding protein. Mol. Immunol. *45*, 1485–1493.

Walker, G.K. (2000). Microbiology of wine-making. In Encyclopedia of Food Microbiology, Robinson, R.K. ed. (Elsevier, Amsterdam, the Netherlands), pp. 2306–2310.

Walker, L.M., Bowley, D.R., and Burton, D.R. (2009). Efficient recovery of high-affinity antibodies from a single-chain fab yeast display library. J. Mol. Biol. *389*, 365–375.

Wan, L., Cai, H.W., Yang, H., Lu, Y.R., Li, Y.Y., Li, X.W., Li, S.F., Zhang, J., Li, Y.P., Cheng, J.Q., and Lu, X.F. (2008). High-level expression of a functional humanized single-chain variable fragment antibody against CD25 in *Pichia pastoris*. Appl. Microbiol. Biotechnol. *81*, 33–41.

Wanchoo, A., Lewis, M.W., and Keyhani, N.O. (2009). Lectin mapping reveals stage-specific display of surface carbohydrates in *in vitro* and haemolymph-derived cells of the entomopathogenic fungus *Beauveria bassiana*. Microbiology *155*, 3121–3133.

Wang, H., Nakata, E., and Hamachi, I. (2009). Recent progress in strategies for the creation of protein-based fluorescent biosensors. Chembiochem. *10*, 2560–2577.

Wang, Q., Li, L., Chen, M., Qi, Q., and Wang, P.G. (2007). Construction of a novel system for cell surface display of heterologous proteins on *Pichia pastoris*. Biotechnol. Lett. *29*, 1561–1566.

Wang, Y., Xuan, Y.J., Zhang, P., Jiang, X., Ni, Z.H., Tong, L.J., Zhou, X.S., Lin, L.P., Ding, J., and Zhang, Y.X. (2009). Targeting expression of the catalytic domain of the kinase insert domain receptor (KDR) in the peroxisomes of *Pichia pastoris*. FEMS Yeast Res. *9*, 732–741.

Wang, Z., Zhang, L.Y., and Tan, T.W. (2010). Efficient screening a high glutathione-content mutant of *Saccharomyces cerevisiae* by flow cytometry. Process Biochem. *45*, 1168–1171.

Warnock, D.W. (2007). Trends in the epidemiology of invasive fungal infections. Nippon Ishinkin. Gakkai Zasshi. *48*, 1–12.

Watson, J.V., Chambers, S.H., and Smith, P.J. (1987). A pragmatic approach to the analysis of DNA histograms with a definable G1 peak. Cytometry *8*, 1–8.

Weaver-Feldhaus, J.M., Lou, J., Coleman, J.R., Siegel, R.W., Marks, J.D., and Feldhaus, M.J. (2004). Yeast mating for combinatorial Fab library generation and surface display. FEBS Lett. *564*, 24–34.

Weigert, C., Steffler, F., Kurz, T., Shellhammer, T.H., and Methner, F.J. (2009). Application of a short intracellular pH method to flow cytometry for determining *Saccharomyces cerevisiae* Vitality. Appl. Env. Microbiol. *75*, 5615–5620.

Wen, F., Esteban, O., and Zhao, H.M. (2008). Rapid identification of CD4+T-cell epitopes using yeast displaying pathogen-derived peptide library. J. Immunol. Meth. *336*, 37–44.

White, M.A., Riles, L., and Cohen, B.A. (2009). A systematic screen for transcriptional regulators of the yeast cell cycle. Genetics *181*, 435–446.

Wozniak-Knopp, G., Bartl, S., Bauer, A., Mostageer, M., Woisetschlager, M., Antes, B., Ettl, K., Kainer, M., Weberhofer, G., Wiederkum, S., *et al.* (2010). Introducing antigen-binding sites in structural loops of immunoglobulin constant domains: Fc fragments with engineered HER2/neu-binding sites and antibody properties. Prot. Eng. Des. Sel. *23*, 289–297.

Yang, C.Z., Gong, W.Y., Lu, J.T., Zhu, X.J., and Qi, Q.G. (2010). Antifungal drug susceptibility of oral *Candida albicans* isolates may be associated with apoptotic responses to Amphotericin B. J. Oral Pathol. Med. *39*, 182–187.

Yang, G.Y., and Withers, S.G. (2009). Ultrahigh-Throughput FACS-based screening for directed enzyme evolution. Chembiochem *10*, 2704–2715.

Yue, L., Chi, Z., Wang, L., Liu, J., Madzak, C., Li, J., and Wang, X. (2008). Construction of a new plasmid for surface display on cells of *Yarrowia lipolytica*. J. Microbiol. Methods 72, 116–123.

Zhang, J., Campbell, R.E., Ting, A.Y., and Tsien, R.Y. (2002). Creating new fluorescent probes for cell biology. Nat. Rev. Mol. Cell Biol. 3, 906–918.

Zhang, L.B., Leibowitz, M.J., and Zhang, Y. (2009). Antisense oligonucleotides effectively inhibit the co-transcriptional splicing of a *Candida* group I intron *in vitro* and *in vivo:* Implications for antifungal therapeutics. FEBS Lett. *583*, 734–738.

Zheng, S., Zhang, Y., Tong, T., Cui, C., and Sun, J. (2010). Dominance of yeast in activated sludge under acidic pH and high organic loading. Biochem. Eng. J. *52*, 282–288.

Zhou, C., Jacobsen, F.W., Cai, L., Chen, Q., and Shen, W.D. (2010). Development of a novel mammalian cell surface antibody display platform. MAbs 2, 508–518.

Zhou, Y., Kajiyama, S., Itoh, K., Tanino, T., Fukuda, N., Tanaka, T., Kondo, A., and Fukui, K. (2009). Development of an enzyme activity screening system for beta-glucosidase-displaying yeasts using calcium alginate micro-beads and flow sorting. Appl. Microbiol. Biotechnol. *84*, 375–382.

The Application of Flow Cytometry to the Study of Lactic Acid Bacteria Fermentations

8

Imelda A. Doolan, Martin G. Wilkinson and Dara K. Hickey

Abstract

This chapter will focus on lactic acid bacteria (LAB) and their importance in the production of a wide range of fermented foods as both viable and non-viable organisms. A comparison of traditional microbiological methods used to study LAB will be outlined as well as examples of how flow cytometry (FCM) can be successfully employed to analyse LAB in viable and non-viable states during a fermentation process such as cheese manufacture. This chapter also deals with some of the difficulties encountered with FCM analysis of certain food samples and how these issues may be overcome. The potential applications of FCM to the study of LAB in both academic and applied research issues are also discussed.

Introduction

Lactic acid bacteria (LAB) are a broad group of Gram-positive, catalase-negative, non-sporing rods and cocci which are usually non-motile and are facultatively anaerobic. LAB are divided into six major genera: *Lactococcus, Lactobacillus, Leuconostoc, Pediococcus, Streptococcus and Enterococcus.* LAB, especially members of the genera *Lactococcus, Lactobacillus, Leuconostoc* and *Pediococcus* are the main bacteria used as starter cultures in the fermentation of milk, cheeses, yogurts, meats and vegetables. During food fermentations LAB starter cultures generate lactic acid as their major end product of carbohydrate metabolism in food substrates such as milk, meat or vegetables (Aguirre and Collins, 1993). Overall, their metabolism can be classified as either homofermentative, where lactic acid represents 95% of the total end products or heterofermentative, in which acetic acid, ethanol and carbon dioxide are produced in addition to lactic acid (Bamforth, 2005). Starter cultures can also produce specific metabolites of economic interest such as acids, alcohol, enzymes, antibiotics and carbohydrates (Giraffa, 2004).

An LAB starter preparation may be defined as an inoculum containing large numbers of viable micro-organisms, which may be added to accelerate a fermentation process initially through acidification and thereafter through activity of their cell wall-associated or intracellular enzymes (Bunthof *et al.*, 2001). The addition of starter culture facilitates improved control over the initial phase of a fermentation process and the predictability of the end products obtained (Holzapfel and Schillinger, 2002). Starter LAB strains are generally selected for their technological, sensory and, more recently, probiotic properties with consideration for aspects such as acidification profile, organoleptic, textural and preservative properties. Before being added to substrates in industrial fermentation processes, LAB strains are themselves grown in large-scale fermentations and the cells harvested and preserved by either freezing or freeze-drying into various formats such as powdered or frozen pellets (Rault *et al.*, 2007, 2010). Frozen or freeze-dried ready-to-use starter cultures are known as either direct vat set (DVS) or direct vat inoculation (DVI) and are highly concentrated preparations (10^{10}–10^{12} CFU/g) containing highly active cultures (Cogan, 1996; Hansen, 2002). However, the activity of the industrial starter cultures in the fermentation is dependent on the ability of cells to survive the processing steps used to produce the DVI/DVS cultures. The freezing step, in particular, may cause cellular osmotic stress, dehydration, molecular and membrane damage resulting in a loss in viability and acidification activity (Rault *et al.*, 2010). Harvesting time and fermentation conditions are also key factors that strongly influence these properties. During fermentation, starter cultures are subjected to the build-up of inhibitory compounds such as lactic acid which can also affect their cellular physiological state (Amrane and Prigent, 1999; Cachon *et al.*, 1998; Rault *et al.*, 2008). A build-up of lactic acid may result in full inhibition of growth and subsequent inhibition of acid production after inoculation causing delayed fermentations, extended production times and inconsistent product quality. Control of industrial fermentation processes therefore requires knowledge of both the physiology and metabolic properties of starter cultures (Giraffa, 2004) as these are critical for end product applications, such as cheese manufacture or production of probiotic cultures (Bunthof *et al.*, 2001a; Bunthof *et al.*, 2001b). The main fermentation conditions that affect the physiological status of starter LAB strains include pH, temperature or salt concentration. Flow cytometry (FCM) has recently been applied to gain a deeper insight into the metabolic activity and cell physiology of starter LAB during fermentation (Fig. 8.1). FCM when combined with fluorescence measurements enables rapid monitoring of the dynamic modifications of the cellular state during fermentation/acidification phase and also in the ripening/maturation phase. Bacteriophage infections which can destroy cellular viability and acidification properties can also be monitored by FCM which can enhance process control over dairy fermentations (Díaz *et al.*, 2010). Michelsen *et al.* (2007) developed a method to study bacteriophage infection in *Lactococcus lactis* using FCM. The method was based on the detection of populations of cells with low density cell walls which appear late in the infection cycle and may be detected by FCM using light scatter versus fluorescence after staining with Mithramycin (a GC-specific stain) and Ethidium Bromide. This method allows for rapid and early detection of phage-infected bacteria and a method which can be performed in real time, increasing the possibility of a successful fermentation.

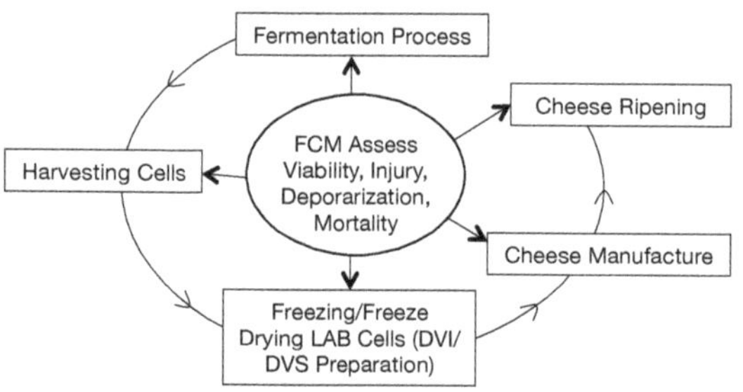

Figure 8.1 Application of flow cytometry for physiological analysis of LAB during production, storage and application of direct vat set (DVS) or direct vat inoculation (DVI) cultures.

Traditional methods used to monitor LAB and a comparison with flow cytometry methodology

The quality of any given dairy starter culture is generally measured in terms of its viability and acidification activity (Rechinger and Siegumfeldt, 2002). Microbial viability can be defined in terms of: the ability to reproduce, the degree of vitality (defined as being metabolically active), or extent of membrane integrity (Nebe-von-Caron and Badley, 1995). Reproductive (living) cells are deemed to be able to proliferate, whereas non reproductive cells (dead) cannot. Vital cells are generally considered to be metabolically active while intact cells possess a cytoplasmic membrane with selective permeability (Bunthof *et al.*, 1999). Methods for viability assessment are based on evaluating the basic requirements required for microorganisms to survive including: (i) an intact cytoplasmic membrane; (ii) active transcription of DNA and translation of RNA; (iii) generation of energy for maintenance of metabolism, biosynthesis of proteins, nucleic acids, polysaccharides, and other cell components; and, (iv) growth and multiplication (Mason *et al.*, 1986). Membrane integrity, vitality, and the ability of a culture to reproduce depends on both environmental conditions and the physiological status of the cell at the point of investigation (Bunthof *et al.*, 1999).

The colony count is the traditional standard method by which to evaluate starter culture viability. In terms of dairy fermentations and cheese production, traditional methods of starter enumeration involve viable cell count (VCC) on selective media such as L-M17, LBS or MRS agars. In the dairy industry, acidification and heat tolerance are routinely investigated in assays such as the Pearce activity test (Pearce, 1969) which simulates the process temperatures involved in Cheddar cheese manufacture. The Pearce test is frequently undertaken in addition to plate counting to determine the performance of dairy starter cultures (Bunthof *et al.*, 1999). However, plate counts require relatively long incubation times (up to 3 days) and yield no information regarding the degree of heterogeneity existing within bacterial populations or the physiological state of the cells. Population heterogeneity refers to the presence of cells in various states such as live, permeabilized/damaged or dead. Fig. 8.2 shows an example of the application of FCM used to monitor loss of membrane potential during the Pearce activity test. Microorganisms which do not form colonies because they are dead, sublethally injured, viable but non-culturable (VBNC), dormant or inactive are generally undetected by standard plate count methods (Breeuwer and Abee, 2000). These non-viable, damaged or permeabilized subpopulations are potentially highly important in fermented foods such as cheese or fermented meats as they may comprise the particular subpopulations which contribute the most to the release or leakage of intracellular enzymes into the food matrix thereby determining the rate of ripening and flavour forming potential of the product (Sheehan *et al.* 2005; Doolan and Wilkinson, 2009; Doolan *et al.*, 2014; Kilcawley *et al.*, 2012). The accuracy of the viable plate count method for starter cultures may also be influenced by the degree of cell clumping, inhibition by neighbouring cells and composition of the growth media (Mason *et al.*, 1986). The use of FCM not only allows the determination of viable cell counts but also allows for enumeration of damaged and dead cell subpopulations. Absolute counts of each of the subpopulations can be achieved through the inclusion of defined numbers of fluorescent beads, e.g. Trucount beads™ (BD Biosciences, San Jose, CA, USA) with the bacterial samples, or through the use of particular software capabilities.

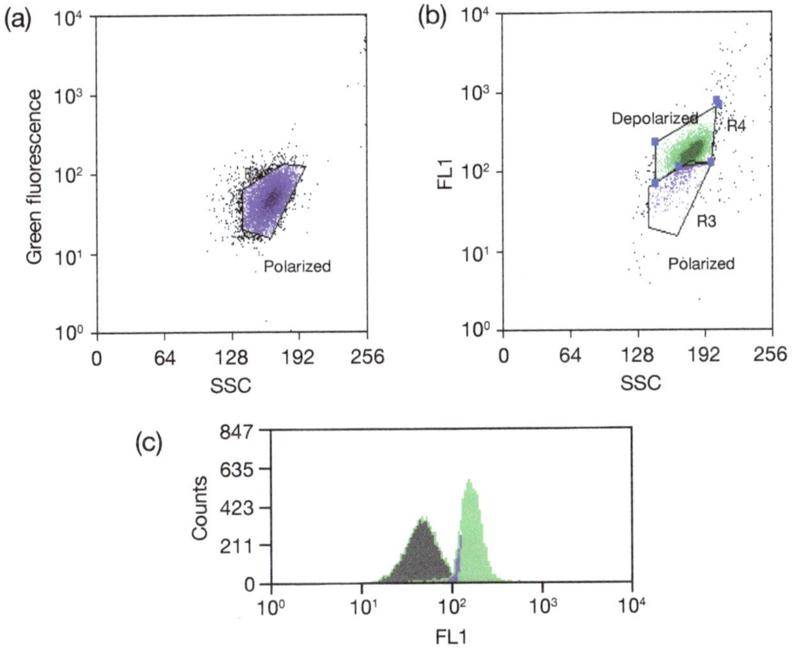

Figure 8.2 (a) LAB cells stained with DiBAC$_4$(3), (b) depolarization of membrane potential during exposure to Pearce activity test, and, (c) shift in FL1 fluorescence (Doolan and Wilkinson, unpublished data).

This is achieved when the volume of the sample fluid going through the cytometry system can be accurately measured and hence this information is subsequently used to calculate the absolute numbers of events/cells per ml. Current methods to study starter cell permeability or autolysis during maturation or storage involves using a combination of plating to determine viability/loss in viability rates as well as monitoring of the activity of various released intracellular enzyme markers from both viable and non-viable cells including: Lactate dehydrogenase (LDH), Pep X or Pep N peptidases. This combination of methodologies while useful does not permit direct cellular analysis or discrimination of cells which are either permeabilized or completely autolysed. An outline of the limitations of traditional methods for LAB cell viability, permeabilization and autolysis are compared with flow cytometric methods for assessment is shown in Fig. 8.3.

Flow cytometry (FCM)

Fluorescence measurements combined with FCM have been reported for assessment of viability and vitality of LAB. The main advantages of FCM include an ability to rapidly count individual cells (Bunthof and Abee, 2002) with simultaneous monitoring of both structural and functional cellular properties such as membrane integrity or intracellular enzyme activity using light scattering and/or specific fluorescent probes in a multi-parameter assay. A range of fluorescent dyes are available for evaluating specific physiological properties of bacteria including: activity, intracellular pH, membrane integrity, and viability (Bunthof *et al.*, 2001a). A number of reports have outlined the use of FCM to demonstrate the presence of viable but not cultivable (VBNC) subpopulations of LAB (Ananta *et al.*, 2004; Ben Amor *et al.*, 2002; Lahtinen *et al.*, 2006; Papadimitriou *et al.*, 2006; Sunny-Roberts and Knorr, 2008). Ananta *et al.* (2004) assessed the effects of exposure to high pressure on *Lactobacillus*

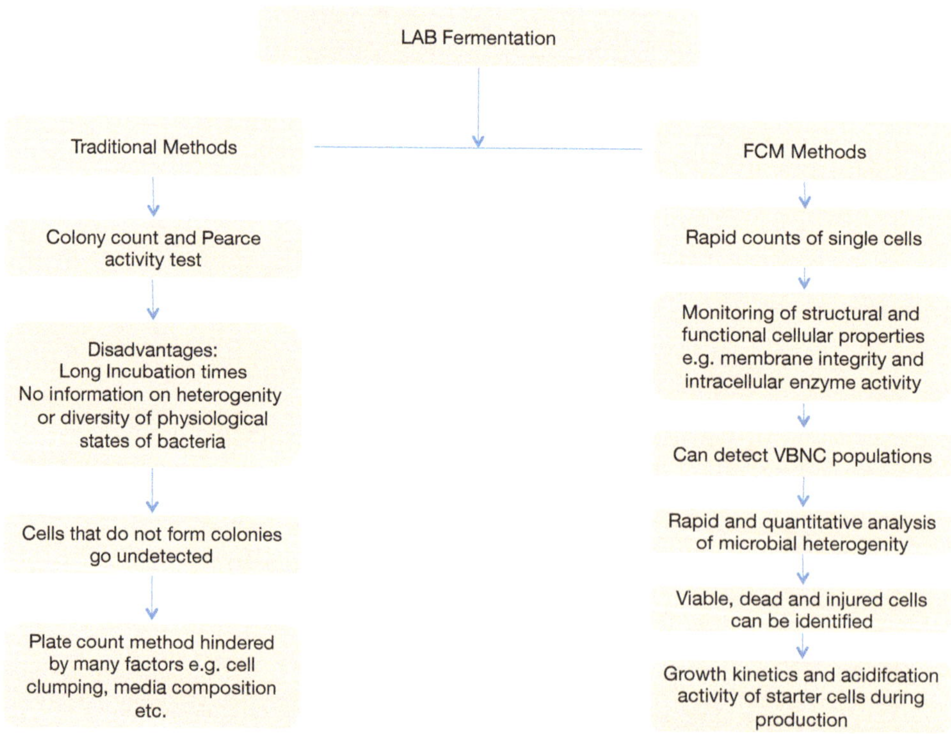

Figure 8.3 Comparison of traditional enumeration and physiological analysis of starter LAB with flow cytometric analysis.

rhamnosus GG using FCM. These researchers noted a growing body of evidence to suggest the existence of a proportion of metabolically active bacteria that cannot be cultured by traditional means. VNBC subpopulations which may fail to grow as visible colonies on agar plates may be particularly important in terms of starter bacterial enzyme contribution to flavour formation reactions. Lahtinen *et al.* (2006) evaluated the intrinsic properties of so called dormant probiotic bacteria by applying FCM viability assays. In this study, plate counting and four culture-independent cytometric assays were used to determine the viability of three probiotic strains during storage. During storage (cells were maintained at 4°C and subjected to weekly viability assays) the strains showed a reduction in plate counts but were able to maintain intracellular esterase activity, an intact cytoplasmic membrane while maintaining pH gradients between intracellular and extracellular matrices. However, the FCM assays classified uncultivable cells as being metabolically active and stress resistant.

Specific examples of flow cytometry and LAB analysis

Multi-parameter FCM has been used to analyse the extent of viability, mortality, injury, and depolarization of LAB cells at various times during the starter cell production process (Rault *et al.*, 2008). FCM and fluorescent probes have assessed the following properties; membrane integrity, intracellular esterase activity, cellular vitality, membrane depolarization, and intracellular pH values (Table 8.1).

Live/dead assays were carried out using dual staining of samples with carboxyfluorescein diacetate (cFDA) or propidium iodide (PI) to differentiate viable, dead, or stressed cells during fermentation (Rault *et al.*, 2008). To assess the viability of LAB, cells were stained using cFDA a non-fluorescent precursor (prefluorochrome) which readily diffuses across intact cell membranes. When cleaved by non-specific intracellular enzymes a positively charged green fluorescent compound, carboxyfluorescein (cF), is generated (Mason *et al.*, 1986; Rault *et al.*, 2007). The carboxyfluorescein (cF) has a maximum excitation wavelength of 492 nm and an emission wavelength of 517 nm (Rault *et al.*, 2007). Retention of cF within the cell indicates an intact membrane and functional cytoplasmic enzymes (Ben Amor *et al.*, 2002; Bunthof *et al.*, 2001a; Nappy and Edwards, 1994).

Permeabilized or dead cells were monitored by uptake of the nucleic acid dye Propidium iodide (PI). This probe can only enter cells with compromised membranes, and binds to DNA and RNA to form a red fluorescent DNA-complex having a maximum excitation wavelength of 535 nm and an emission wavelength of 617 nm. (Ben Amor *et al.*, 2002; Rault *et al.*, 2007). The extent of exclusion of PI by cells can also be an indication of viability (Bunthof *et al.*, 2001a).

Bis (1.3 dibutylbarbituric acid) trimethine oxonol (DIBAC$_4$(3)) has been used to assess the degree of membrane depolarization of LAB cells. In bacteria, membrane potential is usually generated by the extrusion of H^+ ions by the H^+-ATPase or the electron transfer chain (Mason *et al.*, 1986). Oxonol is a negatively charged molecule which can enter depolarized or dead cells and bind to lipid-rich compounds (Ben Amor *et al.*, 2002). It can be excited by 488-nm wavelength and emits green fluorescence, thus depolarized cells are indicated by a shift in green fluorescence.

Cellular activity of LAB strains can be monitored by measuring the ability of cells to release intracellular cF upon energization by lactose addition (Breeuwer and Abee, 2000; Rault *et al.*, 2008). LAB labelled with the non-fluorescent precursor cFDA retained cF when not energized, but rapidly extruded this probe after addition of lactose (Bunthof *et al.*, 2000). However if the ATP driven pump becomes damaged cF will be retained within the cell (Volkert *et al.*, 2008).

The fluorochrome 5(6)-carboxyfluorescein diacetate succinimidyl ester (cFDA-SE) has also been used in combination with FCM to monitor the intracellular pH (pHi) of LAB strains during fermentation (Rault *et al.*, 2009). During growth of acid-tolerant LAB strains, pH$_i$ decreases along with a decrease in external pH (pHex) in order to maintain a constant pH gradient rather than a constant pHi. The pHi of LAB cells has been determined by staining cells with cFDA-SE, measuring the fluorescence intensity with FCM

Table 8.1 Applications of FCM to the study of lactic acid bacteria

LAB culture	Fluorescent stains	Parameters measured	Reference
Lactobacillus delbrueckii subsp. *bulgaricus* CFL1	cFDA/PI	Live/dead analysis	Rault *et al.* (2008)
Lactococcus lactis subsp. *lactis* ML3	DiBAC (4) 3	Membrane potential	Bunthof *et al.* (2000)
Lactococcus lactis subsp. *lactis* ML3	cFDA	Reproduction and acidification capacity of stressed ML3 cells	Bunthof *et al.* (2000)
Lactobacillus bulgaricus	CFDA-SE	Intracellular pHi	Rault *et al.* (2008) Siegumfeldt *et al.* (2000)
Lactobacillus bulgaricus	5-(and 6) carboxyfluorescein diacetate	Intracellular enzyme activity	Riis *et al.* (1995)
Lactobacillus delbrueckii	cFDA/PI	Cryotolerance	Rault *et al.* (2007) Rault *et al.* (2010)
Lactobacillus rhamnosus GG	CFDA/PI	Physiological effects high pressure treatments Esterase activity	Ananta *et al.* (2004) Volkert *et al.* (2008)
Lactococcus lactis subsp. *cremoris* AM2 *Lactococcus lactis* subsp. *cremoris* HP *Lactococcus lactis* subsp. *lactis* 303	PI/Syto 9	Lysis in cheese starter cultures Chemical attenuation of starter cultures	Bunthof *et al.* (2001b) Sheehan *et al.* (2005) Doolan and Wilkinson (2009)
Oenococcus oeni	CF/PI	Membrane integrity	Graça da Silveira *et al.* (2002)
Lactococcus lactis, *Lactobacillus helveticus* and *Leuconostoc mesenteroides*	cFDA, TOTO-1, PI	Effect of exposure to deconjugated bile salts on probiotic cultures	Bunthof *et al.* (2001a)
Bifidobacteria lactis, *Lactococcus lactis*	5-(and 6) carboxyfluorescein diacetate, Syto 24, polyclonal antibody	Rapid and specific enumeration of viable *Bifidobacterium* in dairy products	Geng *et al.* (2014)

and comparing data with calibration curves established for pH-equilibrated cells with known pHi values (Siegumfeldt *et al.*, 2000). Rault *et al.* (2008) found that during the exponential growth phase of *Lb. delbrueckii* in a 2-l bioreactor, viable cell counts of *Lb. delbrueckii* assessed by PI/cFDA using a Cyflow cytometer equipped with a specific volumetric counting system gave data similar to plate counts. However, during the stationary phase, viable cell counts on the cytometer remained stable while cultivability on MRS agar declined. This particular finding may indicate that during stationary phase some cells enter a non-culturable state but remain both viable and active as determined by flow cytometry. Using flow cytometry combined with cFDA, PI, and DIBAC$_4$(3) staining, Rault *et al.* (2008) demonstrated that a high percentage (89.9%) of *Lb. delbrueckii* cells remained viable with intact and polarized membranes present throughout both the exponential and stationary phases, despite a continuous decrease in pH of the media and a decrease in viable cells as determined by plate counting. These authors also demonstrated that the physiological state of the cells during fermentation was strongly affected by pHi and pH gradient.

FCM was also used (Rault *et al.*, 2009) to study the influence of pH on the physiology of *L. bulgaricus* CFL1 during fermentation controlled at either pH 6.0 or pH 5.0. Differences were noted in mortality as measured by membrane damage, viability by (enzymatic activity), membrane depolarization and intracellular pH.

Bunthof *et al.* (2000) studied the exposure of *L. lactis* subsp. *lactis* ML3 to various stressors including: heating, freezing, conjugated or deconjugated bile salts. These treatments were used to reflect the environmental stresses that may occur during production of LAB starters and during storage of fermented end products. Following exposure to stressors, cF labelling indicated a difference between FCM profiles and viable plate counts, therefore a two-step assay was developed. *Lc. lactis* cells were shown to be easily labelled with cF by incubation with the non-fluorescent precursor carboxyfluorescein diacetate (cFDA). The labelled cells retained cF when not energized, but extruded the probe rapidly upon lactose addition, most likely mediated by an ATP driven transport system. Combining cF labelling with lactose-energized efflux appeared a more reliable indicator for reproduction and acidification capacity of stressed *L. lactis* subsp. *lactis* ML3 cells, and FCM data corresponded well with plate counts. This FCM method represents a combined assay (use of a non-fluorescent precursor) which may be a good rapid indicator for reproduction and acidification capacity of stressed cells with potential for LAB physiological research in dairy applications. The cited studies demonstrate the potential for FCM and multi parameter analysis for process improvement for industrial starter production and improved definition of process conditions (harvesting time and pH) to generate culture preparations in the desired or optimum physiological states (Rault *et al.* 2009).

Starter performance

Following production, starter cultures are normally stored as frozen concentrates or as a freeze-dried powder (Volkert *et al.*, 2008). Both preservation processes can affect the integrity of the cell structure due to interactions between cell structures and ice crystals (McGann, 1978). Rault *et al.* (2010) suggested that the long term performance of frozen LAB cells could be predicted by early determination of physiological parameters such as a low acidification activity and monitoring by FCM following harvesting of cells. Several studies have therefore utilized FCM as a technique for predicting cryotolerance of LAB cultures during the starter production process.

Riis *et al.* (1995) reported the use of FCM for predicting the fermentation performance of frozen and stressed frozen cultures of *Lb. delbrueckii* subsp. *bulgaricus*. These researchers used FCM analysis of 5- (and-6)-carboxyfluorescein diacetate stained cells and an acidification power (AP) test to predict fermentation performance of cultures of *Lb. delbrueckii* subsp. *bulgaricus* and compared them with traditional fermentation tests. Frozen or freeze dried cultures stressed by exposure to 5°C for 2 hours or 20 hours were examined. FCM was able to differentiate between freeze dried or frozen cultures but did not appear to be able to differentiate cultures based on their fermentation abilities.

Multiparametric FCM was used to evaluate the cryotolerance of four *Lb. delbrueckii* strains (Rault *et al.*, 2007; Rault *et al.*, 2010). A combination of cFDA and PI probes were used to quantify viable, dead or injured subpopulations before and after freezing, and throughout one month storage at −80°C. This methodology enabled differentiation of strains according to their susceptibility to freezing or frozen storage. Resistant strains had <10% of dead cells after one month storage at −80°C.

FCM has been used to determine the impact of three different freezing methods on the physiology of the probiotic culture *Lb. rhamnosus GG* based on intracellular esterase activity (cFDA) membrane integrity (PI) or cell vitality (cF extrusion activity) (Volkert *et al.*, 2008). *Lb. rhamnosus* GG was analysed by FCM before and after freezing in either phosphate buffered saline or 20% (w/v) skim milk powder, respectively. The freezing methods applied were static air freezing at −30°C for 3 hours, high pressure shift freezing at −21°C and 210 MPa or spray freezing at −30°C. After each treatment all samples were stored at −20°C for one day before analysis. FCM clearly demonstrated that various freezing methods had differing impacts on the physiological characteristics of the cells while the growth media also influenced survival rates. Higher survival rates were achieved when 20% (w/v) RSM was used as a protecting media. This study highlighted the importance of monitoring cell vitality (metabolically active) in combination with cell viability.

FCM has been used to understand the influence of the initial physiological state of cells as defined by fermentation pH and harvesting time on the cryotolerance of *Lb. bulgaricus* CFL1 (Rault *et al.*, 2010). The probes cFDA, PI and cFDA-SE were used to monitor intracellular esterase, membrane integrity during growth with and without pH control, and during subsequent freezing and frozen storage of harvested cells at −20°C for up to 5 months. Cultivability was monitored by plate counting and acidification using the Cinac system, respectively. The Cinac system continuously monitors the acidification activity of multiple LAB strains by measuring the pH evolution over time. While intracellular esterase activity and membrane integrity remained relatively stable throughout frozen storage, cultivability and acidification activity declined. FCM, in combination with traditional viability measurements, indicated that freezing or frozen storage affected various physiological properties of the cells.

Spray drying is another technology used for the production and preservation of LAB cultures. However, the high air temperatures (outlet temperatures from the drying chamber can reach up to 100°C), during the passage of the bacteria through the drying chamber may negatively affect viability and therefore their activity in end use industrial applications such as cheese or yogurt manufacture (Streit *et al.*, 1979). FCM in combination with dual staining by cFDA and PI was used to evaluate sites where cellular injury may have occurred during spray-drying of *Lb. rhamnosus* GG in skim-milk based preparations (Ananta *et al.*, 2005). The fluorescence profile of the spray–dried bacteria indicated that the viable population decreased as the air outlet temperature increased from 70°C to 100°C with an increase in the dead/comprised membrane population. A strong correlation was found between the survival rates of *Lb. rhamnosus* as determined by plate counting and the percentage of cF-stained cells indicating that cells with membrane damage became non-viable and did not grow on solid agar. These authors concluded that membrane damage was the underlying mechanism for increases in cell death noted during spray drying (Ananta *et al.*, 2005).

Wine flavour is also influenced through LAB activity which carry out the important malo-lactic fermentation (MLF), significant in modifying the final flavour profile of the product. The control of MLF is an important feature of the technology of modern commercial wine production (Graça da Silveira *et al.*, 2002). When *Oenococcus oeni* starter cells are inoculated directly into wine, this can lead to significant cell mortality and potentially a failure of the fermentation process. In wine production and for MLF to be achieved correctly, one or more steps of reactivation and adaptation of starter cultures to the conditions are required to enhance the survival of these cultures when inoculated into wine. The ethanol and acid resistances of *O. oeni* strains are therefore considered crucial for survival in wine (Graça da Silveira *et al.*, 2002; Lafon-Lafourcade *et al.*, 1983). Graça da Silveira *et al.* (2002) studied the membrane integrity of ethanol stressed *Oenococcus oeni* cells using FCM. The effect of ethanol on these cells was studied based on the ability of labelled cells to efficiently retain cF and take up propidium iodide (PI). FCM analysis of both ethanol-adapted and non-adapted cells after dual staining revealed three main subpopulations: cF stained intact cells; cF and PI stained permeable cells, and, PI stained damaged cells. FCM showed that the subpopulation of cells retaining membrane integrity was three times larger when cells were grown in the presence of ethanol, prior to inoculation into wine, reflecting the protective effect of ethanol adaptation on cells.

Cheese manufacture and FCM applications

Lactic acid starter bacteria have a dual role during cheese production: (1) rapid acid production in milk during growth by viable cells in the cheese vat during manufacture from conversion of milk lactose to lactic acid and (2) release of intracellular peptidases from permeabilized viable or non-viable cells or from non-viable autolysed cells during ripening to generate flavour components, mainly lower molecular mass peptides and free amino acids (Bunthof *et al.*, 2001(b)). Generally, intracellular LAB enzymes access their substrates in the cheese matrix via autolysis or permeabilization of cells following a loss in viability (Bunthof *et al.*, 2001a; Niven and Mulholland, 1998; Wilkinson *et al.*, 1994). The loss in viability of

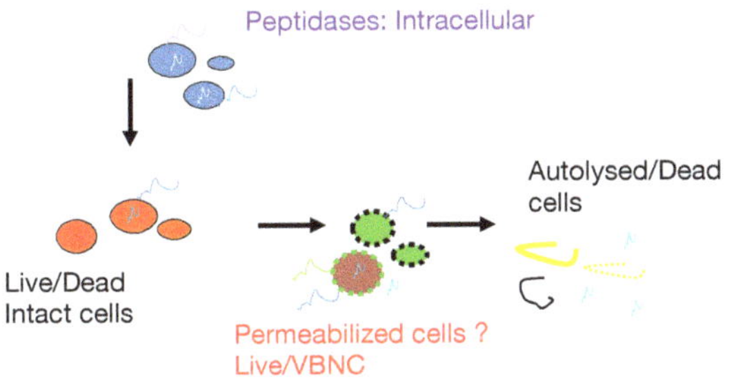

Figure 8.4 Physiological pathways affecting intracellular enzyme release by LAB during cheese ripening.

starter LAB occurs initially during manufacturing in the cheese vat as a strain-specific response to process conditions such as cooking temperature or added salt levels. Thereafter, in the early stages of ripening up to day 42, this progressive loss in viability is accompanied by varying degrees of cell wall and cell membrane damage (Crow *et al.*, 1995; Wilkinson *et al.*, 1994). Consequently, the cheese environment during ripening appears to be populated by a range of subpopulations of intact, viable, non-viable and partially or fully disintegrated cells (Sheehan *et al.*, 2005) (Fig. 8.4). There is much industrial and academic interest in understanding the relationship between intracellular enzyme release/accessibility from the various subpopulations of starter LAB during ripening as a means to accelerate or modify the flavour generation process. In particular, it is important to try and assess the optimal balance between intact, permeabilized and autolysed subpopulations required for optimal flavour development from each particular LAB starter strain. These individual LAB subpopulations present in cheese during various stages of ripening may correspond to an enzymatic balance between cell envelope proteinases (CEP) from intact cells, a mixture of CEP and intracellular peptidases from permeabilized cells or primarily intracellular peptidases released from autolysed cells (Sheehan *et al.*, 2006). The importance of permeabilized subpopulations is that they may function as metabolic cell factories by maintaining a higher degree of intracellular structures required for more complex flavour bio-transformations compared with completely autolysed cells (see Fig. 8.5) (Bunthof *et al.*, 2001b; Doolan and Wilkinson, 2009; Sheehan *et al.*, 2005). Currently, cell autolysis and permeabilization in cheese during ripening is monitored by the decrease in viable counts of starters on selective agars, and release of intracellular markers indicating cell structural damage including: DNA, and intracellular enzymes, such as phospho-β-galactosidase, lactate dehydrogenase, (LDH) or peptidases such as Pep X or Pep N in aqueous extracts of cheese samples. Bunthof *et al.* (2001b) monitored permeabilization of LAB strains where cheese conditions were mimicked by incubating cells in a stabilizing buffer with high protein, potassium and magnesium levels which stabilizes cells and protects them from autolysis. When non-stabilizing conditions were used, addition of a high concentration of the cell wall hydrolysing

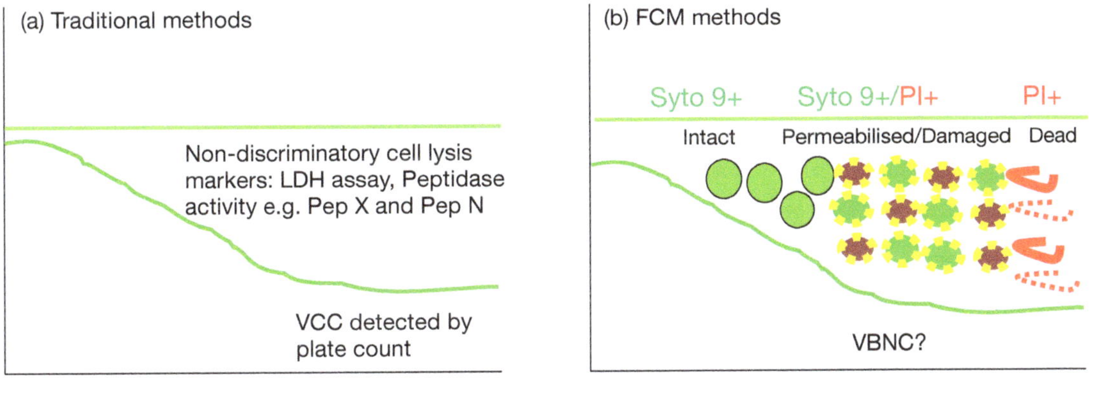

Figure 8.5 Diagram illustrating the use of traditional and flow cytometric methods for monitoring the physiology of cheese starter cultures using various parameters.

enzyme mutanolysin caused complete disruption of the cells and resulted in a decrease in the total numbers of cells and a release of cytoplasmic LDH. However, when cells were held in the stabilizing buffer, mutanolysin also caused membrane damage but cells disintegrated at a much lower rate. FCM analysis indicated that the stabilizing buffer appeared to protect permeabilized cells as indicated by a high number of PI positive, labelled damaged cells.

Sheehan *et al.* (2005) developed FCM methodology to directly differentiate, identify and enumerate intact or permeabilized LAB in the aqueous phase of Cheddar cheese expressed hydraulically as cheese 'juice' during ripening. Strains of LAB starters were stained using the Live/Dead *Bac*Light bacterial viability kit and data were correlated with traditional viable cell counts and release of intracellular autolytic markers LDH or Pep X. Permeabilized cells were determined by FCM from a comparison with control profiles obtained from LAB strains treated with isopropyl alcohol or hexadecyltrimethylammonium bromide (CTAB) to generate discrete dead or permeabilized subpopulations, respectively. Three main subpopulations could be clearly differentiated by FCM corresponding to intact, CTAB and isopropyl alcohol-treated cells. Using this methodology the influence of Cheddar cheese cooking temperature, where the temperature of the curds and whey are raised gradually from 32 to 38.5°C over a 35-minute period, on the permeabilization, autolysis and intracellular enzyme release from two lactococcal starter strains *Lactococcus lactis* subsp. *cremoris* AM2 or *Lactoccus lactis* subsp. *lactis* HP in broth and in Cheddar cheese juice during ripening were compared. Strain AM2 is known to have low viability in cheese and exhibits strong autolysis with elevated intracellular peptidase release and consequently generates high levels of free amino acids giving good quality cheese flavour. In contrast, strain HP is known to have a high viability in cheese, is poorly autolytic with low levels of released intracellular peptidases with consequent production of low levels of free amino acids with inferior flavour. This study was undertaken to evaluate whether the above differences were also reflected in permeabilization properties and intracellular peptidase release in cheese during ripening. In broth systems simulating the Cheddar manufacturing process, FCM data indicated that both strains had a high proportion of permeabilized subpopulations at the end of the cooking stage. However, in response to the temperature shift from 32 to 38.5°C during Cheddar manufacture the highly autolytic *L. lactis* AM2 strain contained a higher proportion of permeabilized cells. It was also shown that an increase in permeabilized cells was accompanied by an increase in the level of accessible intracellular enzyme activities. In juice extracted over ripening from 20 kg blocks of Cheddar cheese made with both strains, FCM detected subpopulations of intact, or permeabilized cells. Permeabilized cell numbers increased during ripening for both strains; at day 1, permeabilized cells accounted for 47% of total cells in *L. lactis* AM2 cheese juice compared with 13% in *L. lactis* HP cheese juice and at day 35, permeabilized cell types accounted for 95% of the total cells in *L. lactis* AM2 cheese juice compared with 49% in *L. lactis* HP cheese juice. Despite the increase in levels of permeabilized cells for both strains, an increase in accessible intracellular enzyme activity was observed only for the highly autolytic strain *L. lactis* AM2 (Sheehan *et al.*, 2005). This study highlighted the ability of FCM to rapidly identify strain related differences in permeabilization properties of lactococcal starters within product samples and relate these differences to intracellular enzyme release and flavour generation potential of these strains during cheese ripening. Doolan and Wilkinson (2009) also used FCM in combination with PI/Syto 9 staining to monitor the effects of various attenuation methods on cell permeability and accessibility of intracellular peptidases Pep X or Pep N in *L. lactis* cheese starter cultures. In this study, three *Lactococcus lactis* cheese starter cultures were exposed to various cell attenuation methods to release intracellular peptidases including sonication, chemical treatments, heat-shock and freeze-shock. Treated cells were subsequently assessed for changes in cell permeability and accessibility to the intracellular peptidases Pep X and Pep N. Generally, for all three strains, as the duration of sonication increased, the levels of intracellular peptidases in the cell free extract increased. For the other treatments, cell pellets of strains were prepared and treated with isopropyl alcohol (IPA), sodium dodecyl sulphate (SDS), hexadecyltrimethylammonium bromide (CTAB), heated over the range 45–65°C or freeze-shocked using liquid nitrogen. The effects of various treatments on cell permeability were assessed using FCM after staining with the Live/Dead® BacLight™ viability kit. Treatment of cell pellets with CTAB again generated the unique highly permeabilized subpopulation previously noted by Sheehan *et al.*, (2005) which had the highest levels of accessible intracellular enzyme activity. Treatment with SDS also resulted in highly permeabilized cells with enhanced accessibility to Pep X or Pep N. The findings in this study are significant for the artificial production of permeabilized *L. lactis* cell fractions having enhanced accessibility to intracellular enzymes for accelerated ripening of Cheddar cheese (Doolan and Wilkinson, 2009).

More recently, FCM has been applied to monitor and evaluate the addition of commercially available enzyme systems for the acceleration of Cheddar cheese ripening. In this study, three exogenous enzyme preparations were added either to the cheese milk or to cheese at the salting stage. In this instance, FCM was employed initially to monitor the partitioning of starter bacteria between the curd and whey during manufacture and was used thereafter to monitor the profile of live, dead or permeabilized/damaged cells present throughout ripening. Doolan *et al.* (2014) concluded that no significant differences could be observed between the numbers of live cells present in both the control and experimental cheeses. Pre-screening of the preparations had identified that an AHC50 peptidase preparation contained primarily permeabilized/dead cells which may have been generated as a result of an attenuation treatment (Kilcawley *et al.*, 2012). However in cheeses that had been treated with the AHC50 enzyme preparation (Commercial preparation from Danisco, Dange-Saint-Roman, France) (added with salt), FCM profiles indicated that levels of dead/permeabilized cells were significantly higher in the salted and pressed whey samples indicating that a majority of the added cells were lost immediately after addition representing an economic cost (Doolan *et al.*, 2014). FCM was also used to monitor the profile of the cells throughout ripening (to day 112) and to compare the levels of live, permeabilized or dead cells in the control and experimental cheeses. Significant differences were observed for numbers of live cells at day 112 in the cheese to which the preparation AM317 (Danisco, France) was added. This is an enzyme preparation derived primarily from *L. lactis* and was added with the cheese milk. Using FCM, the authors concluded that the addition of AM317 may have stimulated cell viability through carry-over of residual substrate (Kilcawley *et al.*, 2012).

Probiotic bacteria

Probiotics are described as 'live microorganisms, which when administered in adequate amounts confer a health benefit to the host' (FAO/WHO, 2001). In addition to being used in dairy fermentations, several LAB species are also used as probiotics. Viability at the point of consumption by the consumer is essential for their efficacy, therefore the accurate determination of viability is of fundamental importance for probiotics (Doherty *et al.*, 2010). In addition to their potential health enhancing effects, for any LAB strain to be selected as a probiotic, they must possess a degree of resistance to technological processing, survival during the shelf-life of the product with resistance to gastric and bile acids (Bunthof *et al.*, 2001a). FCM has been shown to be useful in evaluating the effect of relevant stress conditions on the viability of probiotic bacteria.

Three fluorescent probes cFDA, PI, and TOTO-1 were evaluated for their usefulness for discriminating live or dead cells of nine LAB species by FCM. Exponential-phase cells were used as positive controls while heat-killed cells (70°C for 10 minutes) were used as negative controls (Bunthof *et al.*, 2001a). FCM data indicated that cFDA successfully labelled live cells of all species, and did not label dead cells. However, cF fluorescence intensity differed among the LAB species when using the same standard protocol which may reflect strain related differences in cell permeability affecting the diffusion of cFDA, or strain-related differences in intracellular esterase activity or specificity. When evaluated as a counter stain for cFDA, TOTO-1 appeared to be better than PI at discriminating between intact or damaged cells. TOTO-1 is a yellow fluorescent dimeric cyanine dye and because it binds nucleic acids it is excluded by cells with an intact membrane (Bunthof *et al.*, 2001a). For some LAB species, peaks of PI-labelled cells and non-labelled cells in FCM fluorescence histograms were unresolved and some live LAB species exhibited labelling by PI. It was suggested that the larger molecular mass of TOTO-1, its high fluorescence enhancement, or a lower partitioning into the membrane may influence the trends noted by FCM (Bunthof *et al.*, 2001a).

In the same study (Bunthof *et al.*, 2001a) examined the effect of deconjugated bile salts or hydrochloric acid on the viability of three LAB species, *Lactococcus lactis*, *Lactobacillus helveticus*, and *Leuconostoc mesenteroides* using FCM and staining with cFDA followed by TOTO-1 counterstaining. FCM data were compared with plate count enumeration. Exposing cells to different concentrations of bile salts resulted in membrane damage which was also indicative of culturability. Decreases in populations recovered on plate counts were in agreement with decreases in cF labelled cells and increases in TOTO-1 labelled cells (Bunthof *et al.*, 2001a). Plate count data indicated that when the cells were exposed to hydrochloric acid solutions with pH values as low as 3.0 a decrease in culturability was not observed. However at pH 2.0, there was a dramatic decrease in survival. FCM data indicated that above pH 3.0 the cell membrane remained intact while at pH 2.0, membrane damage may have occurred.

FCM with cell sorting has been used to establish a direct relationship between an FCM viability assay for *Lactococcus lactis* cells using cF and TOTO-1 labelling with viable plate counts. Initially, the live (cF positive) and dead (TOTO-1 positive) regions were defined using mixtures of exponential-phase or heat-killed cells. Labelled or non-labelled subpopulations of mixtures of treated cells were then sorted and plated. When the mixtures of live/dead cells were stained with cFDA only, the sorted cF-labelled subpopulation had high culturability while the non-labelled subpopulation did not generate colonies on plates. When the mixtures were stained with TOTO-1, the labelled subpopulation was not culturable on plates, whereas the non-labelled subpopulation had a high culturability. The authors also suggest that as a result of stresses imposed during the cell sorting process ~20% of *Lactococcus lactis* cells did not recover and grow on solid agar plate counts (Bunthof *et al.*, 2001a). Geng *et al.* (2014) recently developed a rapid and specific analytical method for the enumeration of viable Bifidobacteria in dairy products containing a mix of lactic acid bacteria. These researchers used a polyclonal antibody against *Bifidobacterium animalis* subsp. *lactis* for specific detection by flow cytometry. The antibody used in combination with the viability probe 5,6-carboxyfluorescein diacetate allowed for a species-specific method enumerating viable cells within 2 hours compared to the standard ISO method which takes 72 hours. This particular study showed that FCM could be a rapid solution for probiotic analysis in dairy products (Geng *et al.*, 2014).

Sample preparation

Pure populations of LAB are easily detected by FCM when they are suspended in a broth or a buffer system. However, when 10% reconstituted milk (10% RSM) is used as the growth medium, distinct separation of bacterial cells can be problematic. Gunasekera *et al.* (2000) reported that the application of FCM to microbial analysis of milk is hampered by the presence of milk proteins and lipid particles that can non-specifically bind to fluorescent stains and thus interfere with staining and detection of bacteria. Gunasekera *et al.* (2000) further reported that the treatment of milk by centrifugation to remove lipids was insufficient for detection of bacteria and the most critical barrier for FCM analysis of milk appeared to be the presence of protein and lipid globules. These researchers developed a rapid (<1 hour) FCM assay for detection of bacteria in milk based on enzymatic treatment with either proteinase K or Savinase (a commercial protease) plus Trition X-100. Bunthof and Abee (2002) also reported an optimized milk clearing procedure for FCM analysis of bacteria in milk and investigated the viability of bacteria in dairy starters and probiotic products. In this study a commercially available (ENLITEN Kit, Promega) milk clearing solution was used for both extraction of bacteria and reduction in size of milk particles and was also optimized in respect of the signal to noise ratio. FCM enumerations were found to be accurate to a concentration of 10^5 cells per ml while the level of cells suspended in milk was high and viability appeared to be unaffected by the clearing treatment. Three functional subpopulations could be distinguished in milk: culturable cells, intact and metabolically active cells which were non culturable, and permeabilized cells. Plate counts for cleared samples of untreated cells showed a strong correlation with FCM data ($r > 0.99$).

Doherty *et al.* (2010) reported the use of a cell extraction and digestive pre-treatment to release *Lb. rhamnosus* GG from a dense protein matrix prior to enumeration by FCM. Encapsulated *Lb. rhamnosus* GG was released from the protein matrix by homogenization, followed by double digestion of the protein matrix using Proteinase K at a concentration of 12% (v/v) for 10 minutes at 37°C. Cell extraction and protein digestion of samples was performed at pH 7 in phosphate buffer. Protein clarification by digestion generated a 200-fold reduction in the sample debris and was reported to be a requirement to ensure good correlation between FCM and plate count data. It was also shown that working at pH 7 reduced protein adhesion to *Lb. rhamnosus* GG cells through negative repulsive charges. This extraction mechanism was claimed to remove cells encapsulated in the protein network without adversely affecting probiotic metabolism or viability.

Extraction of cells from a cheese can prove more difficult due to the inherent nature of the cheese matrix. During Cheddar cheese production removal of the whey leads to a compact casein matrix and also results in the fat globules becoming much closer together and appearing as separate entities within the cheese curd matrix. However FCM analysis of cheese has been achieved using two extraction methods: hydraulic pressure (32 MPa) to generate a cheese 'juice' (Sheehan *et al.*, 2005), or aqueous extraction of cells using 2% trisodium citrate to dissolve the protein, thereafter incorporating a refrigerated centrifugation step to remove supernatant fat and filtration of the defatted extract through a 35 μm filter. The latter method was used for the estimation of levels of subpopulations of live, permeabilized or dead cells during cheese manufacture and subsequently for analysis of cheese samples throughout ripening (Kilcawley *et al.*, 2012).

Conclusion

FCM has been successfully applied to the study of LAB fermentations enabling rapid and reliable information to be obtained regarding the physiological states of bacteria during fermentation when compared to the traditional microbiology methods. In comparison to traditional methods, FCM allows for simultaneous multi-parametric data beyond viability to be assessed during various stages of the fermentation process. FCM is a useful tool for identifying subpopulations of cells which are metabolically active and impact on key aspects of maturation such as intracellular enzyme release and flavour generation reactions. FCM also has the ability to provide real time cell count data of absolute cell numbers generated during the early stages of fermentations. Even before fermentation has commenced, FCM can be used to screen strains for their performance and hence enable assessment of suitability of novel strains for use within a particular fermentation process. It is evident from this chapter, that FCM has already been successfully applied to study LAB fermentations and offers the food and allied industries a novel process control method by which to monitor LAB performance.

References

Aguirre, M., and Collins, M.D. (1993). Lactic acid bacteria and human clinical infection. J. Appl. Microbiol. *75*, 95–107.

Amrane, A., and Prigent, Y. (1999). Differentiation of pH and free lactic acid effects on the various growth and production phases of *Lactobacillus helveticus*. J. Chem. Technol. Biotechnol. *74*, 33–40.

Ananta, E., Heinz, V., and Knorr, D. (2004). Assessment of high pressure induced damage on *Lactobacillus rhamnosus* GG by flow cytometry. Food Microbiol. *21*, 567–577.

Ananta, E., Volkert, M., and Knorr, D. (2005). Cellular injuries and storage stability of spray-dried Lactobacillus rhamnosus GG. Int. Dairy J. *15*, 399–409.

Bamforth, C.W. (2005). Food, Fermentation and Micro-organisms (John Wiley and Sons LTD, Oxford, UK).

Ben Amor, K., Breeuwer, P., Verbaarschot, P., Rombouts, F.M., Akkermans, A.D.L., De Vos, W.M., and Abee,

T. (2002). Multiparametric flow cytometry and cell sorting for the assessment of viable, injured, and dead bifidobacterium cells during bile salt stress. Appl. Environ. Microbiol. *68*, 5209–5216.

Breeuwer, P., and Abee, T. (2000). Assessment of viability of microorganisms employing fluorescence techniques. Int. J. Food Microbiol. *55*, 193–200.

Bunthof, C.J., and Abee, T. (2002). Development of a flow cytometric method to analyze subpopulations of bacteria in probiotic products and dairy starters. Appl. Environ. Microbiol. *68*, 2934–2942.

Bunthof, C.J., van den Braak, S., Breeuwer, P., Rombouts, F.M., and Abee, T. (1999). Rapid fluorescence assessment of the viability of stressed *Lactococcus lactis*. Appl. Environ. Microbiol. *65*, 3681–3689.

Bunthof, C.J., van den Braak, S., Breeuwer, P., Rombouts, F.M., and Abee, T. (2000). Fluorescence assessment of *Lactococcus lactis* viability. Int. J. Food Microbiol. *55*, 291–294.

Bunthof, C.J., Bloemen, K., Breeuwer, P., Rombouts, F.M., and Abee, T. (2001a). flow cytometric assessment of viability of lactic acid bacteria. Appl. Environ. Microbiol. *67*, 2326–2335.

Bunthof, C.J., van Schalkwijk, S., Meijer, W., Abee, T., and Hugenholtz, J. (2001b). Fluorescent method for monitoring cheese starter permeabilization and lysis. Appl. Environ. Microbiol. *67*, 4264–4271.

Cachon, R., Antérieux, P., and Diviès, C. (1998). The comparative behavior of *Lactococcus lactis* in free and immobilized culture processes. J. Biotechnol. *63*, 211–218.

Cogan, T.M., ed. (1996). Dairy Starter Cultures (VCH Publishers, Inc., New York, NY, USA).

Crow, V.L., Coolbear, T., Gopal, P.K., Martley, F.G., McKay, L.L., and Riepe, H. (1995). The role of autolysis of lactic acid bacteria in the ripening of cheese. Int. Dairy J. *5*, 855–875.

Diaper, J.P., and Edwards, C. (1994). The use of fluorogenic esters to detect viable bacteria by flow cytometry. J. Appl. Microbiol. *77*, 221–228.

Díaz, M., Herrero, M., García, L.A., and Quirós, C. (2010). Application of flow cytometry to industrial microbial bioprocesses. Biochem. Eng. J. *48*, 385–407.

Doherty, S.B., Wang, L., Ross, R.P., Stanton, C., Fitzgerald, G.F., and Brodkorb, A. (2010). Use of viability staining in combination with flow cytometry for rapid viability assessment of *Lactobacillus rhamnosus* GG in complex protein matrices. J. Microbiol. Methods *82*, 301–310.

Doolan, I.A., and Wilkinson, M.G. (2009). Comparison of the effects of various attenuation methods on cell permeability and accessibility of intracellular enzymes in Lactococcus lactis strains. I. Dairy J. *19*, 215–221.

Doolan, I.A., Nongonierma, A.B., Kilcawley, K.N., and Wilkinson, M.G. (2014). Partitioning of starter bacteria and added exogenous enzyme activities between curd and whey during Cheddar cheese manufacture. Int. Dairy J. *34*, 159–166.

Food and Agricultural Organization of the United Nations (FAO)/World Health Organization (WHO) (2001). Report of a joint FAO/WHO expert consultation of evaluation of heath and nutritional properties of probiotics in food including powder milk with live lactic acid bacteria. Cordoba, Argentina.

Geng, J., Chiron, C., and Combrisson, J. (2014). Rapid and specific enumeration of viable Bifidobacteria in dairy products based on flow cytometry technology: A proof of concept study. Int. Dairy J. *37*, 1–4.

Giraffa, G. (2004). Studying the dynamics of microbial populations during food fermentation. FEMS Microbiol. Rev. *28*, 251–260.

Graça da Silveira, M., Vitória San Romão, M., Loureiro-Dias, M.C., Rombouts, F.M., and Abee, T. (2002). Flow cytometric assessment of membrane integrity of ethanol-stressed *Oenococcus oeni* cells. Appl. Environ. Microbiol. *68*, 6087–6093.

Gunasekera, T.S., Attfield, P.V., and Veal, D.A. (2000). A flow cytometry method for rapid detection and enumeration of total bacteria in milk. Appl. Environ. Microbiol. *66*, 1228–1232.

Hansen, E.B. (2002). Commercial bacterial starter cultures for fermented foods of the future. Int. J. Food Microbiol. *78*, 119–131.

Holzapfel, W.H., and Schillinger, U. (2002). Introduction to pre- and probiotics. Food Res. Int. *35*, 109–116.

Kilcawley, K.N., Nongonierma, A.B., Hannon, J.A., Doolan, I.A., and Wilkinson, M.G. (2012). Evaluation of commercial enzyme systems to accelerate cheddar cheese ripening. Int. Dairy J. *26*, 50–57.

Lafon-Lafourcade, S., Carre, E., and Ribéreau-Gayon, P. (1983). Occurrence of lactic acid bacteria during the different stages of vinification and conservation of wines. Appl. Environ. Microbiol. *46*, 874–880.

Lahtinen, S.J., Ouwehand, A.C., Reinikainen, J.P., Korpela, J.M., Sandholm, J., and Salminen, S.J. (2006). Intrinsic properties of so-called dormant probiotic bacteria, determined by flow cytometric viability assays. Appl. Environ. Microbiol. 72, 5132–5134.

McGann, L.E. (1978). Differing actions of penetrating and non-penetrating cryoprotective agents. Cryobiology *15*, 382–390.

Mason, C.A., Hamer, G., and Bryers, J.D. (1986). The death and lysis of microorganisms in environmental processes. FEMS Microbiol. Lett. *39*, 373–401.

Michelsen, O., Cuesta-Dominguez, Á., Albrechtsen, B., and Jensen, P.R. (2007). Detection of bacteriophage-infected cells of *Lactococcus lactis* by using flow cytometry. Appl. Environ. Microbiol. 73, 7575–7581.

Nebe-von-Caron, G., and Badley, R.A. (1995). Viability assessment of bacteria in mixed populations using flow cytometry. J. Microsc. *179*, 55–66.

Niven, G.W., and Mulholland, F. (1998). Cell membrane integrity and lysis in *Lactococcus lactis*: the detection of a population of permeable cells in post-logarithmic phase cultures. J. Appl. Microbiol. *84*, 90–96.

Papadimitriou, K., Pratsinis, H., Nebe-von-Caron, G., Kletsas, D., and Tsakalidou, E. (2006). Rapid assessment of the physiological status of *Streptococcus macedonicus* by flow cytometry and fluorescence probes. Int. J. Food Microbiol. *111*, 197–205.

Pearce, L.E. (1969). Activity tests for cheese starter cultures. New Zealand J. Dairy Tech. *4*, 246–250.

Rault, A., Béal, C., Ghorbal, S., Ogier, J.-C., and Bouix, M. (2007). Multiparametric flow cytometry allows rapid assessment and comparison of lactic acid bacteria viability after freezing and during frozen storage. Cryobiology 55, 35–43.

Rault, A., Bouix, M., and Béal, C. (2008). Dynamic analysis of *Lactobacillus delbrueckii* subsp. *bulgaricus* CFL1 physiological characteristics during fermentation. Appl. Microbiol. Biotechnol. *81*, 559–570.

Rault, A., Bouix, M., and Beal, C. (2009). Fermentation pH influences the physiological-state dynamics of *Lactobacillus bulgaricus* CFL1 during pH-controlled culture. Appl. Environ. Microbiol. 75, 4374–4381.

Rault, A., Bouix, M., and Béal, C. (2010). Cryotolerance of *Lactobacillus delbrueckii* subsp. *bulgaricus* CFL1 is influenced by the physiological state during fermentation. Int. Dairy J. *20*, 792–799.

Rechinger, K.B., and Siegumfeldt, H. (2002). Rapid assessment of cell viability of *Lactobacillus delbrueckii* subsp. *bulgaricus* by measurement of intracellular pH in individual cells using fluorescence ratio imaging microscopy. Int. J. Food Microbiol. 75, 53–60.

Riis, S.B., Pedersen, H.M., Sørensen, N.K., and Jakobsen, M. (1995). Flow cytometry and acidification power test as rapid techniques for determination of the activity of starter cultures of *Lactobacillus delbrueckii* ssp. *bulgaricus*. Food Microbiol. *12*, 245–250.

Sheehan, A., O'Loughlin, C., O'Cuinn, G., FitzGerald, R.J., and Wilkinson, M.G. (2005). Cheddar cheese cooking temperature induces differential lactococcal cell permeabilization and autolytic responses as detected by flow cytometry: implications for intracellular enzyme accessibility. J. Appl. Microbiol. *99*, 1007–1018.

Sheehan, A., O 'Cuinn, G., FitzGerald, R.J., and Wilkinson, M.G. (2006). Proteolytic enzyme activities in Cheddar cheese juice made using lactococcal starters of differing autolytic properties. J. Appl. Microbiol. *100*, 893–901.

Siegumfeldt, H., Rechinger, K.B., and Jakobsen, M. (2000). Dynamic changes of intracellular pH in individual lactic acid bacterium cells in response to a rapid drop in extracellular pH. Appl. Environ. Microbiol. *66*, 2330–2335.

Streit, K., Ruegg, M., and Blanc, B. (1979). Influence of water activity on the growth of Lactis and propionic acid bacteria and the dependence on the solute added to the culture media. Milchwissenschaft, Milk Sci. Int. *34*, 459–462.

Sunny-Roberts, E.O., and Knorr, D. (2008). Evaluation of the response of *Lactobacillus rhamnosus* VTT E-97800 to sucrose-induced osmotic stress. Food Microbiol. 25, 183–189.

Volkert, M., Ananta, E., Luscher, C., and Knorr, D. (2008). Effect of air freezing, spray freezing, and pressure shift freezing on membrane integrity and viability of *Lactobacillus rhamnosus* GG. J. Food Eng. *87*, 532–540.

Wilkinson, M.G., Guinee, T.P., O'Callaghan, D.M., and Fox, P.F. (1994). Autolysis and proteolysis in different strains of starter bacteria during Cheddar cheese ripening. J. Dairy Res. *61*, 249–262.

Flow Cytometry for Rapid Microbiological Analysis of Drinking Water: From Science to Practice – An Unfinished Story

9

Thomas Egli and Stefan Kötzsch

Abstract

For many years routine microbiological analysis of drinking water has been dominated by plating methods. Established by Koch more than a century ago, the world-wide standards include both indicator organisms for faecal contamination as hygiene parameters (*Escherichia coli*, coliforms, enterococci), and the 'heterotrophic plate count' (HPC) as a general parameter of microbiological water quality. These methods are slow, requiring 1–10 days before data become available. In particular, the HPC has been criticized over the past 30 years as underestimating the real number of microbial cells by two or three orders of magnitude. Flow cytometry (FCM) offers a range of options to detect cells quickly, and we have recently explored the potential of a number of FCM-based methods for rapid (<1 hour) microbiological analysis of drinking and surface waters. These include total cell counting (TCC), the ratio of small/big cells (low nucleic acid (LNA)–high nucleic acid (HNA) ratio), community fingerprints, assessing cell viability and physiological state (all in 15 minutes), or detection and quantification of selected pathogens (in <2 hours). Determination of TCC and the LNA:HNA-ratio are now officially accepted and recommended parameters for freshwater analysis in Switzerland. Determination of the TCC by flow cytometry is already routinely used by a number of Swiss and Dutch drinking-water suppliers. FCM has been proven to allow fast and realistic assessment of water treatment processes and rapid quantification of microbial regrowth and bio(in)stability in drinking-water distribution networks and in public and private buildings. In addition, online monitoring of TCC has already been reported and will be commercially available shortly. The potential for applications of FCM-based methods in the field of drinking-water analysis is still far from being fully exploited.

Setting the scene

Established drinking-water microbiology and its methods

The relationship between consumers and producers of drinking water with microbes has always been – and still is – biased. On the one hand, to obtain good and safe drinking water, for probably more than 4000 years – first by experience, later based on scientific knowledge – mankind has relied on the purifying activity of microbes and filtering effect during passage through soil or sand (Baker and Taras, 1981), or deliberate application of microbiological processes, such as either rapid sand, slow sand, or granular activated carbon filters, and employing microbial cells attached to these particles to remove growth-supporting carbonaceous compounds, nitrate removal using denitrifying bacteria, or metal (iron, manganese) oxidation processes catalysed by microbial cells. On the other hand, since early times many different methods have been used to make water safe for consumption, including various techniques that eliminate, or at least inactivate, microbes, such as boiling, dipping hot pieces of copper or silver into water, exposure of water to sunlight, and cloth filtering. Nowadays, chlorination, ultraviolet (UV)C light, chemical oxidation processes and ultrafiltration are used for this purpose. Even today this bias is, to some extent, reflected in the two opposite strategies used to produce and distribute drinking water: Whereas one approach is to ensure 'sterility' of the water during distribution right to the consumer's tap by adding disinfectants, the other approach relies on using water that contains so few nutrients that no microbial regrowth can occur (i.e. it is 'biologically stable') during distribution. The latter can be achieved by either tapping 'microbiologically cleaned' water from groundwater resources, or by producing a high quality (groundwater-like) drinking water employing microbiological processes, such as slow sand filtration (van der Kooij *et al.*, 1999).

It was only in the late 17th century that microbes became 'visible' with the help of Antonie van Leeuwenhoek's microscopes (1675). Even though there were hints for a role of organisms invisible to the naked eye in the spread of diseases, it then took a further 200 years until the first specimen, the anthrax-causing *Bacillus anthracis*, was isolated in pure culture in 1876 by Robert Koch. At that time, isolation of pathogenic microbes was to a significant extent driven by the regular appearance of waterborne diseases, particularly cholera and typhoid. Both, *Salmonella* Typhii (by Eberth and Gaffky) and *Vibrio cholera* (by Koch and co-workers) were isolated in 1884. A year earlier, Koch had presented a cultivation method on solid media that allowed the assessment of, as he stated, 'the number of cultivable bacterial cells in soil, air and water' (Koch, 1883), which later became known as the 'heterotrophic plate count' (HPC) method.

Based on this information some of the principles for microbiological drinking-water quality assessment, including the essential methods required, were established. The cornerstones were: (i) that one important pathway of distribution of waterborne pathogens was due to contamination of water by faeces of infected persons or animals; (ii) that instead of the difficult isolation of bacterial pathogens one can use the harmless, easy-to-cultivate intestinal bacterium *E. coli* to indicate possible contamination of water with faeces, which is today known as the 'indicator concept' (Péré, 1891; Schardinger, 1892); and, (iii) that the number of heterotrophic, aerobic colony-forming microbes (heterotrophic plate count) is a general measure of water pollution, and, as a rule, 'good' water contains <100 colonies/ml, a number that can be achieved by soil passage or slow sand filtration (Koch, 1893; Frankland and Frankland, 1894).

These principles remained virtually unchanged for a century, with *E. coli* (0/100 ml) and HPC (<200–500/ml, depending on country) as the two central microbiological parameters. Although a number of other microbiological parameters have been suggested as additional or even better indicators of the hygienic state of drinking water (for an overview see Payment *et al.*, 2003; Köster *et al.*, 2003), routine testing still relies on plating for indicators, such as *E. coli*. In Switzerland, for example, enterococci (0/100 ml) are used in addition to *E. coli* (lately also recommended in the EU), and the HPC is also still required by law.

The 'method dilemma' in microbiological water analysis

The driving role of OECD and WHO

In the early 1990s the first author became involved in the activities of the OECD Working Party on Biotechnology as a Swiss Delegate, and issues of biological wastewater treatment and water reuse became an important part of the agenda (OECD, 1996). After joining forces with the WHO, and including hygiene and health-related points in the discussion, it was quickly realized that in routine water analysis the use of cultivation-dependent, slow plating methods was one of the critical points in restricting reaction time, and that faster methods were necessary in this area (OECD, 1999). In a common effort, the methodological issues concerning detection of microbes and viruses in drinking water were summarized and possible solutions were outlined (OECD/WHO, 2003). Discussions in the two international bodies can be boiled down to a few simple methodological questions, namely: which methods allow improved pathogen detection, what is the role and usefulness of the HPC method, and generally, which methods allow relevant results to be obtained much quicker.

With respect to pathogen detection, either immunological or nucleic-acid-based polymerase chain reaction (PCR) methods were favoured; other fast techniques, including flow cytometry and confocal laser-scanning immunofluorescence microscopy were considered to be both very tedious and too expensive (OECD/WHO, 2003).

Concerning the second microbiological parameter that had been essential for almost a century in routine water analysis, the HPC method, became down-rated and crossed off the list of hygiene-relevant parameters by the WHO, the EU, the US, and many other countries (Bartram *et al.*, 2003; WHO, 2008), with no real substitute proposed to fill the resulting gap. The HPC was now considered to be 'of little value as an index of pathogen presence but can be useful in operational monitoring as a treatment and disinfectant indicator, where the objective is to keep numbers as low as possible' (WHO, 2008).

So, microbiological analysis of fresh- and drinking water was left with counting of *E. coli* as the only generally accepted hygiene-related routine method, a number of suggested potential index and indicator organisms were listed as needing 'further research' (OECD, 2008), and the HPC stigmatized as being of no use as an index, and of little use as an indicator. All this with no substitute for HPC in sight! This dilemma is reflected, for example, in the regulations of the EU, where routine analysis of drinking water does not include HPC, but for drinking water in containers analysis for *E. coli*, enterococci, *Pseudomonas aeruginosa* (all 0/250 ml) and also HPC at 22 and 37°C (< 100/ml and <20/ml, respectively) is required (EU 98/83/EC).

The need for speed

The general issue that was addressed again and again was related to speed. At present, all legally established routine methods are cultivation-based. Although long-term experience has shown that they have mostly served their purpose, their disadvantages and occasional failure have also become obvious (Huq *et al.*, 2000; Payment *et al.*, 2003; Köster *et al.*, 2003; Figueras and Borrego, 2010). However, their main disadvantage is that all of them take considerable time until the result becomes available. The fact that routine microbiological analysis of drinking water requires between roughly one (*E. coli*) and 10 days (for some modifications of the HPC method) has generated calls for faster methods.

Molecular methods

Much hope was placed on molecular, usually nucleic-acid/PCR-based methods, for the fast detection of microbial pathogens and viruses. However, analytical chemistry-based methods for the identification and quantification of cellular components, as well as enzymatic activity-, antibody/immuno-based techniques, either on their own, or in combination with laser scanning fluorescence microscopy or flow cytometry (see below), have been proposed (for more details consult Table 8.6 in Köster *et al.*, 2003). Although many such non-cultivation-based methods have been extensively tested, none of them have been adopted for routine testing of drinking water (Girones *et al.*, 2010; Cabral, 2010; Figueras and Borrego, 2010). This is in contrast to the food industry and medical diagnosis where such methods are commonly used in routine testing of organisms of hygienic importance and pathogens (Emmadi *et al.*, 2011; Hoorfar, 2011; Dwivedi and Jaykus, 2011).

The challenges in drinking-water analysis originate from the fact that, in contrast to food and medical diagnosis, low numbers of target organisms have to be detected in a large volume of water, such that concentration steps are usually necessary. This severely restricts the number of potential methods (Medema *et al.*, 2003; Köster *et al.*, 2003).

Flow cytometry-based methods: out of the hospital into the aquatic environment

Over the last 30 years, flow cytometry (FCM)-based rapid detection and analysis of cells, after staining with fluorescent dyes or markers, has developed into a robust and widely applied technique in routine medical analysis, diagnosis and research. Many different options are available to stain both cellular components as well as biochemical activities of cells (Fig. 9.1) However, commercially available flow cytometers were originally designed for large eukaryotic cells, so that the detection of small 'starved' bacterial cells from environmental samples was difficult and usually

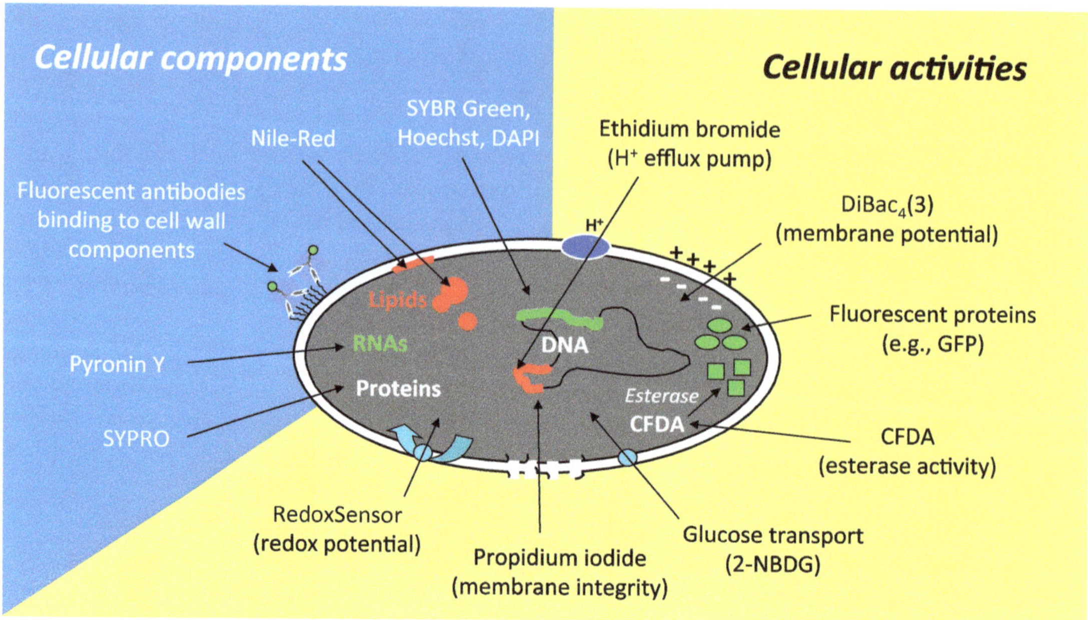

Figure 9.1 Some staining options used frequently in fluorescent microscopy and also regularly applied in flow cytometry (adapted from Kötzsch *et al.*, 2012).

required adjustments (Shapiro, 2003; Davey and Kell, 1996; Müller and Nebe-von-Caron, 2010).

The first studies on bacterial cells date back to the late 1970s and were focused on the one hand on determining growth, composition and physiological state in the laboratory, and on the other hand on detection of immuno-labelled pathogens from pure cultures (reviewed in Shapiro, 2003; Porter *et al.*, 1997; Gasol and del Giorgio, 2000; Vives-Rego *et al.*, 2000). However, advances made over the past 20 years with respect to the hardware, staining and software have continuously improved to allow the detection of ever smaller particles down to viruses (Weinbauer, 2004; Wang *et al.*, 2010).

The first application of FCM for quantifying and characterizing aquatic microbial consortia appear to be the reports by Don Button's group who studied cell size, filterability, DNA content and growth kinetics of marine and freshwater bacterial communities (Robertson and Button, 1989; Button and Robertson, 1989; Robertson *et al.*, 1998). Staining procedures from epifluorescence microscopy were often adapted to FCM, and from the outset results were compared with those obtained by direct microscopy counting or plating (Pinder *et al.*, 1990; Monfort and Baleux, 1992; Troussellier *et al.*, 1993; Porter *et al.*, 1995; Yamaguchi and Nasu, 1997). Numerous studies followed, including aspects of viability (dormancy, starvation, 'VBNC'), physiological state, enzyme activities, cellular constituents, toxicity, or community composition and dynamics (see reviews by Gasol and del Giorgio, 2000; Vives-Rego *et al.*, 2000; Bergquist *et al.*, 2009; Czechowska *et al.*, 2009; Hammes and Egli, 2010; Wang *et al.*, 2010). Hence, particularly in the field of marine microbiology, FCM is well established as an analytical technique and it is increasingly being applied to freshwater microbiology.

A very interesting observation originating from cytometric studies on bacterioplankton in marine waters was described by Button's group shortly after beginning to employ FCM in their microbiological studies (Button and Robertson, 1989; Robertson and Button, 1989; Robertson *et al.*, 1998). They reported the occurrence of two clusters of cells that had distinctly different properties with respect to their DNA-staining and light-scattering signal. One of the clusters contained cells that stained well and also appeared to be large, whereas the second cluster consisted of small cells staining only faintly with obviously little cellular DNA; today, the two fractions are referred to as 'high nucleic acid' (HNA) content and 'low nucleic acid' (LNA) content cells. Such clustering has been observed in most aquatic ecosystems, including fresh- and drinking water and represents to some extent a 'fingerprint' of the microbial community (Gasol and del Giorgio, 2000; Bouvier *et al.*, 2007; Hammes and Egli, 2010).

First attempts to use FCM in the drinking-water area

As far as the authors are aware, all initial attempts to employ FCM-based methods in the drinking-water area were focussed on pathogen detection. The first reports appeared in the early 1980s on the detection of *Legionella* cells stained with fluorescently labelled antibodies (Ingram *et al.*, 1982; Tyndall *et al.*, 1985). In subsequent years attempts were focused on the rapid detection of various pathogens using FCM (Phillips and Martin, 1988; Pinder and McClelland, 1994; Nexmann *et al.*, 1997). However, detection limits were not low enough to allow it to compete with other methods (Phillips and Martin, 1988) and, therefore, more attention was given to laser scanning fluorescence microscopy techniques where cells from a large volume can be concentrated on a filter for detection (e.g. Pyle *et al.*, 1999; Reynolds and Fricker, 1999; Baudart *et al.*, 2002; Lisle *et al.*, 2004).

Remarkably little attention was given to the quantification of the natural 'microbial flora' in raw and drinking water using FCM-based methods. Reports on its occurrence and behaviour starting from the used raw water, throughout the treatment train and during distribution of the finished water are largely lacking. This is rather surprising, not only because counting of bacterial cells in natural waters had already been established in ecological research (see above), but also in the light of Koch's effort to assess 'general' microbiological quality of drinking water by developing the HPC method. As far as the authors are aware, the few examples can be counted on the fingers of one hand. Developing a simple, micro-fluidic device, Sakamoto and colleagues reported fast detection of bacterial cells in potable water (Sakamoto *et al.*, 2005, 2007; Yamaguchi *et al.*, 2011), whereas Hoefel and his group used conventional FCM for enumeration of activity-stained bacterial cells in raw and chlorinated drinking water and compared data with the HPC method (Hoefel *et al.*, 2003). The latter research group later also enriched active cells by fluorescence-activated cell sorting (FACS) to characterize the composition of the microbial community through a water treatment train and the distribution system (Hoefel *et al.*, 2005a,b). Another example comes from Block and his group who investigated the effect of chlorination on natural bacteria in drinking water using a live/dead-staining procedure (Phe *et al.*, 2005).

The obvious lagging behind (not to say disinterest) of the research community in exploring FCM-based methods for their usefulness for the microbiological analysis of drinking-water systems is not easy to understand. Explanations and justifications for the verdict 'FCM is good for research, but not for practice' are manifold and include:

- that it is difficult to challenge cultivation-based procedures 'cast in stone' by legislation;
- that the focus of hygienic concepts was always on pathogens and indicator organisms, especially as the deficiencies of the HPC method were recognized and, consequently, was abandoned as a hygiene-relevant parameter (Bartram *et al.*, 2003);
- the fact that early attempts to detect pathogens using FCM failed because of insufficient sensitivity;
- the enthusiasm for molecular methods that had promised broad pathogen-screening and detection on chips for the near future;
- the 'belief' (supported by HPC data) that good groundwater-like drinking water is virtually sterile – even without adding a disinfectant – and contains little or no alive microbial cells;
- the concept of solving the problem by killing all microbes when adding a disinfectant residual;
- the idea that the system is microbiologically not very interesting because it is extremely oligotrophic with only a few and mostly uncultivable microbes;
- the general conception that FCM is much too expensive and is a piece of equipment that is demanding to operate and not sufficiently robust in practice;

- and last but not least, the rather conservative nature of the drinking-water legislators and practitioners relying on their experience 'that established methods had worked and provided safety for more than 100 years'.

Whatever the reasons were, it has now become obvious that flow cytometry offers numerous advantageous options for microbiological drinking-water analysis.

Developing FCM-based methods in the sheltered lab with an eye on application

The decision for taking the FCM option

In 2003, the first author was given the opportunity to set up a new drinking-water microbiology research group at Eawag. From the lively discussions within WHO, OECD and in the literature it seemed that a molecular, PCR-based chip technology was close to a breakthrough which would allow detection and quantification of target organisms, including many pathogenic microbes and viruses, simultaneously and quickly (Köster *et al.*, 2003). Hence, the 'research niche' regarding method development for the detection of pathogens in drinking water using molecular methods seemed rather busy and saturated. For this reason, it was decided not to go into (molecular) pathogen detection method development but, instead, to focus on two central but broad topics, namely (i) identification of the nature of damage occurring in microbial cells during disinfection measures, quantifying cell injury and investigating survival, repair and re-growth of injured cells, this – whenever possible – not only at the population but also at the single cell level; and (ii), to study the survival and growth ability of bacterial pathogens in the nutrient-poor drinking-water environment and their competitive behaviour within the naturally present microbial flora.

In the first area, Eawag was at that time strongly promoting 'solar disinfection' (SODIS) of drinking water, a simple method intended for use in remote areas of developing countries at a household level, where hygienically unsafe water is exposed in PET-bottles to intense sunlight for a day. Investigating the molecular mechanisms of damage occurring in bacterial cells during sunlight exposure appeared to be a particularly attractive research area, as little was known at that time with respect to the nature of damage brought about by solar radiation. At the same time there were several reports on transient inactivation and later re-growth of the indicator bacterium *E. coli* and enteric pathogens after exposure to sunlight. Therefore, total cell counting using flow cytometry, together with different viability stains and combined with classic methods such as plating (still the gold standard for assessing viability), or adenosine-tri-phosphate (ATP) content, were considered to give a better understanding of the processes occurring.

Regarding the second area, microbial growth and competition at low concentrations, quantification of available carbon, so called 'assimilable organic carbon' (AOC), and its influence on growth kinetics also appeared to be an interesting area, as AOC was reported to be the most important parameter in determining the extent of bacterial re-growth in non-chlorinated drinking-water distribution systems (van der Kooij *et al.*, 1982; van der Kooij, 1992).

For both areas, sensitive monitoring of growth and properties of individual cells at very low cell densities was required and was the real experimental challenge.

This prompted us to purchase our first flow cytometer, as this technique was reported to allow recording of cell numbers at low densities and to enable investigation of cellular activities and physiological states at the population as well as the single cell level, similar to, but faster than fluorescence microscopy methods.

First 'warming-up exercises' in FCM: exploring its potential, including some surprises

Initial results using established flow cytometry cell-counting protocols (Gasol and del Giorgio, 2003) demonstrated (or rather confirmed) that suspended, free-living bacterial cells were present in large numbers, typically in the range of 10^4–10^6 cells/ml, not only in all natural freshwaters and raw waters employed as sources of drinking water, but also in all tap-water samples and bottled still mineral waters tested (Fig. 9.2). The range of total cells detected using flow cytometry was particularly wide in groundwaters, where some, whenever sampled, contained only a few thousand cells/ml, whereas others reached as much as half a million (Kötzsch and Sinreich, 2014), probably depending on meteorological conditions, soil usage and other unknown factors. Cell concentrations in water samples collected from rivers and lakes considered 'clean' were usually one order of magnitude higher. Tap water distributed without disinfectant residual and bottled drinking water contained roughly one hundred thousand cells/ml; also, bottled still mineral water was in the same order of magnitude. We demonstrated that most of these cells were active but unable to grow on HPC media (Hammes *et al.*, 2008; Berney *et al.*, 2008). *Per se*, this was not new or particularly surprising as microscopy- and FCM-based investigations had already shown the presence of cell densities in this range for many aquatic, oligotrophic marine and freshwater systems (see literature cited in ZoBell and Grant; 1943; Gasol and del Giorgio, 2000; Staley and Konopka, 1985), as well as in chlorinated and untreated drinking water (Hoefel *et al.*, 2003, 2005a,b), and in bottled still mineral water (Leclerc and Moreau, 2002).

Learning by doing: developing an improved AOC assay

From earlier, in particular Dick van der Kooij's, work (van der Kooij *et al.*, 1982; van der Kooij, 1992; Morita, 1988,), it was known that in many oligotrophic natural systems, including high-quality freshwater and drinking water, (re-)growth of heterotrophic microbial flora is mostly limited by available carbon and energy sources (method-dependent referred to as 'assimilable organic carbon', AOC, or 'biologically degradable organic carbon', BDOC). At that time, the gold standard to quantify available AOC, developed by van der Kooij and co-workers, was based on quantifying regrowth of bacterial cells using the plate count method after inoculation of pasteurized water samples with pure cultures of (cultivable) strains isolated from freshwater. Although elegant, it was evident that a pure culture would be able to use only a small (and AOC composition-dependent) fraction of the total carbon that was actually present. Therefore, a number of attempts were made to modify and improve this method (LeChevallier *et al.*, 1993; Kaplan *et al.*, 1993; Escobar and Randall, 2000). When contacting scientists at a number of water works and research laboratories who used Dick van der Kooij's method, the general consensus was that this AOC assay was

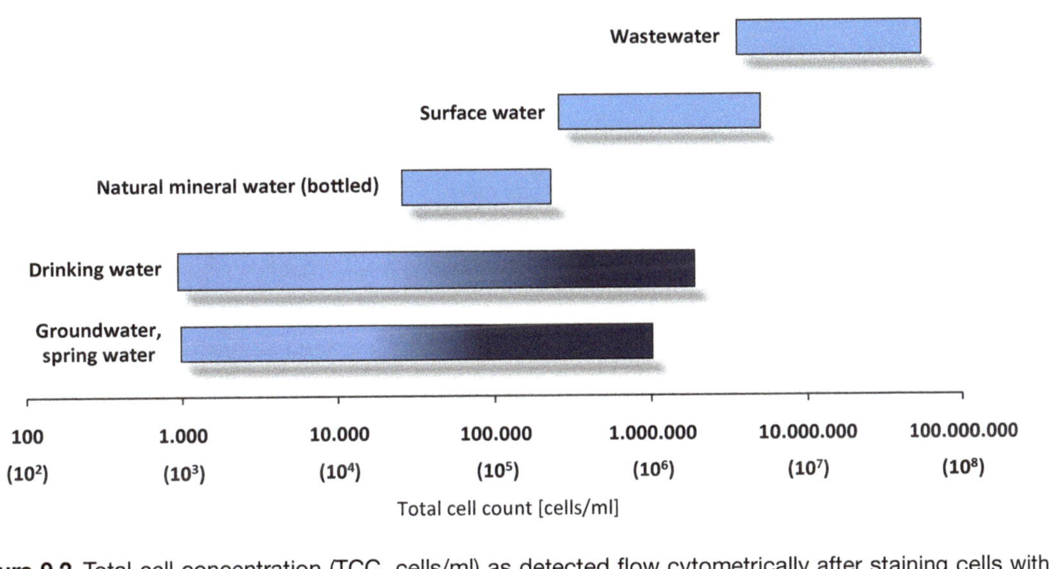

Figure 9.2 Total cell concentration (TCC, cells/ml) as detected flow cytometrically after staining cells with SGI and detection with a 488 nm blue laser. The dark-blue parts indicate unusually high (but observed) TCC concentrations in drinking, spring and groundwater (adapted from Kötzsch and Sinreich, 2014).

tedious and difficult to perform. Also, at that time our colleagues from the Zurich Waterworks (WVZ) made an attempt to establish the Dutch AOC assay in their laboratory, and, because of the difficulties experienced, they had great interest in a simpler method. Therefore, in lively exchanges with the practitioners from WVZ, we set out to improve the assay by employing natural microbial consortia instead of the recommended strains *Pseudomonas* P17, and/or *Aquaspirillum* NOX (van der Kooij *et al.*, 1982; van der Kooij, 1992, 2000), with the idea that microbial consortia 'ideally adapted' to growth in freshwater would produce a better indication of the AOC actually available in a sample. It was obvious that flow cytometric total cell-counting was very well suited to enumerate and quantify growth much quicker and with less effort than determination of cell number by plating.

When developing this modified AOC assay (Hammes and Egli, 2005) we made a number of (to us eye-opening) observations. For example, when preparing 'adapted/acclimatized' mixed inocula from various natural waters and storing them in the fridge for later inoculation of AOC assays with the same (or at least a similar) microbial flora, we observed a long persisting 'vitality' and 'longevity' of microbial cells present in all of the waters that we analysed. This fact was subsequently supported with various FCM-adapted 'viability' test procedures (Berney *et al.*, 2007, 2008). A natural microbial flora derived from surface or drinking water inoculated into heat-sterilized river water, grown to stationary phase and then stored in the refrigerator at 4°C for more than four months, lost virtually none of its growth potential. When re-inoculated into sterilized or pasteurized natural water and incubated at 30°C, these cells displayed hardly any observable lag phase and commenced growth almost instantaneously reaching doubling times in the range 1–3 hours with growth yields similar to those observed previously with the freshly prepared inocula (Hammes and Egli, 2005), i.e. from each µg of AOC approximately 1×10^7 cells were formed (van der Kooij *et al.*, 1982).

Even more surprising was that various pathogens, including *Vibrio cholerae* O1 Ogawa Eltor (Fig. 9.3), *Escherichia coli* O157, *Pseudomonas*

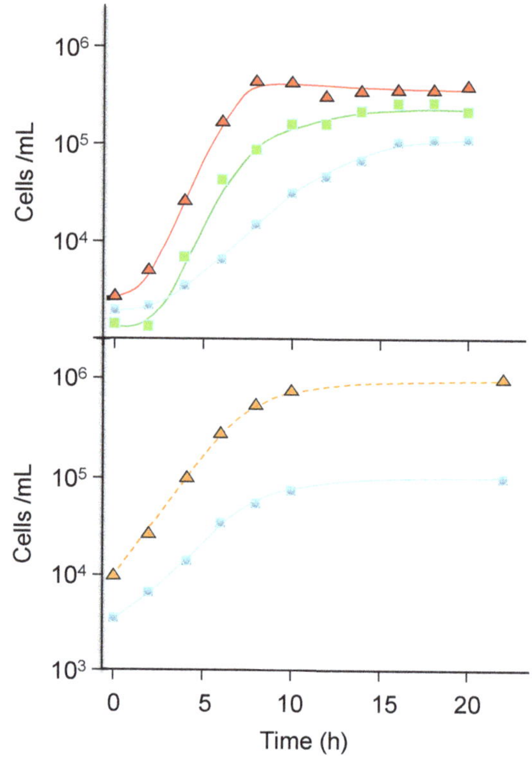

Figure 9.3 Growth of *Vibrio cholerae* O1 Eltor in sterilized and pre-filtered surface-water from eutrophic Swiss lake (Greifensee). A) Growth in original lake-water (blue dots) and in the same lake-water amended with 5 g/l of NaCl (green squares). For comparison, growth of the original microbial lake-water flora is shown (red triangles). The maximum specific growth rates reached in the exponential growth phase were approximately 0.40 $hour^{-1}$, 0.84 $hour^{-1}$and 0.84 $hour^{-1}$, respectively. B) Growth of *V. cholerae* in competition with the indigenous lake-water flora. At time 0 hours, the batch culture was inoculated with the same cell number of both microbial flora and *V. cholerae* (5×10^3 cells/ml). Mathematical modelling indicated that the maximum specific growth rates reached during competition were 0.34 $hour^{-1}$ and 0.50 $hour^{-1}$ for *V. cholerae* and the lake-water flora, respectively. Symbols: *V. cholerae* (blue dots), total cell concentration (yellow triangles). Lake-water was heat-sterilized and filtered (0.22 µm pore-size) before use. All experiments were performed at 30°C, and the initial AOC concentration was in the range of 300 µg/l (±10%). Cells were enumerated flow cytometrically after staining with SGI. In mixed culture experiments *V. cholerae* cells were immunolabelled with a fluorescent, surface-specific antibody. For more experimental details see Vital *et al.* (2007). Adapted from Vital *et al.* (2007).

aeruginosa, and *Listeria monocytogenes*, were also perfectly able to utilize a considerable part of the AOC present and to proliferate in waters from rivers and lakes (Vital, 2010; Vital *et al.*, 2007, 2008, 2010, 2012; Schneebeli, 2012). For example, as visualized in Fig. 9.3, *V. cholerae* grew surprisingly fast reaching doubling times in freshwater in the range from 20 to 45 minutes (with or without added NaCl, respectively). Even when competing with the indigenous lake-water flora it was able to grow, although at a slightly reduced rate, and withdraw AOC and other nutrients from the shared nutrient pool. These results give a first glimpse into the so far much ignored area of bacterial pathogen survival and growth in fresh- and brackish water, and demonstrated that much still has to be learned with respect to their abilities to survive, grow and compete with natural bacterial flora.

Another important, although not really surprising finding was that inocula from different natural waters did not always perform similarly with respect to growth rate and cell yield (Hammes and Egli, 2011). Also included in these tests were microbial floras from a number of commercially available still, natural (not disinfected or filtered) mineral waters. In particular, the flora present in 'Evian', originating from a natural groundwater source with a low DOC concentration and virtually no AOC (Servais *et al.*, 1989), performed very well and reproducibly. As this bottled water originates from a single source and has fairly global distribution, it seemed to us well suited as a standard inoculum that is available to everyone cheaply and without difficulty throughout the world. Indeed, the use of the natural flora present in Evian bottled mineral water as an inoculum for AOC (or for similar purposes) has since been reported by a number of colleagues working in the area.

Many of the details on the development of our AOC assay, control experiments performed, influence of various parameters, such as source and size of inocula, temperature, and comparison with other AOC assays, are compiled a recent report (Hammes and Egli, 2011; see also Hammes and Egli, 2005).

Subjecting practitioners to rapid cell counting (a personal account)

Convincing practitioners and legislators to try out a new and expensive technology

It is definitely much easier for a researcher than for a practitioner or a legislator to accept the (long-known) fact that HPC has always given incomplete information on drinking-water microbiology. To practitioners and legislators, this means a fundamental change in the perception of 'microbiology' in drinking water, which, according to the HPC method is virtually sterile. In Zürich, for example, HPC numbers for the water delivered are generally in the range of 0–20 CFU/ml. Although there has been a long-standing discussion on the usefulness and interpretation of the results obtained with the HPC method (Bartram *et al.*, 2003), it takes some courage to tell the customer that the real number is five to ten thousand times higher! The immediate questions that would be asked were obvious: is this dangerous, what is the role of these 'beasties', and do we really need them to be there? The established HPC method in combination with the search for indicator organisms and pathogens has obviously led to a very restricted (even distorted) view of the microbial presence and activity for all steps of drinking water production and distribution.

Initial discussions with several drinking-water professionals and legislators confirmed that it would be very difficult to replace the HPC method that has been used for more than a hundred years, by something completely new, and particularly by flow cytometry, a technique that was considered expensive, difficult to operate and not available in any water laboratory. However, after initial mouth-watering demonstrations to practitioners of the many possibilities offered by FCM, the new and additional information obtained with this instrument appeared to be very useful. Also it was obvious to them that FCM methods allow a much faster and more realistic assessment of the microbiological processes taking place in the various stages involved in drinking-water treatment and during distribution in the network. Particularly for countries such as Switzerland, Austria, the

Netherlands and Germany, where drinking water is commonly distributed without addition of a disinfectant residual (van der Kooij *et al.*, 1999), FCM-based methods, such as total cell counting, AOC and viability assessment, might provide valuable information on the biological stability of the water produced and its behaviour during distribution. With such a 'tool-box' on hand, drinking-water systems, from the resources to the consumer's tap, could suddenly become an accessible microbiological ecosystem, in which growth, survival and die-off processes can be realistically monitored.

From these discussions we also learned that in the 'tradition-oriented', established drinking-water community, new methods would have to be not only elegant and 'fool-proof', but also officially accepted and compulsory. Hence, to facilitate the acceptance of FCM-based methods by the Swiss authorities (i.e. the Federal Office of Public Health and the Swiss Gas and Water Board), the strategy adopted was to first convince practitioners to test FCM-based methods in their water treatment and supply systems, with the idea to shorten the demanding and time-consuming 'long march through the institutions'. We were fortunate that right from the very beginning our approach was supported by many Swiss practitioners.

A pilot plant to link the bridge from research to practice

During 2003/4, together with drinking-water chemists and engineers from Eawag, a large 'cross-cutting' project (Wave21) was initiated in close collaboration with the Zurich Waterworks (ZWV) and partners from the private, cantonal and governmental drinking-water sector, with a focus on alternative ways of producing drinking water from lake water. Wave21 resulted in a close cooperation with practice and enabled us to evaluate the feasibility of FCM-based methods in real drinking-water treatment and distribution systems of the city of Zürich.

With our Wave21 partners, a drinking-water production pilot plant was set up, including sequential ozonation, granular activated carbon (GAC) filtration and membrane ultrafiltration (separate, modular units that could be flexibly arranged), with the goal of studying the production of drinking water from lake water, resulting in similar quality to that achieved by the existing full scale lake-water treatment system operated by the ZWV (consisting of filtration, pre-ozonation, rapid-sand filtration, followed by ozonation, GAC filtration and slow sand filtration).

The pilot plant performance was monitored regularly using 'classical' as well as 'new' parameters, and TCC, HPC and ATP, determined at the same time and locations were perfectly stable after a three week period of steady-state operation. It took little time to realize that the impact of the three treatment steps on the microbiology in the pilot plant was accurately (and quickly) described by total cell counting (Hammes *et al.*, 2008): ozonation resulted in complete destruction of microbial cells and the generation of AOC (Hammes *et al.*, 2006, 2007); the AOC produced from the oxidative treatment supported considerable re-growth in the GAC filter that followed ozonation; membrane ultrafiltration again physically removed all the cells (Hammes *et al.*, 2008). Although HPC discerned a similar trend as that indicated by FCM-TCC, flow cytometric cell counts were always ca. 10 (after GAC filtration) to 10,000 times (raw water) higher than those obtained by plating, implying that the fraction of cultivable cells detected by the HPC was not constant. Surprisingly, total ATP concentrations did not reflect the cell number but this parameter was strongly influenced by ozonation, as ATP released from cell lysis was rather stable and all ATP detected in the effluent from this treatment step could be attributed to free, dissolved ATP (Hammes *et al.*, 2008, 2010a,b). Hence, it was essential to distinguish between cell-bound and free ATP. Also, after membrane ultrafiltration, free ATP was detected whereas neither cells nor cell-bound ATP was detected.

An important initial message of this explorative study was that, with respect to the microbiological state and performance of standard treatment steps used in drinking-water production trains, flow cytometry-based TCC is a simple parameter which supplies information much more rapidly than any other method. The essential results and knowledge obtained in this pilot plant study were confirmed in a number of subsequent investigations performed on full-scale drinking-water treatment systems by our, and other, research groups (e.g. Egli *et al.*, 2008; Siebel *et al.*, 2008; Hammes *et al.*, 2010a; Vital *et al.*, 2012; Lautenschlager *et al.*, 2013, 2014; Prest *et al.*, 2013; Liu *et al.*, 2013; Lipphaus *et al.*, 2014; Nescerecka *et al.*, 2014); this will be discussed in further detail later.

Lessons learned on both sides

The collaboration within Wave21 was fruitful for both sides. On the one hand, we, the researchers, got to know the most urgent microbiological problems in practice, and in this way we were forced to adapt FCM-based methods to practical needs; it also allowed us to test out a range of different FCM-based methods and to compare the results obtained with other techniques. On the other hand, practitioners were able to experience that the technique was not at all difficult to apply and understand, and produced results (although quite different from those obtained with the established HPC method) which allowed a quick visualization of microbiological events taking place during treatment and distribution. A very encouraging outcome of this collaboration was that our partners from ZWV were the first water company that decided to purchase their own FCM and to establish FCM-based methods in their laboratory. Soon afterwards we were afforded the opportunity to test our methods extensively again, this time at an international level within the EU project 'Techneau'.

Basics – counting cells, big and small

Total cell counting

These first years of working together with the water industry clearly demonstrated that counting microbial cells using flow cytometry was simple, very fast, robust and provided practitioners with an immense amount of meaningful data about their drinking-water production plant from source water through the whole treatment train and distribution network, right to the place of delivery. Hence, the method proved to be very well suited for fast monitoring and for obtaining a quick survey of the presence of microbial flora in such a system. However, the visual inspection of dot plots suggested that one might extract even more information than just cell numbers from the raw data obtained with this technique.

Determination of the total cell number in natural water samples based on the use of DNA-binding fluorescent stains with microscopy has already been used in microbial ecology for a long time (Jones and Simon, 1975; Hobbie *et al.*, 1977; Porter and Feig, 1980; Zimmermann, 1977). These or similar stains have also been applied for counting cells using either laser scanning microscopy (Reynolds *et al.*, 1997; McFeters *et al.*, 1999; Rhamalho *et al.*, 2001; Lepeuple *et al.*, 2004), or flow cytometry (Monfort and Baleux, 1992; Robertson and Button, 1989; Gasol and del Giorgio, 2000). Although FCM-based total cell counting had already been described for aquatic microbes (see earlier), the method required adaptation and optimization, as it had to provide reproducible results for the very small cells present in the 'starved' natural microbial consortia in raw and drinking waters. The same holds for assessment of 'viability' (see later).

In our laboratory we have used SYBR Green I (SGI) for the determination of the TCC and have adopted this strategy to pick up both the (major) green plus the (minor) red fluorescence signal from this dye (Hammes and Egli, 2005; Hammes *et al.*, 2008), instead of the frequently used combination of green fluorescence with a scatter signal. In our experience (and confirmed by round robin testing, see later), this usually leads to a better separation of the small bacterial cells from the background, as natural waters often contain small inorganic particles which interfere with the scatter signal from cells (Hammes and Egli, 2010). Accurate determination of TCC in a sample has turned out to be the central method for most applications in microbiological drinking-water analysis. This parameter proved to be not only very reliable but also able to respond to a range of known (and as yet unknown) factors. Usually, up-shifts in temperature, increased availability of AOC from external or internal sources, or stagnation leading to detectable growth of the indigenous microbial flora.

What you get for free: the ratio of LNA to HNA content cells

Staining cells present in both marine and freshwater samples with SGI or a similar nucleic acid-binding dye, typically reveals two, sometimes more, distinct cell clusters (Fig. 9.4); light scattering and fluorescence signals suggest that the two main clusters contain cells that are small and only weakly fluorescent, and cells that are larger and strongly fluorescing (Robertson and Button, 1989). Accordingly, they are often referred to as 'low nucleic acid' (LNA) content and 'high nucleic acid' (HNA) content cells (Gasol *et al.*, 1999; Bouvier *et al.*, 2007; Hammes and Egli, 2010). Although this must be considered a phenomenological observation, and the presently used terminology may be only temporary, this clustering can be employed as a type of 'fingerprint' of a water sample (discussed later). There has been a lively debate about the properties and

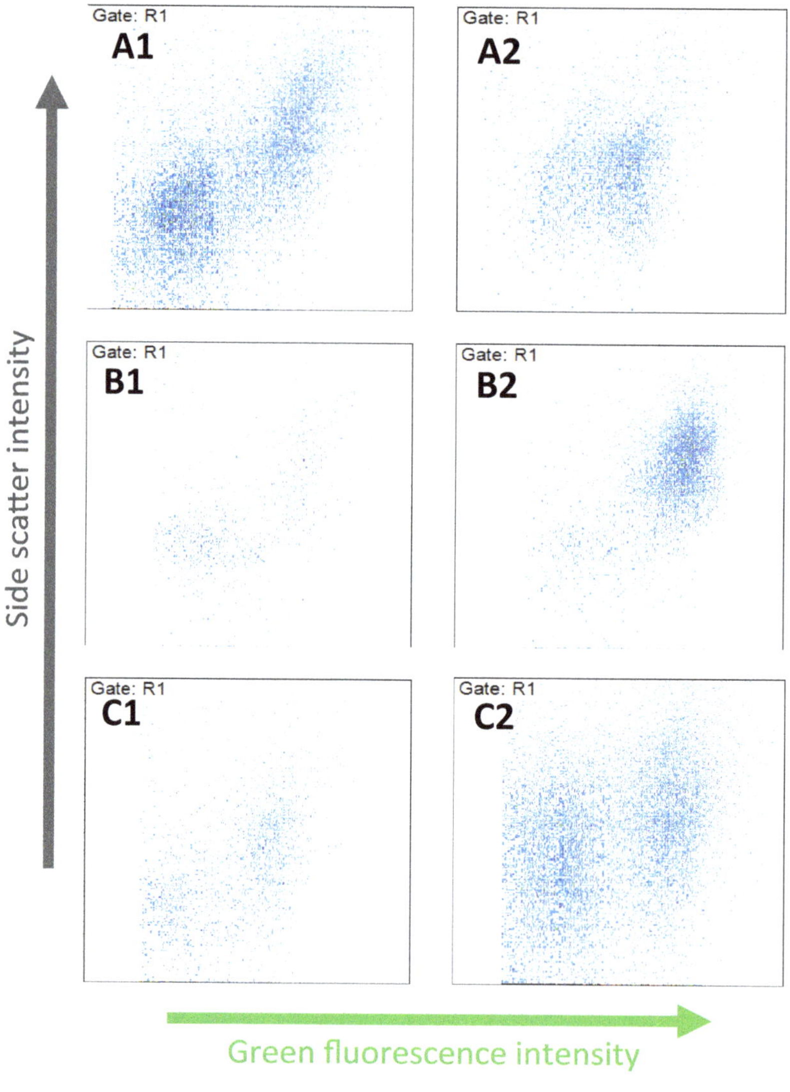

Figure 9.4 Dot-plot representation of fluorescent and side scatter signals of SGI-stained cells present in different natural waters ('fingerprints'). The clusters of so-called low nucleic acid (LNA) and high nucleic acid (HNA) content cells can be recognized. A1, river water; A2, pond water; B1, groundwater; B2, the same groundwater after stagnation and regrowth in an in-house distribution system; C1, spring water (dry period); C2, water from the same spring (after rain event).

viability of the cells within these clusters, particularly those exhibiting LNA properties, which were often proposed to be inactive (starting with Gasol *et al.*, 1995; for a summary see Hammes and Egli, 2010). However, it has recently been shown that both clusters contain active, cultivable cells (Bouvier *et al.*, 2007; Wang *et al.*, 2009).

The ratio of LNA to HNA content cells (for convenience abbreviated to 'LNA:HNA-ratio') can be employed to obtain additional, supporting information on the state of single cells and of communities in sea- and freshwater in general, and for drinking water monitoring in particular. For example, in most natural waters, including drinking water, the LNA:HNA-ratio changes whenever the water is affected in some way, be it from an increase in temperature, sudden availability of nutrients, or other changes; usually it shifts in favour of larger cells due to proliferation of the HNA cell fraction. Furthermore, LNA and HNA cells present in drinking water have been observed to react differently towards changing environmental conditions, chemical disinfection or sunlight (Berney *et al.*, 2007; Wang *et al.*, 2008a, 2009; Ramseier *et al.*, 2011a). Whereas natural waters mainly show several clusters in both the LNA and HNA regions, communities that experienced stagnation or re-growth exhibit one or several clusters of well-fluorescing and -scattering HNA content cells; pure (laboratory) cultures or communities, in which selection has occurred, exhibit one or two single intensive dots in the HNA region. Hence, this parameter combined with the TCC may be useful for water monitoring as shifts to HNA content cells indicate mostly (re)growth due to contamination or changes in temperature or other environmental parameters in the system. Shifts in quantity and quality of cells in defined clusters may be used for routine monitoring and as early warning systems, which indicate that 'something is happening' (or has happened) at a particular control point.

Obtaining an 'official' touch: standardization and validation of FCM-TCC method

The successful application of total cell counting by flow cytometry in practice, its central role, simplicity and usefulness in practical issues, paved the way in Switzerland for establishing FCM-TCC together with the LNA:HNA-ratio, as an alternative, accepted method for the determination of the general microbiological state of drinking water.

Standardization of the method was performed in 2012 under the supervision of the Swiss Gas and Water Board (SVGW). The 18 participating institutions (with 13 flow cytometers and 16 individuals actively taking part) in this standardization programme included major water works in Switzerland, cantonal water laboratories, water research institutes, the Federal Office of Health and several private water-testing laboratories and companies from the water industry in both Switzerland and Germany. During the first two round-robin test series, staining, fixation, dilution buffers and analysis conditions were optimized using a range of waters of different complexity with respect to background and cell clustering. For example, staining at elevated temperatures instead of the commonly used 'room temperature' was found to result in better focussed cell clusters and, hence, easier gating. During this procedure the participants became convinced that not only the TCC, but also the LNA:HNA-ratio was a useful parameter to be included in the new method, as it is obtained at the same time as a side product of total-cell-counting. Standardization of the staining- and gating-procedure allowed fairly reproducible results for this parameter to be extracted (despite the comments made earlier), even in this particular situation where data were collected not only by different personnel but also with different instruments (SLMB, 2012).

For SGI-staining, optimum results were obtained between 35 and 40°C; so, arbitrarily, standard staining conditions were set to 10 minutes at 37 ± 2°C. With this staining protocol, addition of ethylene-diamine tetra-acetate (EDTA) or other additives for improving membrane permeability, as required for certain pure bacterial cultures (Berney *et al.*, 2008), was not necessary. After this optimization the standard deviation (SD) of TCCs from different waters was consistently within the range of ± 10–15% (for 3 μm standard beads the SD was ± 5.7%), and an example is shown in Fig. 9.5. The largest contributions to error were due to individual gating strategies (usually LNA cells disappeared in the

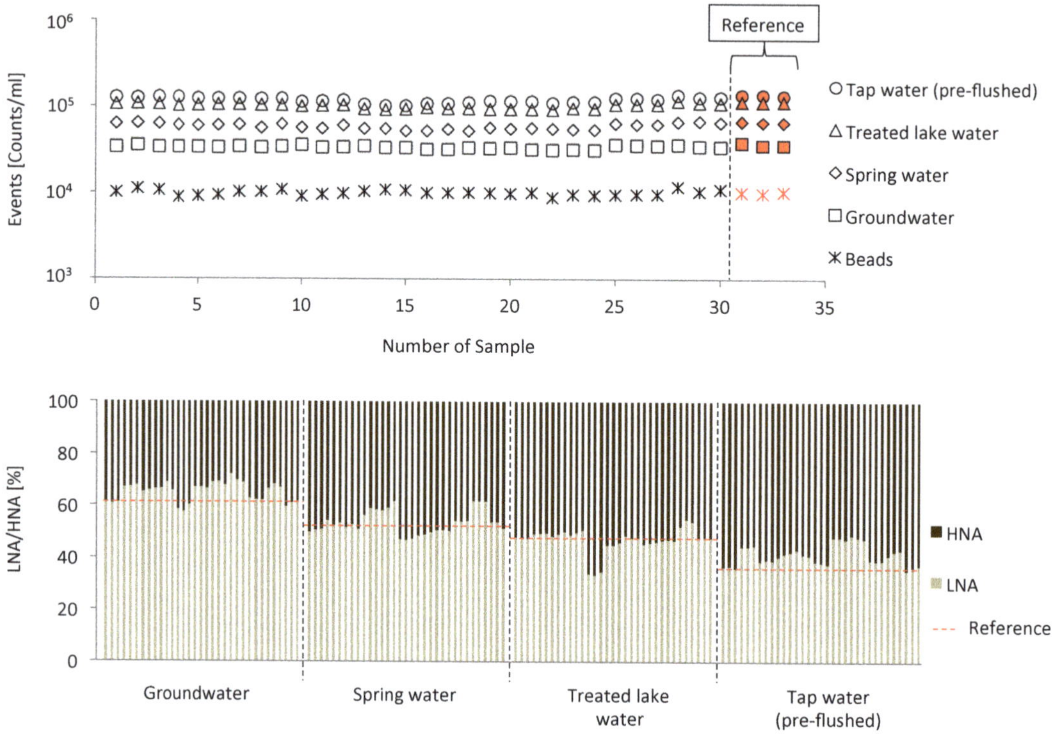

Figure 9.5 Standardization and validation study performed for the flow cytometric determination of the microbial cell number and the ratio of 'small': 'big' cells in freshwater (TCC, and LNA:HNA-ratio). Involved in the study were 11 different persons from eight institutions working on 13 different flow cytometers. Data set in the top panel shown was obtained for glutaraldehyde-fixed samples from four different water samples and standard beads. The RSDs were 8.08, 6.92, 7.97, 4.56, and 6.74% for tap water, treated lake water, spring water and ground water, and standard beads, respectively. Data set in the lower panel shows the results obtained for the %-age of LNA and HNA cells from the same water samples. RSDs for the %-age of LNA cells were 7.81, 13.68, 15.22 and 13.42, respectively. Adapted from SLMB (2012) with permission.

background) or instruments that were not well calibrated. For example, re-analysis of original data sets from the participants at our institute for three 'difficult' water samples resulted in a reduction of SD from ±20–30% to less than ±10%.

After completion of this standardization and validation procedure, the method was included as an alternative method for the determination of the total microbial cell number in drinking water in the Swiss book of foodstuffs (SLMB: Schweizerisches Lebensmittelbuch). A handbook for users that describes the basics of the FCM-TCC method is available (although to date only in German), including a number of useful applications in the drinking water sector and guides for practitioners to interpretation of data obtained with this technique. All of the participants hope that this will facilitate general acceptance and wide-spread application not only of the TCC but also of other FCM-based methods. Hence, the approach to introduce the new methods into legislation by first establishing them in practice and allowing this practical experience subsequently to convince the legislators appears to have worked well in Switzerland.

Although the 'Swiss book of foodstuffs' does not recommend or set a threshold or upper limit for the total microbial cell concentration in (drinking) water, the method can be applied in a comparative manner, i.e. in Switzerland drinking water is commonly distributed without or after treatment with 10,000 in groundwater to 100,000 cells/ml in treated surface water (compare Fig. 9.2). After stagnation in both the distribution and/or household water supply system it may increase to 200,000 to 300,000 cells/ml. Hence, drinking water from the tap that contains a (still 'harmless') microbial flora in the range of 500,000 cells/ml appears to be fairly close to the quality of a good surface water than to a biostable drinking water distributed through a well-operated piping system.

The issue of interpretation of FCM cell numbers and of the LNA–HNA ratio obtained with this new method will be discussed later in more detail. This discussion must be based on data obtained in practical case studies, combined with microbiological reasoning.

Fingerprints and other ways of data representation

Fingerprints

Two-dimensional dot-plotting of FCM data often results in clustering of cells, such as the LNA and HNA cell clusters discussed earlier. More complex clustering is often observed and, as the clustering pattern can be typical for a certain water, they are referred to as 'water fingerprints'. Such 'fingerprints' can be produced by combining various pairs of signals (originating from both fluorescence and scattering signals (Li *et al.*, 1995; Gasol *et al.*, 1999; Bouvier *et al.*, 2007; Sekar *et al.*, 2004; Günther *et al.*, 2009; Wang *et al.*, 2009; Müller and Nebe-von-Caron, 2010) and it should be pointed out that the clusters observed appear to depend on the stains and staining-conditions used, the instrument and its settings, and a number of other possible parameters not yet investigated in detail. Until these issues are properly addressed, results from different studies concerning molecular analyses of cells from certain clusters after cell-sorting (e.g. Günther *et al.*, 2009; Vila-Costa *et al.*, 2012) will remain difficult, if not impossible, to compare. However, for 'fingerprints' collected under identical conditions, cluster analyses can be employed for a rapid characterization of different types of water (see later).

How easy it is to obtain essential information from TCC and fingerprints is exemplified in Fig. 9.6 for two groundwaters that serve as a source of bottled drinking water (Wang *et al.*, 2008a). In this study, the effect of membrane filtration and storage in PET bottles for three weeks on re-growth was investigated using flow cytometry. Even simple visual inspection of the dot plots supplies much information on the character of the two groundwaters, the efficiency of the treatment and the effect of storage. Whereas GW1 contained only around 3000 cells/ml, the TCC concentration in GW2 was roughly one order of magnitude higher (35,000 cells/ml). Filtration through an industrial scale 0.22 μm pore-size membrane filtration unit leads to between 1 and 10% of the cells passing the filter (see also Wang *et al.*, 2007). Storage of the untreated water filled into sterile PET bottles resulted in re-growth of up to 180,000 (GW1) – 130,000 cells/ml (GW2). Surprisingly, the final levels of regrowth in the filtered water was not much different from that of unfiltered water (180,000 and 100,000 cells/ml for GW1 and GW2, respectively), indicating that filtration eliminated a large fraction of the cells, but not the growth-supporting AOC (which questions the use of filtration in the first place). The 'fingerprints' of the two unfiltered groundwaters after storage suggest a balanced re-growth of several clusters of cells primarily in the HNA-region, which was also the case for filtered GW2 (Fig. 9.6H). Storage of filtered GW1 in PET bottles, however, resulted in a thick, single HNA-cluster, which suggests that filtration of this high-quality GW resulted in a selection of 'membrane-passing' bacterial cells leading to regrowth of a virtually 'pure culture' (Fig. 9.6D). Microscopic inspection revealed the dominance of long-thin bacterial cells (*Hylemonella* spp.) in this sample, which are known to primarily pass membrane filters (Wang *et al.*, 2008b). This example illustrates the influence of filtration on the biological stability of water, and suggests that care has to be taken when applying this process to drinking water that is not intended for immediate use.

Pattern recognition, spider diagrams

The patterns of LNA or HNA cells are only one particular form of visualizing FCM data obtained after nucleic acid-staining of natural communities. Today, virtually all of the reported microbiological dot-plot FCM results rely on a visual judgement of clusters, which leaves much room for individual interpretation of the data. However, there are methods and software packages available for cluster analyses (Bakker-Schut *et al.*, 1993; Boddy

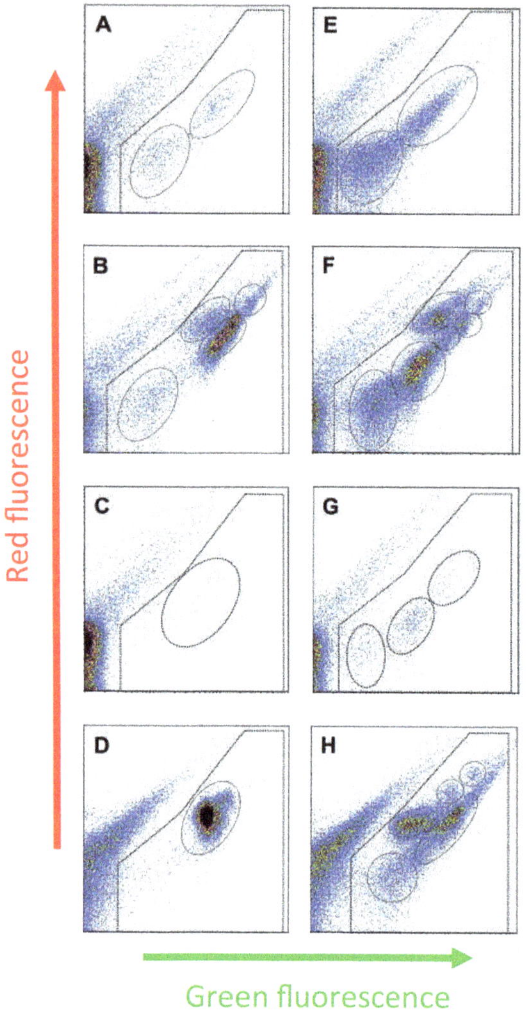

Figure 9.6 Microbial community 'fingerprints' of two different groundwaters before and after membrane-filtration and storage in PET bottles. The two groundwaters (GW1, GW2) were filtered through an industrial-scale 0.22 µm pore-size membrane cartridge, filled into sterile PET bottles and stored for 21 days at 25°C. Samples were stained with SGI. Oval gates visualize the different microbial cell clusters (arbitrarily defined), and microbial cells and background are separated by the dashed gate. The plots A, B, C, and D show 'fingerprints' for GW1, before filtration at day 0, unfiltered after storage, after filtration at day 0, and after filtration and storage, respectively. Plots E-H show the corresponding 'fingerprints' for GW2. Adapted from Wang *et al.* (2008a).

et al., 2001; Bashashti and Brinkman, 2009; De Roy *et al.*, 2012; Koch *et al.*, 2013, 2014), and automated analysis of FCM fingerprints would be particularly helpful for on-line monitoring of drinking water as reported recently (Hammes *et al.*, 2012; Besmer *et al.*, 2014). Automated pattern recognition could probably help to recognize even slight changes occurring in these clusters and, therefore, may be used to trigger early warning-systems. The authors are convinced that there is much yet to come in this area.

For the representation of fluorescence and scatter signals in combination with other cellular properties in net diagrams may be a way of characterizing different waters when more detailed information is not accessible (Fig. 9.7). This method of data visualization was successful in characterizing different surface and drinking waters and of ground- and spring-water sources with respect to their stability throughout the year.

More FCM-based methods with potential for routine monitoring

In addition to the flow cytometric determination of the TCC and LNA–HNA -ratio, both of which are now approved by the Swiss Federal Office of Public Health (SLMB 2012), several other FCM-based methods exist that are not yet ready for routine monitoring, and which are neither standardized nor validated. Despite this, some of them are occasionally employed in practice, or are offered by commercial analytical service laboratories. Here, five different areas will be presented, in which practitioners and regulators are showing great interest and that appear to be suited for routine monitoring. These areas include (i) the use of 'viability' and activity stains, (ii) online detection of cells, (iii) microbial (re)growth potential, (iv) pathogen growth potential and (v) rapid pathogen screening.

Alive or dead? The use of activity stains as a cultivation-independent, fast 'viability test kit'

After establishing the total number of cells in a water sample, one would obviously also like to know how many of them were active and potentially able to grow. As discussed above, this question cannot be answered with traditional plating, as the HPC method largely underestimates the number of viable cells (Bartram *et al.*, 2003).

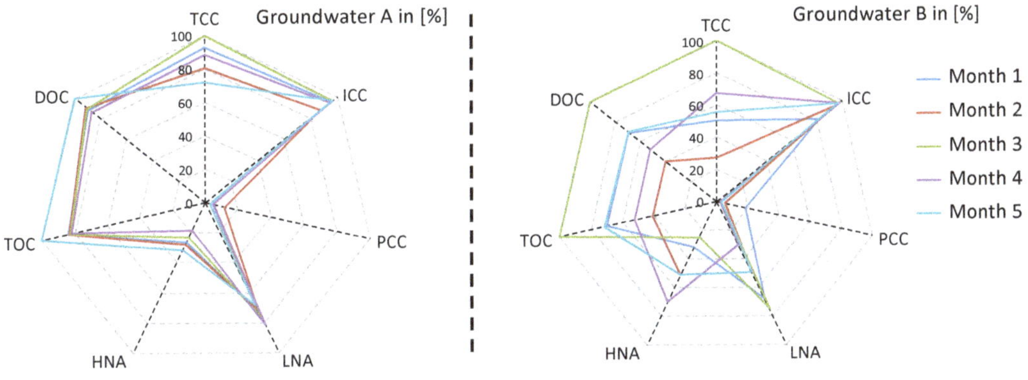

Figure 9.7 Spider diagram of two different groundwaters (A and B), showing the (in)stability during the late winter/spring/summer season 2010. Whereas groundwater A was perfectly stable, groundwater B exhibited considerable fluctuations with respect to chemical and microbiological parameters. Parameters: Total number of cells (total cell count, TCC); cells with intact membranes (intact cell concentration, ICC); cells with damaged membranes (permeabilized cell count, PCC); cells with low or high nucleic acid content (LNA or HNA cells); total organic carbon (TOC); dissolved organic carbon (DOC). Adapted from Kötzsch and Sinreich (2014) with permission.

A number of techniques have been proposed to address the issue of 'viability' and cultivability, however, none of them are able to give a definite answer to 'the central, simple question': is the very cell we observe active and able to grow and reproduce, maybe not right now but under appropriate conditions?

The majority of methods proposed only give a 'less biased' answer, and – at best – shift the question to an advanced level. Those coming very close are difficult and expensive to perform, require much time and are not suited for quick routine testing (e.g. the detection of cells that incorporate radioactively labelled precursors into cellular biomass or the presence of rapidly degrading RNA products (discussed in, for example, Davey, 2011; Fittipaldi *et al.*, 2012). Therefore, fast and cultivation-independent methods for assessing activities of individual cells that correlate with 'viability', e.g. after applying a disinfection procedure, would be very attractive for both drinking-water researchers and practitioners.

Many stains are available that allow assessment of membrane integrity, membrane potential, activity of efflux pumps, esterases, respiration and transport of fluorescent substrate analogues (Fig. 9.1), and a range of these fluorescent dyes were evaluated earlier for assessing different activities of microbial cells by FCM (Vives-Rego *et al.*, 2000; Joux and Lebaron, 2000; Gasol and del Giorgio, 2000; Hammes and Egli, 2010; Fittipaldi *et al.*, 2012). However, most of them were applied only to pure cultures and only few of them were employed to investigate activities of natural microbial flora. As a number of dyes are known to react in a strain-specific manner (discussed, e.g. in Joux and Lebaron, 2000; Hammes and Egli, 2010; Sträuber and Müller, 2010; and Müller and Nebe-von-Caron, 2010), one has to be very cautious when interpreting data obtained for undefined mixed consortia.

With a 'fast FCM-viability test kit' in mind, we set out to investigate different activity/viability stains for their usefulness for freshwater bacterial flora. Our selection criteria were based on previous studies (Berney *et al.*, 2006, 2007) and information from the literature (Hoefel *et al.*, 2003a,b; Phe *et al.*, 2005, Yamaguchi and Nasu, 2007, Pianetti *et al.*, 2005, Yokomaku *et al.*, 2000). To decide whether or not a specific stain provides realistic information on cellular activity/viability similar to some of the established tests, we used as benchmark parameters total cellular ATP content, standard plating, and FCM–TCC combined with the LNA:HNA ratio for estimating the active cellular biovolume (Hammes *et al.*, 2010b). After optimization of staining protocols for natural microbial drinking-water flora, good correlation was obtained between cellular ATP and the stains indicative of membrane integrity (propidium iodide: PI), membrane potential (DiBACC$_4$(3), and esterase activity (carboxyfluorescein diacetate: CFDA), whereas, very little correlation was obtained with both PCA- and R2A-plating (Berney *et al.*, 2008). The good correlation found for cellular ATP and the biovolume of membrane-intact cells (Hammes *et al.*, 2010b) suggested that resting and/or only slowly growing cells maintain on average in their cytoplasm a concentration of 2.95×10^{-9} nmol of ATP/μm^3 of biovolume.

For drinking water treatment and distribution 'viability' stains are of particular interest for a fast and cultivation-independent assessment of kinetics and extent of disinfection processes. In particular for the combination SYTO9/PI (the classic 'live/dead'-staining procedure) but also the SGI/PI a considerable quantity of data is now available in the literature (Berney *et al.*, 2007). A selection of examples is shown in Fig. 9.8. In some cases 'live/dead'-staining was also used in combination with classic, either enzyme activity or cultivation-dependent methods (Bosshard *et al.*, 2009, 2010a). The effects of oxidative chemicals, heat, solvents and sunlight in particular are easily detectable (Berney *et al.*, 2006, 2007, 2008; Bosshard *et al.*, 2009, 2010a,b; Ramseier *et al.*, 2011a; Bigoni *et al.*, 2012).

During exposure to many inactivating factors, a typical transient flow cytometric cellular staining-pattern was observed for selected Gram-negative bacterial indicator bacteria, pathogens (*E. coli*, *Shigella flexneri* and *Salmonella* Typhimurium (Berney *et al.*, 2007), as well as for microbial freshwater flora (Berney *et al.*, 2007; Ramseier *et al.*, 2011a): cells first took up more of the green-fluorescent 'TCC stain' and their green fluorescence intensity continuously increased as membrane damage began; only subsequently, when membranes were seriously damaged, did the cellular red fluorescence become more intensive due to more of the 'dead-stain' PI entering and displacing the 'live-stain'. As a result, cell clusters moved in the two-dimensional green/red-dot plot in a curve-shaped manner, from the 'live' position, through an intermediate high-green/high-red-fluorescing, to the final 'dead' position (Fig. 9.8). This pattern seems to due to the presence of an outer membrane in Gram-negative cells, as the Gram-positive bacterial strain tested (*Enterococcus faecalis*) moved from the 'live' directly to the 'dead' position (Berney *et al.*, 2007).

Unfortunately (but logically, when considering the target and nature of the damage), UVC disinfection is difficult to assess with this technique, for it results in patterns that are hard to interpret as the activity stains tested often responded only after high doses and in a delayed manner to cellular inactivation (assessed by plating).

Interestingly, for the microbial flora present in drinking water and groundwater, a distinct difference in response to chemical disinfection was observed between cells making up the LNA cluster and those clustering in the HNA region (Wang *et al.*, 2008a, 2009; Ramseier *et al.*, 2011a). To most disinfectants the 'big' cells in the HNA cluster appeared to be particularly sensitive, whereas members of the 'small' LNA cluster seemed to survive higher doses before their membrane integrity was affected. Clearly differing properties of 'small' and 'big' cell clusters were also observed in other studies, but results are difficult to compare and the reason for this is not yet known. It might well be necessary to isolate LNA-strains in pure culture, and to cultivate them to sufficiently high cell densities, to allow further biochemical investigations of this phenomenon.

A number of applications of microbial 'viability' assessment in real drinking water distribution systems and in buildings using 'live/dead'-staining techniques have been published recently (some of them will be discussed in more detail later) and more studies are certainly to come. For example, in a very recent study performed in the UK on water samples collected from either chlorinated or chloraminated drinking-water distribution systems it was found that the ratio of membrane-damaged versus intact cells correlated well with the levels of residual disinfectant (Gillespie *et al.*, 2014). The same group reported analogous results for drinking water installations in a public building and several private households (Lipphaus *et al.*, 2014). Also in a recent report on the susceptibility of biological water stability in the distribution network of a large European city, which is fed with chlorine-disinfected drinking water, FCM-'live/dead'-staining was successfully employed for

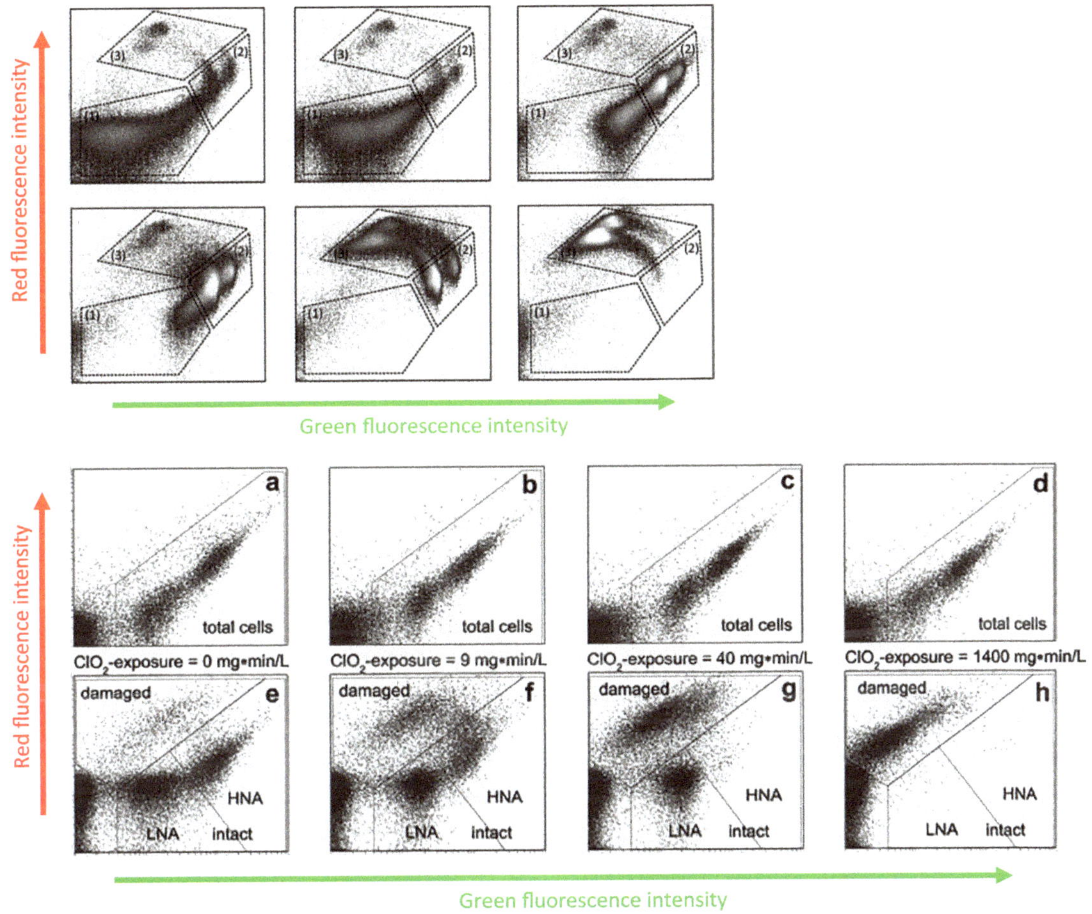

Figure 9.8 Disinfection of microbial cultures can be assessed rapidly by FCM using a combination of 'live/dead'-stains (e.g. SGI/PI, or Syto 9 (SY9)/PI). The top panel shows the effect of increasing doses of artificial UV-A light (from top left to bottom right) on membrane damage to a pure culture of *E. coli*. The three gates indicate (1) intact cells, (2) cells with compromised outer but intact cytoplasmic membranes, and (3) cells with outer and cytoplasmic membrane damaged (from Hammes *et al*., 2012). The lower panel demonstrates the effect of increasing exposure to chlorine dioxide (indicated as cxt-value) on membrane damage to a microbial freshwater flora. The top row (a-d) are dot-plots from SGI-stained cells for assessing the TCC, the bottom row (e-h) show dot-plots from SGI/PI-stained cultures. Note that the small LNA-cells appear to be less sensitive to chlorine dioxide than the larger HNA-cells (from Ramseier *et al*., 2011). In both examples, the cells pass an intermediate state of increased green fluorescence intensity before they move into the enhanced red fluorescence intensity gate.

quickly monitoring the microbiological state of the distributed water (Nescerecka *et al*., 2014).

These studies suggest that flow cytometric assessment of the ratio of membrane-damaged to intact cells may be an easy and fast way for (in the future maybe even online) assessment of the hygienic status of water in distribution networks. Furthermore, they indicate that this can be done not only at the chemical level by monitoring disinfectant residual concentrations, but – similar to FCM–TCC – soon also with online-assessment of microbiological 'viability' parameters.

Online detection of microbial cells

Present situation

Over the past few decades online and real-time monitoring of physico-chemical parameters have become indispensable components in routine monitoring of drinking-water quality at all levels of resource management, production and distribution (Storey *et al.*, 2011; Banna *et al.*, 2014). In particular pH, temperature, conductivity, DOT, TOC and chlorine, have become reliable, accurate and 'affordable' parameters for drinking-water management (Lee *et al.*, 2012), not only for monitoring levels and changes occurring, but also for controlling these parameters. The recent development of fast data transmission and computing systems has made it possible to place robust sensors at almost any place in drinking-water systems and to monitor system performance and processes virtually continuously. Today, such parameters are used frequently for online water quality monitoring and/or as early warning systems for indicating either occasional failures or intentional contamination.

Presently, no sensors are available for determining microbiological parameters online (not to mention real-time detection), early warning for biological risks is deduced from physico-chemical parameters, using (mostly only weakly established) correlations with microbiological data, and from experience. In the literature (Storey *et al.*, 2011; Banna *et al.*, 2014) prospects for the development and commercial availability of online methods for a direct detection of microbiological parameters seem to be judged as being gloomy; however, we see the future considerably brighter.

Approaches to online monitoring of (micro)biological parameters

To allow rapid and online detection of microbial cells, two different approaches have been followed in recent years. On the one hand, the direct detection of cells using optical methods is employed, either without or after staining the cells (Mogensen *et al.*, 2011). On the other hand, indirect detection is achieved by measuring the generation of a product from a specific enzymatic activity after substrate addition.

An example of the latter approach is the detection of β-D-glucuronidase and β-galactosidase activity, which is commonly used for detecting the presence of active cells of *Escherichia coli* and coliforms, the hygiene indicator organisms *per se* (Lendenfeld, 2008; Ryzinska-Paier *et al.*, 2014). A fully automatic and online version is already commercially available (mbOnline-COLIGUARD®). However, this method still requires at least 2–6 hours before a result is obtained (depending on target organism concentration and sampled volume); its sensitivity is sufficient for assessing contaminated surface waters, but it is not yet suited for frequent routine online monitoring of faecal contamination in drinking water.

Other devices with potential for online-detection include the commercially available 'Bactiquant-water' instrument (www.mycometer.com), which allows detection of a range of microbial cells in water based on hydrolase activity, and the interesting report on a hand-held device for *E. coli* detection (Wildeboer *et al.*, 2010) based on the well-known β-D-glucuronidase activity monitored with the fluorogenic MUG substrate. However, the detection limits for both methods are still rather high.

Online FCM

Flow cytometric online detection for cells of pure cultures of yeast and bacteria were first reported for biotechnological applications (Omann *et al.*, 1985, Lindberg *et al.*, 1992; Zaho *et al.*, 1999, Abu-Absi *et al.*, 1993). Also online monitoring of concentrations of autofluorescent phytoplankton and population shifts in surface water were described (Dubblear *et al.*, 1999; Thyssen *et al.*, 2008). The methods have improved over the years and today produce astonishing results (see e.g. Broger *et al.*, 2011; Brogneaux *et al.*, 2013). Even real-time monitoring of cellular properties and activities has been reported recently for selected parameters that do not need staining (Arnoldini *et al.*, 2013).

The first steps in online detection of microbial flora in drinking water were reported only recently (Hammes *et al.*, 2012; Helbing *et al.*, 2012; Besmer *et al.*, 2014). Hammes and co-workers linked-up in series units for automatic sampling, staining and a commercially available flow cytometer to obtain an online device that worked independently at a 5 minute resolution between subsequent samples and delivered results with a 10–15-minute delay. After optimization of staining conditions (Hammes *et al.*, 2012) a number of tests were performed in the laboratory, and the two parameters TCC and 'live/dead' were monitored in pure bacterial cultures and freshwater bacterial flora. The dynamic range for TCC was within 10^3 and 10^6 cells/ml, and it was easily possible to follow changes in TCC concentration (both slow and abrupt) as well as inactivation of an *E. coli* culture with UVA light using the 'live/dead'-staining procedure. The same online device was subsequently used to monitor the release of TCC from AC filters in the WAVE21 pilot plant (see earlier) following a routine cleaning (backwashing) procedure in order to assess the time necessary to re-establish pre-backwash conditions (Helbing *et al.*, 2012).

Very recently, online FCM data, collected over a period of approximately two weeks at 15-minute intervals, were published for the dynamics of TCC and LNA:HNA ratio in drinking water abstracted from a continuously running tap, and for a small creek that experienced two rain events during the measuring period (Besmer *et al.*, 2014). In the latter case, additional online data for physico-chemical parameters were included (T, EC, pH, DOT). The data document a diurnal cycling of the microbiological parameters together with some of the physico-chemical parameters. In this study, the best correlation was observed between TCC and electric conductivity. As total microbial cell number has not been monitored previously at such a high time resolution, interpretation of this novel set of data is still speculative and much is left to common sense interpretation. However, Besmer and colleagues certainly demonstrated that online and real-time (Arnoldini *et al.*, 2014) monitoring of microbiological parameters in surface and drinking water systems is possible over an extended period of time; they also give us a taste of what is to come when monitoring of TCC and maybe also microbial activity parameters will become routine parameters in environmental and drinking-water analysis.

Need for conceptually new FCMs for online monitoring in practice

So far, commercially available flow cytometers equipped with sampling- and staining-robots have been employed for online monitoring. However, in our opinion, instruments still have to be developed that allow a continuous, automatic online monitoring. Such machines have to work fully independently for months without human interference, and may be positioned in places that could be difficult to access for servicing on a regular basis; full electronic data exchange and control options are also compulsory, which automatically judge and/or signal critical situations where interference is required. This requires the instrument to be small and robust, self-sufficient, and yet affordable. Conventional flow cytometers produce large volumes of toxic waste due to the necessity to stain cells with mutagenic, DNA-binding fluorescent dyes and the sheath-fluid-based technique; therefore, they may have a fairly restricted range for application in practice.

However, it seems that routine monitoring of TCC and ratio of LNA:HNA content cells, as well as fingerprinting will soon be possible for water utilities at a reasonable price. Very recently, Kuenzi and colleagues reported the development of a novel type of sheath-fluid-less flow cytometer that is compact and portable, which will be self-sufficient and able to process up to 3000 samples without interruption. Stains, cleaning liquids as well as the produced waste will be contained in a cartridge that can be replaced by the user. A first demonstration flow cytometer was tested in a collaboration with the local water works (Kuenzi *et al.*, 2014), and data are shown in Fig. 9.9. The demonstration instrument was installed in a springwater collector (before chlorination) that delivers water for the city of Lausanne. During 17 days of uninterrupted, automatic operation without any human interference, the instrument collected over 350 individual samples at hourly intervals, processed them for TCC and LNA:HNA-ratio, and the results were sent immediately to the laboratory. Data from parallel samples, collected manually on working days and analysed in the laboratory with two commercially available standard flow cytometers for TCC and percentage of LNA cells, showed excellent agreement for both parameters with those obtained with the demonstration instrument. The field test fell into an extended dry-weather period, such that a surprising microbiological stability of this particular spring water was documented. During this dry period the concentration of cells gradually dropped initially from around 80,000 cells/ml to 50,000 cells/ml where it seemed to stabilize,

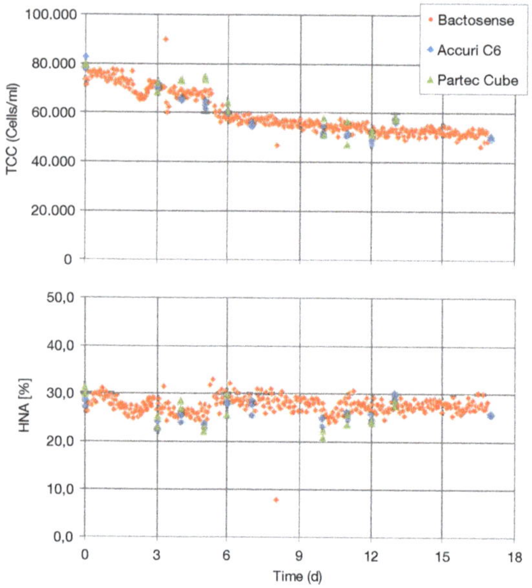

Figure 9.9 Total cell concentration (TCC) and percentage of LNA-content cells (red symbols) monitored online at 1 hour intervals with the BACTOSENSE demonstrator instrument, installed in a spring water collector (before chlorination) of the town of Lausanne (Switzerland). During the entire 17 day period the instrument worked independently without human interference. Samples for comparison with two commercial FCMs were collected manually and analysed in the local water laboratory (blue and green symbols). The continuous downwards shift of the TCC was probably due to an extended period of dry weather as it was also observed in the control samples analysed externally. Data courtesy of '*eauservice*' Lausanne, and S. Kuenzi, EPFL.

whereas the fraction of big HNA-cells remained constant at 26–29% of the TCC.

Hence, the authors are convinced that fully automated flow cytometric online-monitoring of TCC and the LNA:HNA-ratio, as approved in Switzerland (SLMB, 2012), will become feasible and affordable in the very near future. FCM-based 'fingerprints' and the proportion of membrane intact/compromised ('live/dead') cells will be the most likely parameters to follow in due course. Such instruments will enable practitioners for the first time to monitor online microbiological parameters at all levels of drinking-water production and distribution, virtually from the catchment to the consumer's tap. In combination with online physico-chemical parameters it will allow monitoring, optimization and control of the various stages of drinking-water production, from the catchment, through the treatment train and in the distribution network, which should result in savings of both resources and time. We anticipate that this might be the start of a new area of early warning concepts and water safety plans, that are no longer restricted to online physico-chemical parameters (and 'in retrospective' data from traditional plating).

Microbial (re)growth potential (MGP)

Faced with the usually 'annoying' presence of suspended cells or biofilms on surfaces in drinking-water networks, it is obvious that the factors which allow (*vice versa*, restrict) microbial (re) growth in such apparently nutrient-poor environments and which deter us from achieving one of the primary goals of water treatment, namely 'biostability' (van der Kooij *et al.*, 1999) should be considered.

Taking a pragmatic 'practitioner's' approach to this question (and leaving aside all the basic considerations by microbial ecologists of which nutrient might restrict growth of the different classes of microbes; see, for example, Morita, 1988), we can try to quantify the potential extent of growth in a water of interest by collecting a sample, incubating it in a batch assay for a certain time under defined conditions, and determining the cells present before and after incubation (note: without addition of inorganic nutrients, such as P, N, and Fe, to ensure carbon-limited growth as, e.g. done in an AOC assay). FCM is evidently a convenient method for counting cells in such a simplified AOC assay (Hammes and Egli, 2005). Of course, care has to be taken that no nutrients are introduced during handling or released from the surface of the test vessel, that grazing by protozoa is avoided, sufficient oxygen is available, and so on. Although such a microbial growth potential test would require some days of incubation, one might perhaps be able to discriminate water that is very unstable from water that does not allow much re-growth during stagnation at elevated temperature. The fact that the water sample is neither filtered nor otherwise treated has the advantage of using as an 'inoculum' the very microbial flora that is actually present in the water. In some way, such a procedure can be considered as mimicking the stagnation of drinking water in a communal network or an in-house distribution system.

To date, the authors are aware of little data in the open literature that were collected under conditions that come close to such an MGP test. The study that comes closest is probably the one already referred to earlier (Wang *et al.*, 2008a), in which the influence of filtration, bottling and storage on microbial re-growth was investigated. The water samples collected from the two groundwater aquifers (GW1, GW2), filled into X-ray-sterilized PET (or glass) bottles and stored for 21 days at 25°C, exhibited roughly the same MGPs when considering the final cell concentrations reached (1.3 to 1.8×10^5 cells/ml). However, the MGPs expressed as the -fold increase (TCC after incubation to initial TCC) spanned an enormous range from 3- to 60-fold. Generally, the samples from GW1 (TCC $\approx3\times10^3$ cells/ml) exhibited a 20- to 60-fold MGP, whereas the MGP in GW2 samples (initial TCC $\approx3\times10^4$ cells/ml) was approximately 3- to 4-fold only. This suggests that for the incubation conditions tested the concentration of nutrient(s) restricting growth under the test conditions was roughly the same in both aquifers, but the niche, for whatever reasons, was only poorly occupied in GW1.

Microbial growth in communal water distribution systems is common in both drinking water with or without disinfectant residual. In one of the coming sections we will discuss published microbial re-growth (or 'biostability') data from communal water networks and drinking-water distribution systems in buildings; although most of the data were not obtained under well-defined conditions, they reflect the practical side and implications of the MGP test described in this section.

Pathogen growth potential (PGP)

In today's approaches to microbial risk assessment, growth of pathogenic microbes in drinking water is only vaguely considered (OECD, 2001; OECD/WHO, 2003; Signor and Ashbolt, 2006; Smeets *et al.*, 2010). This can be partly traced back to the limited knowledge regarding the factors governing survival and growth of pathogens in natural and man-made environments. For a long time pathogens were considered as 'model copiotrophs' and unable to survive in low nutrient environments (Egli, 2012; Vital *et al.*, 2012a,b). However, numerous reports also exist that document the growth of pathogenic bacteria in water used for personal hygiene or for drinking (see references in Vital *et al.*, 2012a,b). Thus, a better understanding of the growth potential of pathogenic bacteria in aquatic oligotrophic environments is required not only for a holistic approach to quantitative microbial risk assessment (QMRA), but also to improve treatment, storage and distribution of drinking water.

Obviously, FCM-based cell counting, either on its own or in combination with fluorescent immuno-labelling of pathogens, is ideally suited for assessing the growth behaviour of pathogens under environmentally relevant conditions, e.g. when present as a contamination in surface water. The simplest form of such experiments is the direct evaluation of a batch growth cycle of pure culture of a pathogen after inoculation into a sterilized water sample (Vital *et al.*, 2010), which is in essence a modification of the AOC assay developed in our laboratory (Hammes and Egli, 2005). When labelling target cells with dye-linked antibodies, it is also possible to investigate their ability to survive and grow in the presence of competing strains and/or with natural microbial flora. As concentrations of pathogens in such experiments are usually fairly high, there is no need for pre-enrichment of the target cells.

Examples of such PGP experiments have been reported by Vital and colleagues (Vital *et al.*, 2010); they tested growth of *E. coli* O157, *Vibrio cholerae* and *Pseudomonas aeruginosa* in (pasteurized and membrane-filtered) water samples obtained from different treatment steps in drinking and wastewater treatment facilities, and from ozone-treated river water. Surprisingly, in these experiments, PGP did not correlate at all with DOC, and only weak correlation was found with AOC; nevertheless, the results suggest (expectedly) that PGP was dependent on both composition and concentration of the available AOC. Unexpectedly, *Vibrio cholerae* exhibited a behaviour very different from the two other pathogens; whereas the AOC generated from the ozonation treatment steps stimulated PGP of *P.*

aeruginosa and *E. coli* O157, *V. cholerae* appeared to be unable to grow with this oxidized form of carbonaceous compounds. Also, the importance of the final polishing treatment steps, such as slow sand filtration, was clearly demonstrated, as this treatment tremendously reduced the PGP of all tested strains (Vital *et al.*, 2010). Similar experiments had earlier shown that both *V. cholerae* O1 Ogawa Eltor (Vital *et al.*, 2007) and *E. coli* O157 (Vital *et al.*, 2008) are perfectly able to grow in sterile freshwater, and *V. cholerae* was even able to draw from the common carbon pool and grow when in competition with natural microbial freshwater flora (Vital *et al.*, 2007).

Hence, including a PGP assay in judging the quality of the drinking water produced might be an important step in minimizing the risk for pathogen regrowth in drinking water during storage and distribution.

Rapid pathogen screening

Present situation

For routine analysis the monitoring strategy presently practiced to ensure the hygienic safety of drinking water relies heavily on Koch's indicator concept, and this applies to much of the water used for human activities. Screening for pathogens in drinking water is usually done only when either the presence of indicator organisms above a set threshold concentration suggests closer inspection, or in the case of outbreaks. Although a number of reasons support the presently practiced indicator-organism based monitoring strategy, we are occasionally made painfully aware of the fact that an absence of 'indicators' does not necessarily indicate microbiological safety (a summary of model failures can be found in OECD/WHO, 2003). Therefore, the direct search for the presence of microbial and viral pathogens in drinking water has always been on the wish-list of practitioners and regulators, be it either with target-based, or broader, non-target-directed, screening methods. Only recently Germany was the first country to introduce the compulsory routine checking for a particular pathogen, which may stimulate other countries to introduce similar regulations. In Germany, hot-water boilers in buildings containing more than two apartments now have to be checked regularly for the presence of *Legionella* (Bundesgesetzblatt, 2011). Also, humanitarian intervention units in disaster situations would greatly benefit from mobile, rapid pathogen screening methods.

For rapid pathogen-screening reliable and cheap detection methods are required, which are both, sensitive and quantitative. Nucleic acid-and/or antibody-based methods offer a number of exciting options (Köster *et al.*, 2003; OECD/WHO, 2003; Girones *et al.*, 2010; Zhao *et al.*, 2012; Fittipaldi *et al.*, 2012); also an enzyme-activity based method for online detection of indicator organisms (*E. coli*, coliforms) within approximately 6 hours has been reported and is commercially available (Stadler *et al.*, 2013; mbonline, 2014; Ryzinksa-Paier *et al.*, 2014). Nevertheless, a fast, direct microscopic observation of individual target cells after 'ear-marking', if possible in combination with activity stains, would be particularly attractive for practitioners. This option is offered by fluorescence microscopy (FM) and flow cytometry and relies on immuno-based detection of pathogens. For example, confocal laser scanning FM is already widely used in food microbiology and medical diagnostics (Dwivedi and Jaykus, 2011; Hoorfar, 2011). The method can also be performed on drinking water, however, only after appropriate enrichment of 'target particles', which is usually achieved by membrane filtration.

FCM-based methods

The first attempts to employ flow cytometry for the detection of pathogenic microbes goes back to the early 1980s (Ingram *et al.*, 1982; Tyndall *et al.*, 1985), however, at that time, detection limits were not sufficiently low (Phillips and Martin, 1988). Some two decades later, successful rapid detection of *Legionella* using solid-phase cytometry (Aurell *et al.*, 2004; Delgado-Viscogliosi *et al.*, 2005) was reported.

A few years ago we started to explore flow cytometry-based methods for rapid enumeration of microbial pathogens in drinking water. Combining filtration and immunomagnetic-bead separation of fluorescent antibody-labelled cells with flow cytometric detection, as depicted in Fig. 9.10, resulted in a number or protocols for rapid

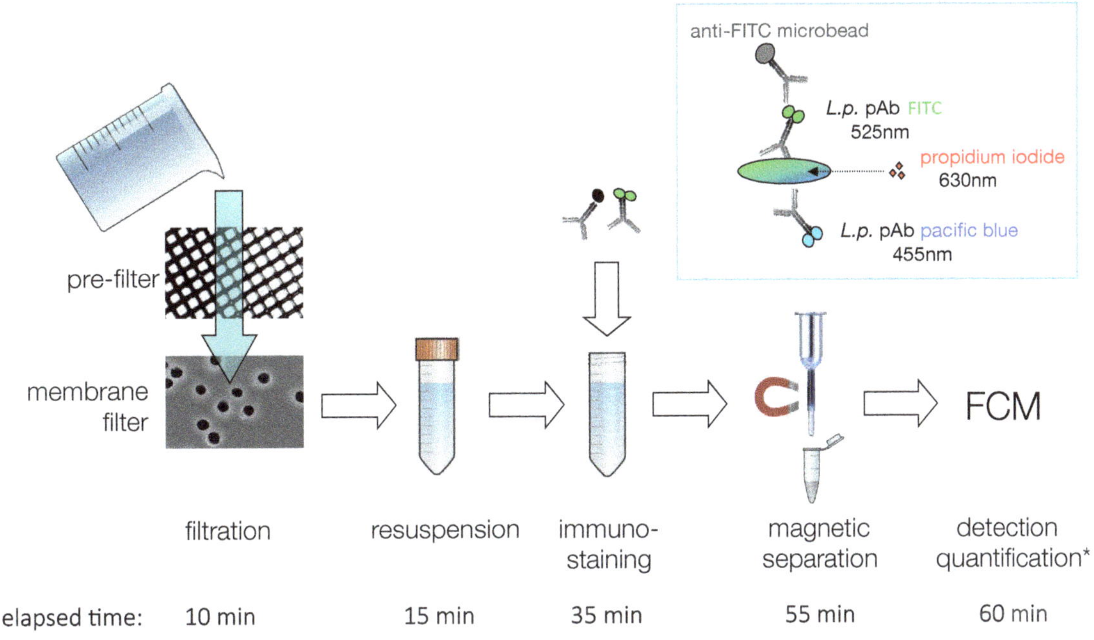

Figure 9.10 Flow sheet for a rapid screening method for pathogens using enrichment on membrane filters, detachment and concentration in a small volume of buffer, labelling of the target organism with immunomagnetic separation and detection in the flow cytometer. In the inset, the labelling concept for *Legionella pneumophila* serogroups 1–12 is depicted. By courtesy of H.A. Keserue, modified from Keserue *et al*. (2012), and Keserue and Egli (2012).

detection and enumeration of the pathogens *Giardia lamblia* and *Cryptosporidium* (oo)cysts, *Legionella pneumophila, Vibrio cholerae, E. coli* O157 and *Listeria monocytogenes* in freshwater and drinking water within 1 to 3 hours (Füchslin *et al.*, 2010, 2012; Keserue *et al.*, 2011; 2012b; Vital *et al.*, 2008; Schneebeli, 2012). Employing in addition the 'viability' dye propidium iodide (live/dead stain), it was possible to simultaneously deduce information on the integrity of membranes and activity of individual cells (Füchslin *et al.*, 2012; Keserue *et al.*, 2012a,b).

For *Giardia lamblia* spp. and *Cryptosporidium* spp. (oo)cysts, the method was recently successfully tested in the field and compared with alternative methods such as fluorescence microscopy and PCR-based detection (Keserue *et al.*, 2012a). In this field study, flow cytometric enumeration of fluorescent antibody-labelled cells required roughly 85 minutes, but resulted in slightly higher numbers than those obtained from microscopic counting, which took about 300 minutes until the result was obtained (Keserue *et al.*, 2012a). This was most probably due to dirt and unspecific material that was deposited on the filters on which cells had been collected, which may have obscured some of the marked cells. In addition, the FCM-based method was considerably quicker than microscopic counting. As frequently reported, PCR was fast (total time to result was approximately 200 minutes) but not reliable as reactions often failed and recovery (established by FCM) was low [30% for *Giardia*, 13% for *Cryptosporidium* (oo)cysts].

As indicated previously, there is much interest at present in a fast and cheap enumeration method for *Legionella pneumophila* cells, as the established plating method (ISO 11731) is slow (10 days are required to obtain a result), and ring trials repeatedly exhibited high interlaboratory volatility. In addition, PCR-based methods, or solid phase cytometry-based detection methods have their difficulties (summarized in Hambsch *et al.*, 2010; Keserue *et al.*, 2012a). The demand for a fast screening method appears to be enormous, as many in-house water distribution systems (and cooling facilities of thermal and nuclear power plants) seem to contain fairly high numbers of this pathogen. We have developed (Füchslin *et al.*, 2010) and adapted a method for quantification of *Legionella pneumophila* cells in drinking water based on immunomagnetic separation, double fluorescent staining and FCM enumeration (Keserue *et al.*, 2012b). The procedure requires two hours and can also discriminate 'membrane-intact' and 'membrane-damaged' cells, with 85% recovery of spiked cells, and a detection limit of 50 total spiked cells (Keserue *et al.*, 2012b). The method has considerable advantages compared with plating, has much potential for rapid routine testing and for obtaining a quick insight into the distribution and occurrence of this pathogen in drinking-water distribution systems.

Examples selected from practice

Introductory comments

Initially, FCM-based methods were proposed to be useful for various 'advanced' practical applications, such as rapid assessment of viability of microbial flora before and after disinfection, or of AOC levels during drinking-water treatment and distribution (Hoefel *et al.*, 2003, 2005b; Phe *et al.*, 2005; Hammes and Egli, 2005; Hammes *et al.*, 2006, 2007). In the meantime, after extensive testing, both in pilot plants and in full scale treatment plants (see for example Hammes *et al.*, 2009) the most simple option that FCM offers, namely counting cells rapidly and reproducibly, has developed into a real 'work-horse method'. In practice, it has proven to be the parameter for producing reliable 'balances and checks' with respect to the presence, (re)growth and die-off (or 'disappearance') of microbial flora during drinking-water treatment and distribution.

Colleagues from several Swiss and Dutch waterworks have communicated that they use flow cytometry as a routine method for a number of applications, as it allows rapid assessment of the state and success of maintenance work done on parts of the system when cleaned, back-flushed, repaired, or replaced, but also of microbial stability and re-growth of drinking water in public distribution networks. Therefore, quick decisions can be made instead of waiting for three or more days until the results from microbiological plating are available (Köster, 2012).

Unfortunately, only a few studies have been published so far by practitioners on such aspects. Nevertheless, by and by, case studies are becoming available from public drinking-water production plants and distribution networks, where the microbiological state of the system has been investigated with flow cytometric methods. They include water works that distribute their water with disinfectant residuals, as well as those that do not disinfect, and results and 'principles' associated with specific treatment steps were essentially confirmed for a number of full scale drinking-water production plants.

In contrast, the few data sets which reported on re-growth of microbial flora in water distribution systems in private and public buildings indicate that this final part of the drinking water distribution system has not obtained sufficient attention. We will focus on a few selected examples that – we feel – demonstrate the power of flow cytometric TCC-LNA:HNA-ratio monitoring for this purpose.

Assessing the various treatment steps in the drinking-water production train

Our initial pilot plant study (Hammes *et al.*, 2008) suggested that information can be obtained very rapidly when monitoring FCM-TCC. Consequently, the know-how obtained with the pilot plant was used to confirm the power of FCM-TCC in practice at the full-scale drinking water treatment plant in Zurich (Hammes *et al.*, 2010a). The plant produces roughly 50% of the drinking water for the city by treating lake water using sequential ozonation and filtration steps and the water is then distributed without addition of a residual disinfectant. The focus in this study was not only on monitoring the performance of the different treatment steps of the production train, but to also inspect the quality of the treated lake water during distribution with the goal of obtaining hard information concerning its biostability (discussed later).

Performed over an 18-month period, concentrations of FCM-TCC, HPC, DOC and AOC were recorded in samples collected twice a month (Hammes *et al.*, 2010a). In addition to conventional parameters (DOC, HPC), microbiologically available carbon (AOC) and FCM-TCC were also monitored over a period of 18 months. Although the raw (lake) water used varied considerably over the year, the subsequent treatment steps exhibited surprising stability. One of the obvious results was that throughout the production a direct correlation was found between the three parameters DOC, AOC and TCC, i.e. substantial growth was always detected when DOC decreased (most probably through the consumption by microbes) and this always correlated with elevated AOC concentrations. The increase in the concentration of TCC in the bulk water phase occurred with a yield in the range of 0.8×10^6 cells/µg C in the first rapid sand filter, and it decreased to roughly 0.2×10^6 cells/µg C in the final slow sand filter (based on decrease in DOC). The former cell yield is only slightly, the latter considerably lower, than the yield of 1×10^7 cells/µg C reported for natural bacterial flora on freshwater AOC (Hammes and Egli, 2005), indicating that the quality of the growth-supporting carbon worsened during treatment. This pattern was not mirrored in the data obtained from the cultivation-dependent plating, and no correlation was found between HPC and FCM–TCC (which is in contrast with the observations made in the pilot plant (Hammes *et al.*, 2008), where data suggest a weak correlation between the two parameters). In the effluent of both ozonation steps, the DOC concentration did not change significantly compared with the inlet concentration, however, it resulted in the formation of AOC. This AOC was consumed in the subsequent treatment step and resulted in growth measured as an increase in the FCM–TCC.

To date, the essential results and information obtained in the first pilot plant study and in the Zurich treatment plant have been confirmed in a number of other investigations performed on full-scale drinking-water treatment systems by our and other research groups (Egli *et al.*, 2008; Siebel *et al.*, 2008; Hammes *et al.*, 2010a; Vital *et al.*, 2012; Lautenschlager *et al.*, 2013; Prest *et al.*, 2013; Liu *et al.*, 2013b).

Rapid monitoring of cleaning and maintenance work

In several Swiss drinking-water facilities one of the most frequently used applications of flow cytometry is that of monitoring the success of the regular maintenance work done on filters and pipe systems.

Rapid assessment of the efficacy and completeness of cleaning pipes after repair or renewal work by flushing with clean water is used routinely by ZWV (Köster, 2012; Helbing *et al.*, 2012). Similarly, distribution strands that are known to have low water exchange are periodically flushed. In such procedures, flow cytometric data are used in a comparative manner. FCM–TCC and LNA:HNA-ratio of the clean water used for flushing are taken as a reference: as soon as cell number and pattern of the water at the end of the strand correspond to that used for flushing, the process can be considered to be complete. As this can be done on-site with portable small FCMs, decisions of whether to continue or to finish cleaning can be made immediately at the site. In particular, for wide-diameter pipes, which are known to be difficult to clean, this has proven to be efficient, reliable and much faster than assessing the cleaning procedure using conventional HPC as reported by Köster and colleagues (Köster, 2012).

Another example is that of backwashing biological filters. It is a normal operational procedure in most waterworks to regularly clean biological filters either by backwashing (GAC filters, sand filters), or by carefully scraping off part of the dirty surface layer (Schmutzdecke) from the sand filters. The cleaned filters then have to first be re-established with respect to their biological performance before they can be integrated back into the actual treatment train. Usually, at ZWV 'clearance for use' for filters is given when the HPC concentration falls below 20 CFU/ml. An example of the former is given in Fig. 9.11 for the backwashing of a GAC filter. The data originate from a large-scale pilot plant at the Zurich Waterworks (WVZ) and include particle counts, FCM–TCC and HPC figures (Egli *et al.*,

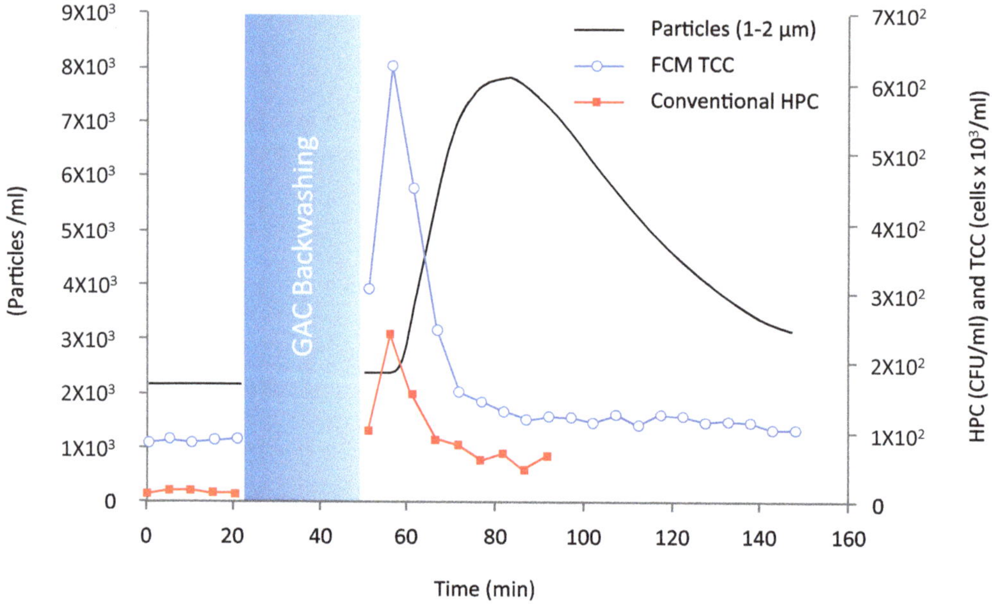

Figure 9.11 Backwashing of GAC filter from large scale pilot plant at WVZ. Results for FCM-TCC, particles (1–2 µm) and conventional HPC data are shown before and after the backwashing procedure. Data demonstrate, firstly, that particles behave differently from microbial cells, and secondly, that (here!) FCM-TCC supplies similar information as the conventional plating; however, FCM–TCC results are available already after 15 minutes, whereas HCPC results are obtained only after three days. Adapted from Egli *et al.* (2011), and with permission from O. Köster and H.P. Kaiser, WVZ.

2011). Backwashing of this GAC filter caused a disturbance during which increased amounts of particles (carbon-fines), and microbial cells were released. Although the plating-results correlate with FCM–TCC numbers here, they are of course only available after three or more days, whereas flow cytometric cell numbers can be determined within 15 minutes (or even online). The data also demonstrate that the signal obtained from the particle counter (1–2 µm) is considerably delayed and obviously not picking up the peak of microbial cells released as they are too small. Hence, in this example, fast FCM–TCC analysis allows a decision to be made within roughly an hour (including both sampling and analysis time) as to when the backwashed filter can be integrated again into the normal treatment train. Similar results were obtained for backwashing of rapid sand filters (Köster, 2012; Helbing *et al.*, 2012). In the future, online FCM–TCC monitoring combined with online turbidity probes will even allow monitoring of such processes almost in real-time.

Monitoring microbial abundance and biological stability during water storage and distribution

The primary goal of any drinking water supplier must be to ensure that no (or only very little) re-growth of the indigenous microbial flora can occur in the water during intermediate storage in reservoirs and/or during distribution (not to speak of opportunistic and real pathogens that might be present in the system). The most common approach to achieve this goal is to distribute the produced water with sufficient disinfectant (usually chlorine, chlorine dioxide or chloramine). A second approach, practiced in some countries with access to sufficient high quality water resources (or, alternatively, sophisticated treatment and distribution facilities), consists of producing and distributing a 'biostable' drinking water, i.e. a water containing virtually no nutrients that might support microbial re-growth (see, for example, Rittmann and Snoeyink, 1984; van der Kooij *et al.*, 1999; van der Kooij, 2000). As availability of growth-supporting reduced carbon compounds is frequently the growth-limiting factor for microbial heterotrophs in freshwater, Dick van der Kooij, in his pioneering work (van der Kooij *et al.*, 1982), defined a 'biostable' water as one that contains less than 10 μg/l AOC.

Although the concept of 'biological stability' has been well known to drinking-water practitioners, few experimental data are available from full-scale plants and networks distributing their water without residual disinfectant, which clearly demonstrated biostability of water during distribution. To experts it is common knowledge that the problems in demonstrating biological stability are due to methodological difficulties (discussed in detail by Liu *et al.*, 2013a). Flawed results were produced primarily due to the use of HPC (as far as the authors are aware, except for Boe-Hansen *et al.*, 2002, the troublesome and time-consuming determination of microbial cell abundance using microscopic cell counting, e.g. after DAPI-staining, was not performed); another reason was the difficult quantification of biologically available carbonaceous compounds either as AOC (according to van der Kooij) or as BDOC (Servais *et al.*, 1989). Both are indirect methods and do not determine microbial growth directly, but indirectly, i.e. by measuring actual AOC or BDOC concentrations and assuming that re-growth does not occur below a certain level of available carbon. Considering these difficulties, and defining 'biostability' in a pragmatic sense, namely, that neither an increase in cell number nor in biomass should occur during handling and distribution, FCM-based cell counting, together with the LNA:HNA-ratio, is ideally suited for a convenient and rapid assessment of biostability of drinking water once fed into a distribution network.

Distributing non-disinfected water

As far as the authors are aware, the first investigation employing the FCM-TCC method for addressing biostability during distribution in a full-scale drinking water treatment and supply system was performed in the city of Zurich (Hammes *et al.*, 2010a). After a final 'polishing', consisting of slow sand filtration, the finished water passed through a buffering reservoir (referred to as R1, located on the plant premises) before it was fed into the distribution network. Another reservoir (R2) was situated 2.4 km from R1. Samples were collected from R1 and two points in the network (DN1 and DN2) at distances of 3.4 and 8.5 km from R1. The average hydraulic retention time of the water sampled was calculated as 29 hours and 43.5 hours, respectively, and the average temperature increased from 6.15 to 8.8°C. In this distribution strand, FCM–TCC increased slightly but significantly during distribution (0.897×10^5 cells/ml at R1, to 1.00×10^5 and 1.09×10^5 cells/ml at DN1 and DN2, respectively), which corresponds to an overall increase of approximately 21%. Correspondingly, DOC and AOC decreased, and DOC consumption was on average 20–30 μg C/l. Although these differences appear minute, a paired student t-test suggested they were significant and indicated that the carbon disappearing was transformed into biomass with a re-growth yield in the range of $0.5–1 \times 10^6$ cells/μg C.

Despite the slight increase in cell number observed in the network, the data suggested a high biological stability of the water during storage and distribution (HPC concentrations allowed no assessment as they were on average 1 CFU/ml and varied in all samples between 0 and 5 CFU/ml). Here it was shown clearly (and for the first time) that FCM-based cell-counting can be used to adequately and quickly demonstrate biological stability of drinking-water in a distribution system. The data also document that FCM–TCC is a very sensitive parameter that can be employed as an indicator of microbial regrowth processes in water, as minute and statistically insignificant differences in concentration of carbon are transposed into significant changes in cell concentration.

In well-maintained networks that distribute non-chlorinated water treated with slow sand filtration as a final polishing step, and therefore usually contain TCC concentrations in the range of 1×10^5 cells/ml, hardly any increase in TCC was reported (Hammes *et al.*, 2010a; Vital *et al.*, 2012b; Lautenschlager *et al.*, 2013; Liu *et al.*, 2013b; Prest *et al.*, 2014). Not surprisingly, also in well-maintained and -managed distribution networks temperatures have to remain low in order to avoid microbial re-growth; for example, Liu and colleagues reported that regrowth in their essentially biostable distribution system was only observed in parts where temperatures rose above 15°C (Liu *et al.*, 2013b).

Distribution with disinfectant residuals

An excellent study of what can happen when biostability is not achieved was reported recently for the drinking-water distribution network in the city of Riga, Latvia (Nescercka *et al.*, 2014). The water is produced in three major and three minor treatment plants, chlorinated and fed into the distribution system with 0.4–0.5 mg/l residual chlorine. The 49 sampling points (fire hydrants) were selected such that a broad coverage of the whole network was obtained and that zones both close to and distant from the treatment plants were included. In addition to physico-chemical parameters, the microbiological parameters HPC, FCM–TCC, membrane compromised/intact cells, free and cellular ATP were determined. The chlorinated water produced at the three major plants contained between 2 and 5% of membrane-intact cells (between 1×10^4 and 2×10^4 cells/ml), demonstrating that disinfection at the plant was effective. However, in half of all samples from the distribution system more than 10×10^4 intact cells per ml were found compared with the effluent water from the same water treatment plant. This corresponds to a 10- or more-fold increase in intact cell count (ICC) concentration. This unmistakably indicated massive microbial regrowth and, hence, high biological instability of this drinking-water distribution network. One major reason for this instability is certainly the high levels of organic carbon in the water used and distributed (average concentrations of TOC in the three major plants were between 3 and 9 mg TOC/l) causing an extensive consumption of chlorine, such that mean free chlorine concentration in half of all samples was below 0.17 mg/l.

In their work, Necerecka and colleagues have convincingly demonstrated that also for water works distributing chlorinated drinking water, flow cytometric methods, in particular TCC and ICC, are able to provide a rapid and reliable description of the general microbiological state and local microbial re-growth in full-scale drinking-water distribution networks. With respect to parameter correlation, good correlation was reported in this work between intracellular ATP and membrane-intact cell count ($R^2 = 0.77$), whereas poor correlation of standard HPC with both intact cells ($R^2 = 0.18$) and intracellular ATP ($R^2 = 0.11$) was observed.

Microbial re-growth in buildings – opening Pandora's box?

Generally, the local drinking-water supplier is responsible for the quality and safety of the drinking water delivered to the customer up to the point where water enters the building and passes the water meter; beyond this point it is the owner of the building who has to take over responsibility. Consequently, in contrast to catchment, treatment and distribution, only very limited information exists on the general microbiological quality of drinking water at the consumer's tap.

Nevertheless, it has been amply documented that microbial re-growth occurs in drinking-water distribution systems, particularly in the summer months; this not only when non-disinfected drinking water is produced and distributed, as is done in some European countries, but also when water is distributed with disinfectant residuals (see Allen *et al.*, 2004). Hence, the survey published by Pepper and colleagues, who during a 1-year period tracked the concentration of cultivable microbes in the water supply system of Tucson, Arizona, from a groundwater well, through treatment, right to the consumer's tap (using HPC and other cultivation-dependent methods), was certainly disturbing but not surprising (Pepper *et al.*, 2004). They observed that from catchment and in the distribution system levels, levels of HPC bacteria remained low, treatment and disinfection worked and the water was delivered to the consumer with roughly 50 CFU/ml, i.e. well below the threshold of 500 CFU/ml recommended in the USA. However, in samples collected from various houses and taps, it had increased on average to more than 3000 CFU/ml. They concluded that 'the major source of bacteria ingested by the average consumer in Tucson originates from bacteria within the household distribution system or the household tap, rather than from source waters or the distribution system' (Pepper *et al.*, 2004).

In an attempt to understand microbial re-growth in in-house water distribution systems supplying drinking water without disinfectant residuals, we investigated the microbiological state of water collected from selected taps in our Institute using conventional plating (HPC) combined with FCM–TCC, LNA–HNA ratio, membrane-intact cells, and ATP (Siebel *et al.*, 2008). Samples were collected during normal working days (at one hour intervals), after stagnation overnight, and during a weekend, in order to observe fluctuations throughout a day, and the influence of regular and non-regular usage. The pattern observed was similar for all taps: FCM–TCC concentration increased 2–3-fold during stagnation at night and over weekends. Depending on user frequency, the concentration of TCC decreased rapidly but continuously in the morning (usually between 8 and 10 am) and remained constant throughout the day, before increasing again during the night. This pattern was reflected by ATP (also 2- to 3-fold change) and by the colony count (ca. 10–20-fold), which was between 1 and 10% of the FCM–TCC. Considering that the finished water leaves the treatment plant with typically less than 20 CFU/ml, the in-house concentrations measured in the range of 500–1000 CFU/ml during the day, and 6000–8000 CFU/ml after overnight stagnation, suggest heavy bacterial re-growth, also when assessed by plating.

The results were confirmed in a more detailed study extended to households of the local community, and focusing on water from taps after stagnation and subsequent flushing, and by also including a comparison with water sampled directly from the communal drinking water network (Lautenschlager *et al.*, 2010). Also here, judged from FCM–TCC and HPC data, re-growth was observed to a very similar extent as in our Institute's taps. After flushing for five minutes, corresponding to approximately 30 l of water, the values decreased to those found in network samples (also based on physico-chemical parameters such as temperature, conductivity, pH, TOC and nitrate concentration). In contrast to the first study, an increase in cellular biovolume of the HNA-cell fraction and their average ATP-per-cell concentrations was reported, indicating that the observed increase in TCC was due to growth and not cell fragmentation. The study was supplemented with a DGGE analysis of the bacterial flora community in the first litre of stagnated water from a tap and in network water samples. DGGE also documented that stagnation sometimes led to a dramatic change of the composition in the microbial flora of many in-house systems. Although this does not necessarily imply a hygiene risk it points to an astonishing lack of knowledge with respect to microbial regrowth in in-house water installations. We do not know as yet whether or not this is due to carbon from the bulk water, caused by nutrients migrating from materials used in in-house installations, or to cells detaching from biofilms adhering to the pipe surfaces.

Obtaining such data is not difficult. The simplicity with which information can be obtained with flow cytometric methods with respect to the state and microbiological behaviour is visualized in Fig. 9.12 for a drinking-water installation in a newly built 13-flat house that was not yet fitted with taps, shower heads and so on (Kötzsch and Egli, 2013). Stainless steel was used for all vertical piping, horizontal pipes were made of PE-Xc. After intensive initial flushing, the water was left to stagnate for 15 hours and samples were then collected from 60 sampling points and analysed with FCM for TCC, LNA:HNA-ratio and HPC (plus TOC, temperature and other parameters). The water supplied by the local water authorities was untreated groundwater of excellent quality (TCC ≈ 8000 cells/ml; LNA ≈ 70%; HPC = 0 CFU/ml). Stagnation led to a dramatic increase in FCM-TCC concentration between one and two orders of magnitude, and a dominance of HNA cells (>90%) at most points. After 10 minutes of flushing, most parameters returned to the levels found in incoming water from the local drinking-water supply system, except for a few points. To pick out a few interesting results: right after the point of entry the highest TCC concentrations were observed in samples obtained from four points, which could be traced back to a heavily contaminated filter. Flushing brought the TCC back down to normal levels (Fig. 9.12). Also obvious were a number of sampling points where neither the TCC nor the LNA:HNA-ratio returned to normal levels after flushing (the exact reason for increased microbial growth at these points could not be explored in more detail). Most interesting was that HPC was not able to

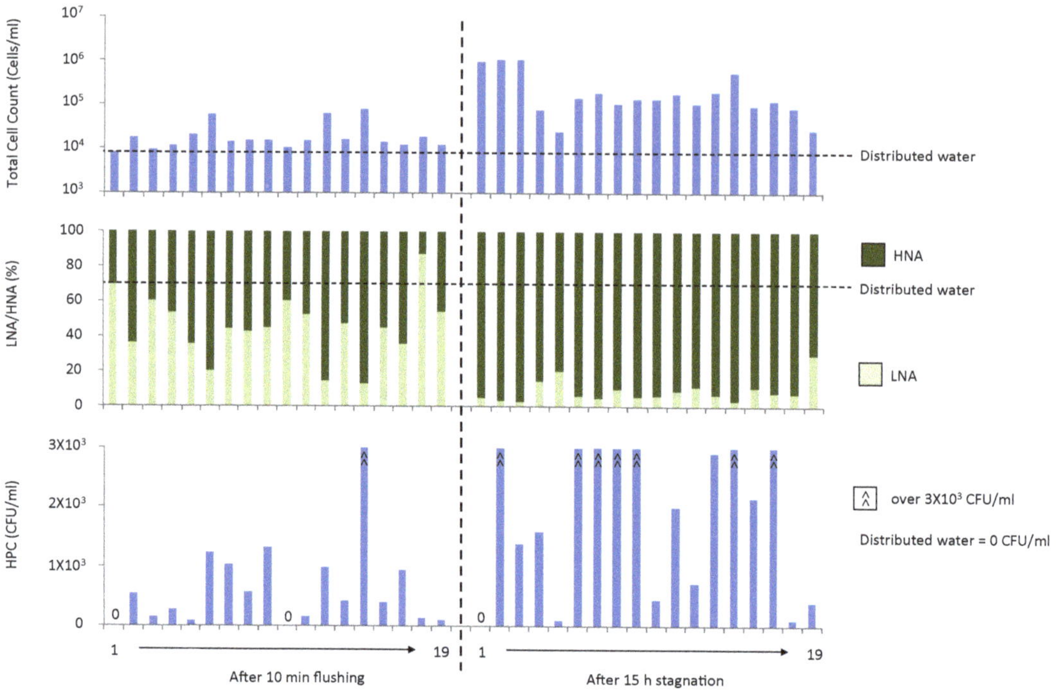

Figure 9.12 Effect of stagnation and flushing on the microbial abundance in the bulk water phase of a drinking water installation in a newly built 13-flat house. After extensive flushing, the water was left to stagnate overnight. Samples were collected from different points and analysed with FCM for TCC, LNA:HNA-ratio and with conventional HPC. A huge increase of FCM–TCC concentration as a result of stagnation can be recognized, paralleled by a shift in the LNA:HNA-ratio to larger cells. At most sampling points, after 10 minutes of flushing, FCM-TCC and LNA:HNA-ratio returned to the values equivalent to those found in incoming water supplied from the local drinking water network, except for a few points where heavy microbial contamination was detected. It should be pointed out that conventional HPC is not able to identify these points. In total 60 sampling points were analysed, here a selection of 19 points is shown; for the full data set see Kötzsch and Egli (2013).

detect most of the points where TCC indicated massive re-growth; for example, right after the contaminated filter no cultivable cells were detected. Selected sampling points in this building were inspected again two months after the first sampling campaign. Although the re-growth pattern was similar, the results indicated a slow improvement of the situation, probably because of continuously reduced migration of growth-supporting compounds from the (plastic) pipe material.

Consistent with these observations is the recently published work by Lipphaus and colleagues (Lipphaus *et al.*, 2014) who investigated the effect of stagnation on re-growth in a system receiving chlorinated water in a medium-sized building. Also here, despite the presence of residual chlorine, massive regrowth was detected using FCM–TCC, particularly at infrequently used taps and after long stagnation periods (supporting the work of Pepper *et al.*, 2004). In addition, the authors also monitored the ratio of intact cells using the 'live/dead'-staining method and found that the number of intact cells increased with stagnation time and decreased again after flushing (as one would expect in water containing disinfectant residuals).

Hence, increasing information demonstrates that biological stability and regrowth is not so much a problem in well-maintained drinking-water distribution networks, rather problematic issues may start within buildings. Some of the most obvious causes for this are smaller diameter plastic pipes, improper installation, increased temperatures and extended stagnation periods. Also, the presently available data indicate that massive re-growth in buildings is not linked to the question of disinfection or no disinfection, but that biological instability is a general problem related to the design of our in-house water installations.

This clearly calls for more information, not only in the field of microbial re-growth in water installations in buildings, but also for new methods that allow for better and faster assessment of materials used in domestic water installations with respect to their properties and potential to support microbial growth, both at the surface and in the bulk water phase. All of the presently established procedures for testing materials in contact with drinking water for their migration and biofilm formation properties are tedious to perform and take a long time until a result is obtained. First attempts to improve this situation have been reported recently (Bucheli-Witschel *et al.*, 2012; Kötzsch and Egli, 2012, 2013).

Outlook and some final comments

On FCM for routine testing

The time has definitely passed when flow cytometry was referred to as an exotic technique that required much more research before it could be applied to drinking-water microbiology (National Academy of Science, 2006). FCM is used in practice already in a few European countries and has pushed the door wide open for a better, and in some areas a new understanding of the role of microbes in drinking-water treatment processes and during distribution. To date, FCM is the only fast method that allows realistic and reliable microbiological 'balances and checks' to be obtained, and this with incredible sensitivity. When supported by additional methods, such as HNA:LNA-ratio, fluorescence fingerprint analysis, cell-associated ATP, or live/dead-staining, practitioners now have a range of cultivation-independent methods available for rapid routine analysis of their drinking-water production and distribution systems.

Currently, small, transportable and robust instruments that serve the needs of practice are commercially available. On-site use of such instruments allows, within a day, to obtain information on small water distribution networks, in-house water installations, or even on water distribution systems in small communities. Flow cytometric online determination of TCC, LNA/HNA-ratio and ICC as reported recently (e.g. Besmer *et al.*, 2014) makes the technique even more attractive. However, so far, these instruments are neither suitable nor robust enough for routine analysis. Practitioners are waiting for a fully automated, small, robust and reasonably priced instrument that can be installed transiently or permanently in places that are difficult to access. A recent report (Kuenzi *et al.*, 2014, and see above) suggests that development of an instrument that might satisfy most of these requirements is on the way.

On rapid pathogen screening

The rapid search for waterborne pathogens, both microbial cells and viruses, suffers from a number of difficulties that have to be overcome. Technically, one of the main challenges is the fact that pathogens are usually present in very low concentrations and are outnumbered by a factor of 10^6 or more by the microbial flora; hence a target cell has to be detected and identified within a large background of cells. This requires techniques that allow enrichment of target cells before or after selective marking. Concentration is usually achieved by filtration procedures; some of these allow sampling of as much as 10–100 l of water. After collection of the cells (and other particles) target cells are 'ear-marked' for detection, either on the filter directly, or after re-suspension in a small volume, using cell-surface-specific fluorescent antibodies (see Figs. 9.1 and 9.10). All in all, flow cytometry-based pathogen detection and quantification, directly or after immunomagnetic enrichment, is no longer restricted by insufficient sensitivity or robustness (Keserue *et al.*, 2012); rather, the success of this method is largely dependent on the binding-properties, affinity, and/or selectivity of the antibodies employed. Unfortunately, from our experience, batches of commercially sold antibodies, even from the same producer, may vary considerably with respect to quality, which makes preliminary quality checks necessary.

Open and closed niches – the role of a healthy microbial flora

For a long time, drinking-water distribution networks – except for unwanted, slimy biofilms on the inside of some pipe surfaces – have not been considered as microbial ecosystems. Drinking water-microbiology was concerned mainly with biofilms and survival or growth of obligate and opportunistic pathogens, and of indicator organisms (see e.g. Szewzyk *et al.*, 2000; Berry *et al.*, 2006; Schoenen, 2011). Only recently, was it recognized that these systems contain an active natural microbial (mostly bacterial) flora, larger in number than that found in most oligotrophic ground- and spring-waters (Poitelong *et al.*, 2011). Molecular, fluorescent-labelling and -staining methods have brought life to this system, and, as outlined here, flow cytometric methods have added speed at the single cell analytical level.

Our perception of the ecological role of the microbial flora was, and still is, rather ambivalent. Whereas we appreciate its 'purifying' activity during soil passage and in groundwater aquifers, and stimulate its growth and activity in biological filters in the drinking-water treatment train, we usually regard its presence in the finished drinking water and during its distribution with much suspicion. This is probably the reason for clinging on to HPC threshold and tolerance numbers in most countries (to cite only two: Bundesgesetzblatt, 2011; HyV, 2008).

The literature and data discussed in the various sections of this review concerning the abundance of microbial cells, their regrowth in, and 'biological stability' of drinking water, points out the difficulty of judging and valuing water quality by using rigid numbers and/or only by considering the -fold increase. For example, it was repeatedly observed that in drinking water of the highest quality, originating from groundwater aquifers with low TCC and DOC concentrations, the relative (-fold) increase was substantially higher than that in excellent drinking water produced from surface water (as done in a number of cities in Switzerland and the Netherlands). As long as data sets are still scarce it may be useful to apply an upper 'threshold' for the maximum acceptable FCM–TCC or ICC (probably in the range of $3–5 \times 10^5$ cells/ml) in combination with a 'stable' fingerprint (and/or other parameters that might prove better suited for this purpose) for judging the quality of a drinking water during storage and distribution based on MGP assays and re-growth observed in full-scale distribution networks.

Consequently, production and distribution of water with a healthy and harmless competitive natural microbial drinking-water flora, i.e. with an already 'well-occupied' niche, appears to offer less opportunity for growth of intruders and of opportunists that are part of the community. In this sense, production steps that physically remove or inactivate the microbial flora – but do not at the same time eliminate growth supporting carbonaceous compounds – are bound to produce a biologically unstable and 'vulnerable' water. This can be true for insufficient chemical disinfection, membrane-filtration, or UVC-disinfection (as reported in, for example, Wang *et al.*, 2008a; Nescerecka *et al.*, 2014; Lipphaus *et al.*, 2014). Hence, as distribution networks are not (and cannot be kept) sterile, we must fill again as quickly as possible microbiological niches that we have opened with harmless drinking-water flora (or consume the produced water immediately).

On interpretation of FCM–TCC data

Since Koch introduced and applied his plating technique in 1883 to assess general water quality (Koch, 1883, 1893) an enormous amount of data has been collected using the traditional HPC method. However, now knowing that this (as any other cultivation-dependent) method allows us to detect only a small fraction of all microbial cells present in an environmental sample, a most obvious question to every drinking-water microbiologist is whether or not there is a correlation between the cell numbers obtained from plating and those determined flow cytometrically. Unfortunately, despite the numerous attempts that have been made, not only for drinking water but also for other ecosystems, to correlate the various techniques for counting cells (total, live/dead, direct viable cell counts, etc.) with the number of cultivable cells, we have been left with no definite answer (see, e.g. Staley and Konopka, 1985; Bartram *et al.*, 2003). Nevertheless, it appears safe to us to accept that in natural untreated waters FCM–TCC numbers usually come fairly close to the 'true' figure of suspended cells (if one takes a pragmatic approach and avoids philosophical questions about what properties a microbial cell should exhibit to be classified as being alive). This statement is supported by numerous microscopy studies combined with micro-radio-autography and activity staining, which suggest that in sea- and freshwater 80–90% of all cells have intact membranes and exhibit some form of metabolic activity, and are, therefore, most probably alive (and may even be able to grow under appropriate conditions).

This implies that in the drinking-water sector one should, as a rule, assume no correlation between results obtained for flow cytometric cell counts with those acquired by the HPC method. Therefore, from a practical point of view, there is much to be said for abandoning the various HPC methods in microbiological drinking-water analysis (but of course not plating to detect *E. coli* or other indicators!) and replacing them with counting particles stained with fluorescent DNA-binding dyes using flow cytometry (or the microscope). This leaves us with the question of how to use and interpret FCM–TCC (or ICC) numbers that we observe in source water, during treatment and distribution in public networks and in buildings. For practitioners and regulators it was fairly convenient to have threshold concentrations (in Switzerland, e.g. < 300 CFU/ml during distribution, <20/ml after treatment) as guidelines for the general water quality and treatment performance.

However, at a closer look, replacing the HPC method with FCM–TCC does not require a paradigm shift. All that is needed for regulators and practitioners is to accept the fact that microbial cells are much more abundant than so far been assumed (e.g. as visualized in Fig. 9.2). In comparison with the present situation this would not change much, as even today HPC-results have to be interpreted with common sense, as 'the test has little value as an index of pathogen presence but can be useful in operational monitoring as a treatment and disinfectant indicator, where the objective is to keep numbers as low as possible' (cited from WHO, 2008; see also Bartram *et al.*, 2003). Hence, 'simple numerical values for an indication of an abnormal change ... (e.g. > 10, > 20, ...> 300 ... CFU/ml at 22°C or 37°C)'... that 'generally serve as triggers to review previous data ... Some have established arbitrary levels of increase ranging from 0.5 log to >2 .3 log increases over previous results' (cited from Bartram *et al.*, 2003). Therefore, we would not go from established HPC threshold values to uncertainty.

Results from FCM–TCC (FCM–ICC) monitoring can be, and are, used today in the same way as suggested by the WHO for results obtained with the HPC method. As discussed in this review, TCC and ICC give realistic information when assessing treatment efficiency and distribution of (drinking) water. If one gets acquainted with the 'new numbers' it becomes obvious that a 10-fold increase in TCC concentration from 50,000 to 500,000 cells/ml is excessive and calls for further investigation (because in the latter case we are getting close to the concentrations found in surface water), whereas, when distributing an untreated groundwater without disinfection a 10-fold increase from 20,000 to 200,000 cells/ml after stagnation may still be considered normal (see all the examples given earlier).

Finally, flow cytometry-based microbiological drinking-water analysis offers not only the rapid determination of cell numbers (soon even online at reasonable cost), but several additional parameters come with it, virtually for free, such as the LNA:HNA ratio and fingerprints (and ICC). All of these supply practitioners with additional information that can be used for making rapid decisions. This could form a solid and science-based background for quantitative microbial risk assessment, water safety plans, and early warning systems, which all require such easy-to-do parameters (see Smeets *et al.*, 2010) for judging process efficiency and monitoring of different stages in water production, including catchment protection, treatment and distribution.

Acknowledgements

Many persons have contributed to this work during the last decade and we would like to thank all colleagues and collaborators who have helped with hands and brains to develop this fascinating project and to build up and establish a new approach to drinking-water microbiology. Particular thanks go to A.J.B. Zehnder and the directorate of Eawag for giving us the freedom to develop new ideas and to purchase our first flow cytometer, as well as to all external partners in practice who have allowed us to test out our methods in reality and have given us access to their facilities, in particular our colleagues from the Zurich Waterworks (WVZ). Also I would like to thank those agencies that have helped to finance various aspects of the work done in our laboratory; this includes: WVZ, IWB Basel, AWWARF, KTI, EU-Techneau, Eawag, the Federal Office of the Environment (BAFU), the Federal Office of Health (BAG), Spiez Laboratory, the Velux Foundation, the Swiss Gas and Water Board (SVGW). We thank H.A. Keserue, rqmicro GmbH, ETH-Zürich, for providing Fig. 9.10. Judith Egan-Shuttler is acknowledged for careful corrections. Furthermore, the first author thanks Claudia for her patience during writing, and for reading of, correcting of, and constructive comments on the various drafts of this review.

References

Abu-Absi, N.R., Zamamiri, A., Kacmar, J., Balogh, S.J., and Srienc, F. (2003). Automated flow cytometry for acquisition of time-dependent population data. Cytometry A *51*, 87–96.

Allen, M.J., Edberg, S.C., and Reasoner, D.J. (2004). Heterotrophic plate count bacteria – what is their significance in drinking water. Int. J. Food Microbiol. *92*, 265–274.

Allen, M.J., Clancy, J.L., and Rice, E.W. (2006). The plain, hard truth about pathogen monitoring. J. Am. Water Work. Assn. *92*, 64–75.

Arnoldini, M., Heck, T., Blanco-Fernandez, A., and Hammes, F. (2014). Monitoring of dynamic microbiological processes using real-time flow cytometry. PLOS ONE *8*(11), e80117.

Aurell, H., Catala, P., Farge, P., Wallet, F., Le Brun, M., Helbig, J.H., Jarraud, S., and Lebaron, P. (2004). Rapid detection and enumeration of *Legionella pneumophila* in hot water systems by solid-phase cytometry. Appl. Environ. Microbiol. *70*, 1651–1657.

Baker, M.N., and Taras, M.J. (1981). The quest for pure water, Vol. 1, 2nd edn. (American Water Works Association, Denver, CO, USA).

Bakker Schut, T.C., De Grooth, B.G., and Greve, J. (1993). Cluster analysis of flow cytometric list mode data on a personal computer. Cytometry, 13, 649–659.

Banna, M.H., Imran, S., Francisque, A., Najjaran, H., Sadiq, R., Rodriguez, M., and Hoorfar, M. (2014). Online driking water quality monitoring: Review on available and emerging technologies. Crit. Rev. Environ. Sci. Technol. *44*, 1370–1421.

Bartram, J., Cotruvo, J., Exner, M., Fricker, C., and Glasmacher, A., eds. (2003). Heterotrophic plate counts and drinking-water safety. the significance of HPCs for water quality and human health (IWA Publishing, on behalf of the WHO, Geneva, London, UK).

Bashashati, A., and Brinkman, R.R. (2009). A survey of flow cytometry data analysis methods. Adv. Bioinform. *2009*, 584603.

Baudart, J., Coallier, J., Laurent, P., and Prévost, M. (2002). Rapid and sensitive enumeration of viable diluted cells of members of the family *Enterobacteriaceae* in freshwater and drinking water. Appl. Environ. Microbiol. *68*, 5057–5063.

Bergquist, P., Hardiman, E., Ferrari, B., and Winsley, T. (2009). Applications of flow cytometry in environmental microbiology and biotechnology. Extremophiles 3, 389–401.

Berney, M., Weilenmann, H.-U., and Egli, T. (2006). A flow cytometric study of vital cellular functions in *E. coli* during solar disinfection (SODIS). Microbiology *152*, 1719–1729.

Berney, M., Hammes, F., Weilenmann, H.-U., Bosshard, F., and Egli, T. (2007). Assessing and interpreting bacterial viability using LIVE/DEAD BacLight TM kit in combination with flow cytometry. Appl. Environ. Microbiol. *73*, 3283–3290.

Berney, M., Vital, M., Hülshoff, I., Weilenmann, H.U., Egli, T., and Hammes, F. (2008). Rapid, cultivation-independent assessment of microbial viability in drinking water. Water Res. *42*, 4010–4018.

Berry, D., Xi, C., and Raskin, L. (2006). Microbial ecology of drinking water distribution systems. Curr. Opin. Biotechnol. *17*, 297–302.

Besmer, M.D., Weissbrodt, D.G., Kratochvil, B.E., Sigrist, J.A., Weyland, M.W., and Hammes, F. (2014). The feasibility of automated online flow cytometry for *in-situ* monitoring of microbial dynamics in aquatic ecosystems. Front. Microbiol. *5*, 265.

Bigoni, R., Kötzsch, S., Sorlini, S., and Egli, T. (2014). Solar water disinfection by a Parabolic Trough Concentrator (PTC): flow cytometric analysis of bacterial inactivation. J. Cleaner Production *67*, 62–71.

Boddy, L., Wilkins, M.F., and Morris, C.W. (2001). Pattern recognition in flow cytometry. Cytometry, *44*, 195–201.

Boe-Hansen, R., Albrechtsen, H.J., Arvin, E., and Jorgensen, C. (2002). Bulk water phase and biofilm growth in drinking water at low nutrient conditions. Water Res. *36*, 4477–4486.

Bosshard, F., Berney, M., Scheifele, M., Weilenmann, H.-U., and Egli, T. (2009). Solar disinfection (SODIS) and subsequent dark storage of *Salmonella typhimurium* and *Shigella flexneri* monitored by flow cytometry. Microbiology *155*, 1310–1317.

Bosshard, F., Bucheli, M., Meur, Y., and Egli, T. (2010a). The respiration chain is the cell's Achilles' heel during UVA inactivation in *Escherichia coli*. Microbiology *156*, 2006–2015.

Bosshard, F., Riedel, K., Schneider, T., Geiser, C., Bucheli, M., and Egli, T. (2010b). Protein oxidation and aggregation in UVA-irradiated *Escherichia coli* cells as signs of accelerated cellular senescence. Environ. Microbiol. *12*, 2931–2945.

Bouvier, T., del Giorgio, P.A., and Gasol, J.M. (2007). A comparative study of the cytometric characteristics of High and Low nucleic-acid bacterioplankton cells from different aquatic ecosystems. Environ. Microbiol. *9*, 250–266.

Broger, T., Odermatt, R.P., Huber, P., and Sonnleitner, B. (2011). Real-time on-line flow cytometry for bioprocess monitoring. J. Biotechnol. *154*, 240–247.

Brogneaux, A., Han, S., Sörensen, J.S., Lebeau, F., Thonart, P., and Delvigne, F. (2013). A low-cost, multiplexable, automated flow cytometry procedure for the characterization of microbial stress dynamics in bioreactors. Microb. Cell Fact. *12*, 100.

Bucheli-Witschel, M., Kötzsch, S., Darr, S., Widler, R., and Egli, T. (2012). A new method to assess the influence of migration from polymeric materials on the biostability of drinking water. Water Res. *46*, 4246–4260.

Bundesgesetzblatt Jahrgang 2011 Teil I Nr. 21. Erste Verordnung zur Änderung der Trinkwasserverordung [Part I Nr. 21. First by-law for a change of the drinking water regulation.] Available online: www.bundesgesetzblatt.de.

Button, D.K., and Robertson, B.R. (1989). Kinetics of bacterial processes in natural aquatic systems based on biomass as determined by high-resolution flow cytometry. Cytometry *10*, 558–563.

Button, D.K., and Robertson, B.R. (2001). Determination of DNA content of aquatic bacteria by flow cytometry. Appl. Environ. Microbiol. *67*, 1636–1645.

Cabral, J.P.S. (2010). Water microbiology. Bacterial pathogens and water. Int. J. Environ. Res. Public Health *7*, 3657–3703.

Czechowska, K., Johnson, D.R., and van der Meer, J.R. (2008). Use of flow cytometric methods for single-cell analysis in environmental microbiology. Curr. Opin. Microbiol. *11*, 205–212.

Davey, H.M. (2011). Life, death, and in between: meanings and methods in microbiology. Appl. Environ. Microbiol. *77*, 5571–5576.

Davey, H.M., and Kell, D.B. (1996). Flow cytometry and cell sorting of heterogeneous microbial populations: the importance of single-cell analyses. Microbiol. Rev. *60*, 642–696.

Delgado-Viscogliosi, P., Simonart, T., Parent, V., Marchand, G., Dollelaere, M., Pierlot, E., Pierzo, V., Menard-Szczebara, F., Gaudard-Ferveur, E., Delabre, K., *et al.* (2005). Rapid method for enumeration of viable *Legionella pneumophila* and other *Legionella* spp. in water. Appl. Environ. Microbiol. *71*, 4086–4096.

De Roy, K., Clement, L., Thas, O., Wang, Y., and Boon, N. (2012). Flow cytometry for fast microbial community fingerprinting. Water Res. *46*, 907–919.

Dubbelar, G.B.J., Gerritzen, P.L., Becker, A.F.R., Jonker, R.R., and Tangen, K. (1999). Design and first results of CytoBuoy: a wireless flow cytometer for *in situ* analysis of marine and fresh waters. Cytometry 37, 247–254.

Dwivedi, H.P., and Jaykus, L.-A. (2011). Detection of pathogens in foods: the current state-of-the-art and future directions. Crit. Rev. Microbiol. 37, 40–63.

Egli, T. (2012). Mikrobiologische Trinkwasseranalyse – Entwicklungen, Stand, Ausblick. Aqua & Gas 5, 14–22.

Egli, T., and Hammes, F. (2010). Neue Methoden für die Wasseranalytik. Gas-Wasser-Abwasser GWA *4*, 315–324.

Egli, T., Berney, M., Hammes, F., and Füchslin, H.P. (2008). New methods for assessing the safety of drinking water. Eawag News *65e*, 20–23.

Egli, T., Hammes, F., Köster, O., and Kaiser, H.-P. (2011). Flow cytometry-based methods for fast microbiological analysis of drinking water and their application in practice. Intern. J. Water Manag. Bluefacts, 70–77.

Emmadi, R., Boonyaratanakomkit, J.B., Selvarangan, R., Shyamala, V., Zimmer, B.L., Williams, L., Bryant, B., Schutzbank, T., Schoonmaker, M.M., Amos Wilson, J.A., *et al.* (2011). Molecular methods and platforms for infectious diseases testing. A review of FDA-approved and cleared assays. J. Molec. Diagnost. *13*, 583–604.

Escobar, I.C., and Randall, A.A. (2000). Sample storage impact on the assimilable organic carbon (AOC) bioassay. Water Res. *34*, 1680–1686.

EU 98/83/EC (1998). Council directive 98 98/83/EU of 3 November 1998 on the quality of water intended for human consumption.

Figueras, M.A., and Borrego, J.J. (2010). New perspectives in monitoring drinking water microbial quality. Int. J. Environ. Res. Public Health 7, 4179–4202.

Fittipaldi, M., Nocker, A., and Codony, F. (2012). Progress in understanding preferential detection of live cells using viability dyes in combination with DNA amplification. J. Microbiol. Meth. *91*, 276–289.

Frankland, P., and Frankland, G.C.T. (1894). Micro-organisms in water: their significance, identification and removal, together with an account of the bacteriological methods employed in their investigation, specially designed for the use of those connected with the sanitary aspects of water supply (Longmans, Green and Co., London, UK).

Füchslin, H.P., Kötzsch, S., Keserue, H.-A., and Egli, T. (2010). Rapid and quantitative detection of *Legionella pneumophila* SG1 applying immmunomagnetic separation and flow cytometry. Cytometry A *77*, 264–274.

Füchslin, H.P., Kötzsch, S., and Egli, T. (2012). *Cryptosporidium* sp. in drinking water of rural sites in Switzerland. Swiss Med. Wkly *142*, w13683.

Gasol, J.M., and Del Giorgio, P.A. (2000). Using flow cytometry for counting natural planktonic bacteria and understanding the structure of planktonic bacterial communities. Sci. Mar. *64*, 197–224.

Gasol, J.M., del Giorgio, P.A., Massana, R., and Duarte, C.M. (1995). Active versus inactive bacteria: size-dependence in a coastal marine plankton community. Mar. Ecol. Prog. Ser. *128*, 91–97.

Gasol, J.M., Li Zweifel, U., Peters, F., Fuhrman, J.A., and Hagström, A. (1999). Significance of size and nucleic acid content heterogeneity as measured by flow cytometry in natural planktonic bacteria. Appl. Environ. Microbiol. *65*, 4475–4483.

Gillespie, S., Lipphaus, P., Green, J., Parsons, S., Weir, P., Juskowiak, K., Jefferson, B., Jarvis, P., and Nocker, A. (2014). Assessing microbiological water quality in drinking water distribution systems with disinfectant residual using flow cytometry. Water Res. *65*, 224–234.

Girones, R., Ferrus, M.A., Alonso, J.L., Rodriguez-Manzano, J., Calgua, B., de Abreu Correa, A., Hundesa, A., Carratala, A., and Bofill-Mas, S. (2010). Molecular detection of pathogens in water – The pros and cons of molecular techniques. Water Res. *44*, 4325–4339.

Günther, S., Trutnau, M., Kleinsteuber, S., Hause, G., Bley, T., Rüske, I., Harms, H., and Müller, S. (2009). Dynamics of polyphosphate-accumulating bacteria in wastewater treatment plant microbial communities detected via DAPI (4′,6′-diamino-2-phenylindole) and tetracycline labelling. Appl. Environ. Microbiol. *75*, 2111–2121.

Hammes, F.A., and Egli, T. (2005). New method for assimilable organic carbon determination using flow-cytometric enumeration and a natural microbial consortium as inoculum. Environ. Sci. Technol. *39*, 3289–3294.

Hammes, F., and Egli, T. (2010). Cytometric methods for measuring bacteria in water: Advantages, pitfalls and applications. Anal. Bioanal. Chem. *397*, 1083–1095.

Hammes, F., and Egli, T. (2011). Development of a simplified biodegradable organic matter quantification technique. Web Report #3128 (Water Research Foundation, Denver, CO, USA).

Hammes, F., Salhi, E., Köster, O., Kaiser, H.P., Egli, T., and von Gunten, U. (2006). Mechanistic and kinetic evaluation of organic disinfection by-product and assimilable organic carbon (AOC) formation during the ozonation of drinking water. Water Res. *40*, 2275–2286.

Hammes, F., Meylan, S., Salhi, E., Köster, O., Egli, T., and von Gunten, U. (2007). Formation of assimilable organic carbon (AOC) and specific natural organic matter (NOM) fractions during ozonation of phytoplankton. Water Res. *41*, 1447–1454.

Hammes, F., Berney, M., Wang, Y., Vital, M., Köster, O., and Egli, T. (2008). Flow-cytometric total bacterial cell counts as a descriptive microbiological parameter for drinking water treatment processes. Water Res. *42*, 269–277.

Hammes, F., Vital, M., Stucki, D., Lautenschlager, K., and Egli, T. (2009). Advances in microbiological methods for drinking water analysis: Flow cytometry, assimilable organic carbon and pathogen growth potential. In TECHNEAU: Safe Drinking Water from Source to Tap, van den Hoven, T. and Kanzner, C., eds (IWA Publishing, London, UK, and New York, NY, USA) pp. 259–171.

Hammes, F., Berger, C., Köster, O., and Egli, T. (2010a). Assessing biological stability of drinking water without disinfectant residuals in a full-scale water supply system. Aqua *59*, 31–40.

Hammes, F., Berney, M., Wang, Y., Vital, M., Köster, O., and Egli, T. (2010b). Measurement and interpretation of microbial adenosine tri-phosphate (ATP) in aquatic environments. Water Res. *44*, 3915–3923.

Hammes, F., Broger, T., Weilenmann, H.U., Vital, M., Helbling, J., Bosshart, U., Huber, P., Odermatt, R.P., and Sonnleitner, B. (2012). Development and laboratory-scale testing of a fully automated online flow cytometer for drinking water analysis. Cytometry A *81*, 508–516.

Helbling, J., Bosshart, U., and Hammes, F. (2012). Online-Flowzytometrie in der Praxis. Von der Platte über die Zelle zur Clusteranalyse. AQUA and GAS *5*, 28–33.

Hewitt, C.J., and Nebe-Von-Caron, G. (2004). The application of multi-parameter flow cytometry to monitor individual microbial cell physiological state. Adv. Biochem. Eng. Biotechnol. *89*, 197–223.

Hobbie, J.E., Daley, R.J., and Jasper, S. (1977). Use of Nuclepore filters for counting bacteria by fluorescence microscopy. Appl. Environ. Microbiol. *33*, 1225–1228.

Hoefel, D., Warwick, L., Grooby, W.L., Monis, P.T., Andrews, S., and Saint, C.P. (2003). Enumeration of water-borne bacteria using viability assays and flow cytometry: a comparison to culture-based techniques. J. Microbiol. Methods *55*, 585–597.

Hoefel, D., Monis, P.T., Grooby, W.L., Andrews, S., and Saint, C.P. (2005a). Profiling bacterial survival through a water treatment process and subsequent distribution system. J. Appl. Microbiol. *99*, 175–186.

Hoefel, D., Monis, P.T., Warwick, L., Grooby, W.L., Andrews, S., and Saint, C.P. (2005b). Culture-independent techniques for rapid detection of bacteria associated with loss of chloramine residual in a drinking water system. Appl. Environ. Microbiol. *71*, 6479–6488.

Hoorfar, J. (2011). Rapid detection, characterization, and enumeration of foodborne pathogens. Acta Pathol. Microbiol. Immunol. Scand. APMS *119*(Suppl. 133), 1–24.

Huq, A., Rivera, I.N.G., and Colwell, R.R. (2000). Epidemiological significance of viable but nonculturable microorganisms. In Nonculturable Microorganisms in the Environment, Colwell, R.R. and Grimes, D.J., eds. (ASM Press, Washington, DC, USA), pp. 301–323.

HyV (2008). Hygieneverordnung des Eidgenössischen Departements des Innern vom 23. November 2005 (Stand am 1. April 2008).

Ingram, M., Cleary, T.J., Price, B.J., Price, R.L. III, and Castro, A. (1982). Rapid detection of *Legionella pneumophila* by flow cytometry. Cytometry *3*, 134–137.

Jones, J., and Simon, B. (1975). An investigation of errors in direct counts of aquatic bacteria by epifluorescence microscopy, with reference to a new method for dyeing membrane filters. J. Appl. Bacteriol. *39*, 317–329.

Joux, F., and Lebaron, P. (2000). Use of fluorescent probes to assess physiological functions of bacteria at single-cell level. Microbes Infect. *2*, 1523–1535.

Kaplan, L.A., Bott, T.L., and Reasoner, D.J. (1993). Evaluation and simplification of the assimilable organic carbon nutrient bioassay for bacterial growth in drinking water. Appl. Environ. Microbiol. *59*, 1532–1539.

Keserue, H.A., Füchslin, H.P., and Egli, T. (2011). Rapid detection and enumeration of *Giardia lamblia* cysts in tap water samples by immunomagnetic separation and flow cytomentric analysis. Appl. Environ. Microbiol. *77*, 5420–5427.

Keserue, H.A., Füchslin, H.P., Wittwer M., Nguyen-Viet H., Nguyen, T.T., Surinkul, N., Koottatep, T., Schürch, N., and Egli, T. (2012a). Comparison of transportable rapid methods for environmental detection of *Giardia* spp. and *Cryptosporidium* spp. (oo)cysts in a Thailand field deployment. Environ. Sci. Technol. *46*, 8952–8959.

Keserue, H.A., Baumgartner, A., Felleisen, R., and Egli, T. (2012b). Rapid detection of total and viable *Legionella pneumophila* in tap water by immunomagnetic separation, double fluorescent staining and flow cytometry. Microb. Biotechnol. *5*, 753–763.

Koch, C., Günther, S., Desta, A.F., Hübschmann, T., and Müller, S. (2013). Cytometric fingerprinting for analyzing microbial intercommunity structure variation and identifying subcommunity function. Nat. Protoc. *8*, 190–202.

Koch, C., Harnisch, F., Schräder, U., and Müller, S. (2014). Cytometric fingerprints: evaluation fo new tools for analysing microbial community dynamics. Front. Microbiol. *5*, 273.

Koch, R. (1883). Über die neuen Untersuchungsmethoden zum Nachweis der Mikrokosmen in Boden, Luft und Wasser. Ärztliches Vereinsblatt für Deutschland, Nr. 137 (Leipzig: Kommissions-Verlag).

Koch, R. (1893). Wasserfiltration und Cholera. Med. Microbiol. Immunol. *14*, 393–426.

Köster, O. (2012). Durchflusszytometrie beim Wasserversorger. Heutige Anwendungen und künftige Bedürfnisse. Aqua & Gas 5, 24–27.

Köster, W., Egli, T., Ashbolt, N., Botzenhart, K., Burlion, N., Endo, T., Grimont, P., Guillot, E., Mabilat, C., Newport, L., *et al.* (2003). Analytical methods for microbiological water quality testing. In Assessing Microbial Safety of Drinking Water. Improving Approaches and Methods (F: OECD Publications, Paris, France), pp. 237–294.

Kötzsch, S., and Egli, T. (2012). Methodenpaket BioMig. Ermittlung des Verkeimungspotentials von Kunststoffen in Kontakt mit Trinkwasser. Aqua & Gas, No 5, 54–62.

Kötzsch, S., and Egli, T., (2013). Kunststoffe in Kontakt mit Trinkwasser. Labortests und Realität. Aqua & Gas No 3, 44–52.

Kötzsch, S., and Sinreich, M. (2014). Zellzahlen zum Grundwasser, Bestimmung mittels Durchflusszytometrie. Aqua & Gas No 3, 14–21.

Kötzsch, S., Alisch, S., and Egli, T. (2012). Durchflusszytometrische Analyse von Wasserproben. Schweizerisches Bundesamt für Gesundheit (BAG), Bern, Switzerland.

Kuenzi, S., Merenda, F., Marclay, S., Watson, D., Bielmann, J., Egli, T., and Ryser, P. (2014). BACTOSENSE – A new and easy-to-use instrument for continuous assessment of microbiological water quality. Poster abstract at: 11th IWA Leading Edge Conference on Water and Wastewater Technologies, 26–30 May, Abu Dhabi, UAE.

Lautenschlager, K., (2014). Abundance and composition of indigenous bacterial communities in a multi-step biofiltration-based drinking water treatment plant. Water Res. *62*, 40–52.

Lautenschlager, K., Boon, N., Wang, Y., Egli, T., and Hammes, F. (2010). Overnight stagnation of drinking water in household taps induces microbial growth and changes in community composition. Water Res. *44*, 4868–4877.

Lautenschlager, K., Hwang, C., Liu, W.-T., Boon, N., Köster, O., Vrouwenvelder, H., Egli, T., and Hammes, F. (2013). A microbiology-based multi-parametric approach towards assessing biological stability in drinking water distribution networks. Water Res. *47*, 3015–3025.

LeChevallier, M.W., Shaw, N.E., Kaplan, L.A., and Bott, T.L. (1993). Development of a rapid assimilable organic carbon method for water. Appl. Environ. Microbiol. *29*, 1526–1531.

Leclerc, H., and Moreau, A. (2002). Microbiological safety of natural mineral water. FEMS Microbiol. Rev. *26*, 207–222.

Lee, A., Francisque, A., Najjaran, H., Rodriguez, M.J., Hoorfar, M., Imran, S.A., and Sadiq, R. (2012). Online monitoring of drinking water quality in a distribution network: a selection procedure for suitable water quality parameters and sensor devices. Int. J. Syst. Assur. Eng. Manag. *3*, 323–337.

Lendenfeld, T.D.S., A-3100 St. Pölten, A.T. (2008). Apparatus for monitoring water for microbial germs. mbOnline GmbH (Steiner Landstrasse 27a, A-3500 Krems, AT). Cited in Ryzinska-Paier *et al.*, 2014.

Lepeuple, A.S., Gioluppe, S., Pierlot, E., and de Roubin, M.R. (2004). Rapid and automated detection of fluorescent total bacteria in water. Int. J. Food Microbiol. *92*, 327–332.

Li, W.K.W., Jellett, J.F., and Dickie, P.M. (1995). DNA distributions in planktonic bacteria stained with TOTO or TO-PRO. Limnol. Oceanogr. *40*, 1485–1495.

Lindberg, W., Ruzicka, J., and Christian, G.D. (1992). Flow injection cytometry: a new approach for sample and solution handling in flow cytometry. Cytometry *14*, 230–236.

Lipphaus, P., Hammes, F., Kötzsch, S., Green, J., Gillespie, S., and Nocker, A. (2014). Microbiological tap water profile of a medium-sized building and effect of water stagnation. Environ. Technol. *35*, 620–628.

Lisle, J.T., Hamilton, M.A., Willse, A.R., and McFeters, G.A. (2004). Comparison of fluorescence microscopy and solid-phase cytometry methods for counting bacteria in water. Appl. Environ. Microbiol. *70*, 5343–5348.

Liu, G., Verberk, J.Q.J.C., and van Dijk, J.C. (2013a). Bacteriology of drinking water distribution systems: an integral and multidimensional review. Appl. Microbiol. Biotechnol. *97*, 9265–9276.

Liu, G., van der Mark, E.J., Verberk, J.Q.J.C., and van Dijk, J.C. (2013b). Flow cytometery total cell counts: A field study assessing microbiological water quality and growth in unchlorinated drinking water distribution systems. BioMed Res. I. *2013*, 595872.

McFeters, G.A., Pyle, B.H., Lisle, J.T., and Broadaway, S.C. (1999). Rapid direct methods for enumeration of specific, active bacteria in water and biofilms. Symp. Ser. Soc. Appl. Microbiol. *85*, 193S-200S.

MbOnline (2014). MbOnline. http://www.mbonline.at/en/ (accessed 9 July 2014).

Medema, G.J., Payment, P., Dufour, A., Robertson, W., Waite, M., Hunter, P., Kirby, R., and Anderson, Y. (2003). Safe drinking water: an on-going challenge. In Assessing Microbial Safety of Drinking Water. Improving Approaches and Methods (F: OECD Publications, Paris, France), pp. 11–45.

Mogensen, C.T., Bentien, A., Lau, M., Hojris, B., Iversen, K., Klinting, M., Winter Berg, T., Agersnap, N., and Valvik, M., (2011). Sensing technologies for global health, military medicine, disaster response, and environmental monitoring: and biometric technology for human identification VIII. Proc. of SPIE *8029*, 80290K-1.

Monfort, P., and Baleux, B. (1992). Comparison of flow cytometry and epifluorescence microscopy for counting bacteria in aquatic ecosystems. Cytometry *13*, 188–192.

Morita, R.Y. (1988). Bioavailability of energy and its relationship to growth and starvation survival in nature. Can. J. Microbiol. *34*, 436–442.

Müller, S., and Nebe-von-Caron, G. (2010). Functional single-cell analyses: flow cytometry and cell sorting of microbial populations and communities. FEMS Microbiol. Rev. *34*, 554–587.

National Academy of Science (2006). Drinking water distribution systems: assessing and reducing risks. National Research Council U.S. (The National Academies Press, Washington, DC, USA).

Nescerecka, A., Rubulis, J., Vital, M., Juhna, T., and Hammes, F. (2014). Biological instability in a chlorinated drinking water distribution network. PLoS ONE 9, e96354.

Nexmann Jacobsen, C., Fremming, C., and Jakobsen, M. (1997). Immunomagnetic separation of *Listeria monocytogenes* for flow cytometric determination of viable cells in liquid. J. Microbiol. Methods *31*, 75–81.

Omann, G.M., Coppersmith, W., Finney, D.A., and Skalar, L.A. (1985). A convenient on-line device for reagent addition sample mixing, and temperature control of cell suspensions in flow cytometry. Cytometry *6*, 69–73.

Organization for Economic Co-operation and Development (OECD) (1996). Biotechnology for Water Use and Conservation: The Mexico '96 Workshop (OECD, Geneva, Switzerland).

Organization for Economic Co-operation and Development (OECD) (1999). Health Policy Brief. Molecular technologies for safe drinking water: results from the Interlaken Workshop, Switzerland, 5–8 July 1998.

Organization for Economic Co-operation and Development (OECD) (2001). Establishing links between drinking water and infectious diseases. Working Party on Biotechnology, Report from Basingstoke Meeting, UK.

Organization for Economic Co-operation and Development (OECD)/World Health Organization (WHO) (2003). Assessing Microbial Safety of Drinking Water. Improving Approaches and Methods (F: OECD Publications, Paris, France).

Payment, P., Waite, M., and Dufour, A. (2003). Introducing parameters for the assessment of drinking water quality. In Assessing Microbial Safety of Drinking Water. Improving Approaches and Methods (F: OECD Publications, Paris, France), pp. 47–77.

Pepper, I.L., Rusin, P., Quintanar, D.R., Haney, C., Josephson, K.L., and Gerba, C.P. (2004). Tracking the concentration of heterotrophic plate count bacteria from the source to the consumer's tap. Intern. J. Food Microbiol. *92*, 289–295.

Péré (1891). Contribution à l'étude des eaux d'Alger. Annales de l'Institut Pasteur 2, 79–91.

Phe, M.H., Dosset, M., and Block, J.-C. (2004). Chlorination effect on the fluorescence of nucleic acid staining dyes. Water Res. *38*, 3729–3737.

Phe, M.H., Dosset, M., Guilloteau, H., and Block, J.-C. (2005). Nucleic acid fluorochromes and flow cytometry prove useful in assessing the effect of chlorination on drinking water bacteria. Water Res. *39*, 3618–3628.

Phillips, A.P., and Martin, K.L. (1988). Limitations of flow cytometry for the specific detection of bacteria in mixed populations. J. Immunol. Meth. *106*, 109–117.

Pinder, A.C., and McClelland, R.G. (1994). Rapid assay for pathogenic *Salmonella* organisms by immunofluorescence flow cytometry. J. Microsc. *176*, 17–22.

Pinder, A.C., Purdy, P.W., Poulter, S.A.G., and Clark, D.C. (1990). Validation of flow cytometry for rapid enumeration of bacterial concentrations in pure cultures. J. Appl. Bacteriol. *69*, 92–100.

Pinto, A.J., Xi, C., and Raskin, L. (2012). Bacterial community structure in the drinking water microbiome is governed by filtration processes. Environ. Sci. Technol. *46*, 8851–8859.

Poitelong, J.-B., Joyeux, M., Welté, B., Duguet, J.-P., and Scott-DuBow, M. (2011). Le réseau de distribution d'eau potable: un écosystème complexe lié à des enjeux de santé publique (The drinking water distribution network: a complex ecosystem related to public health). J. Water Sci. *24*, 383–418.

Porter, K.G., and Feig, Y.S. (1980). The use of DAPI for identifying and counting aquatic microflora. Limnol. Oceanogr. *25*, 943–948.

Porter, J., Deere, D., Pickup, R., and Edwards, C. (1996). Fluorescent probes and flow cytometry: new insights into environmental bacteriology. Cytometry *23*, 91–96.

Porter, J., Deere, D., Hardman, M., Edwards, C., and Pickup, R. (1997). Go with the flow – use of flow cytometry in environmental microbiology. FEMS Microbiol. Ecol. *24*, 93.101.

Prest, E.I., Hammes, F., Kötzsch, S., van Loosdrecht, M.C.M., and Vrouwenvelder, J.S. (2013). Monitoring microbiological changes in drinking water systems using a fast and reproducible flow cytometric method. Water Res. *47*, 7131–7142.

Prest, E.I., El-Chakhtoura, J., Hammes, F., Saikaly, P.E., Loosdrecht, M.C.M., and Vrouwenvelder, J.S. (2014). Combining flow cytometry and 16S rRNA gene pyrosequencing: a promising approach for drinking water monitoring and characterization. Water Res. *63*, 179–189.

Pyle, B., Broadaway, S., and McFeters, G. (1999). Sensitive detection of *Escherichia coli* O157:H7 in food and water by immunomagnetic separation and solid-phase laser cytometry. Appl. Environ. Microbiol. *65*, 1966–1972.

Ramalho, R., Cunha, J., Teixeira, P., and Gibbs, P.A., (2001). Improved methods for the enumeration of heterotrophic bacteria in bottled mineral waters. J. Microbiol. Methods *44*, 97–103.

Ramseier, M.K., Peter, A., Traber, J., and von Gunten, U. (2011a). Formation of assimilable organic carbon during oxidation of natural waters with ozone, chlorine dioxide, chlorine, permanganate, and ferrate. Water Res. *45*, 2002–2010.

Ramseier, M.K., von Gunten, U., Freihofer, P., and Hammes, F. (2011b). Kinetics of membrane damage to high (HNA) and low (LNA) nucleic acid bacterial clusters in drinking water by ozone, chlorine, chlorine dioxide, monochloramine, ferrate(VI), and permanganate. Water Res. *45*, 1490–1500.

Reynolds, D., and Fricker, C. (1999). Application of laser scanning for the rapid and automated detection of bacteria in water samples. J. Appl. Microbiol. *86*, 785–795.

Reynolds, D.T., Fricker, E.J., Purdy, D., and Fricker, C.R. (1997). Development of a rapid method for the enumeration of bacteria in potable water. Water Sci. Technol. *35*, 433–436.

Rittmann, B.E., and Snoeyink, V.L. (1984). Achieving biologically stable drinking water. J. Am. Water. Works Assoc. *76*, 106–104.

Robertson, B.R., and Button, D.K. (1989). Characterizing aquatic bacteria according to population, cell size and apparent DNA content by flow cytometry. Cytometry *10*, 70–76.

Robertson, B.R., Button, D.K., and Koch, A. (1998). Determination of the biomass of small bacteria at low concentrations in a mixture of species with forward light scatter measurements by flow cytometry. Appl. Environ. Microbiol. *64*, 3900–3909.

Ryzinksa-Paier, G., Lendenfeld, T., Correa, K., Stadler, P., Blaschke, A.P., Mach, R.L., Stadler, H., Kirschner, A.K.T., and Farnleiter, A.H. (2014). A sensitive and robust method for automated on-line monitoring of enzymatic activities in water and water resources. Water Sci. Technol. *69*, 1349–1358.

Sakamoto, C., Yamaguchi, M., and Nasu, M. (2005). Rapid and simple quantification of bacterial cells by using a microfluidic device. Appl. Environ. Microbiol. *71*, 1117–1121.

Sakamoto, C., Yamaguchi, N., Yamada, M., Hagase, H., Seki, M., and Nasu, M. (2007). Rapid quantification of bacterial cells in potable water using a simplified microfluidic device. J. Microbial. Methods *68*, 643–647.

Schardinger, F. (1892). Über das Vorkommen Gährung erregender Spaltpilze im Trinkwasser und ihre Bedeutung für die hygienische Beurtheilung desselben. Wiener Klin. Wochenschr. *5*, 403–405.

Schneebeli, H.R. (2012). Occurrence, growth and competitive properties of *Listeria* spp. in the aquatic environment. PhD Thesis ETH Nr. 20763. Swiss Federal Institute of Technology, Zurich, Switzerland.

Schoenen, D. (2011). Mikrobiologie des Trinkwassers (D: Oldenbourg Industrieverlag GmbH, München, Germany).

Sekar, R., Fuchs, B.M., Amann, R., and Pernthaler, J. (2004). Flow sorting of marine bacterioplankton after fluorescence *in situ* hybridization. Appl. Environ. Microbiol. *70*, 6210–6219.

Servais, P., Anzil, A., and Ventresque, C. (1989). Simple method for determination of biodegradable dissolved organic carbon in water. Appl. Environ. Microbiol. *55*, 2732–2734.

Shapiro, H.M. (2003). Practical Flow Cytometry, 4th edn. (Wiley-Liss, New York, NY, USA).

Siebel, E., Wang, Y., Egli, T., and Hammes, F. (2008). Correlations between total cell concentration, total adenosine tri-phosphate concentration and heterotrophic plate counts during microbial monitoring of drinking water. Drinking Water Engin. Sci. *1*, 1–6.

Signor, R.S., and Ashbolt, N.J. (2006). Pathogen monitoring offers questionable protection against drinking water risks: a QMRA (Quantitative Microbial Risk Analysis) approach to assess management strategies. Water Sci. Technol. *54*, 261–268.

SLMB (2012). Determining the total cell count and ratios of high and low nucleic acid content cells in fresh water using flow cytometry. Method 333.1. In Swiss Food Book, Swiss Federal Office of Public Health. Bern, Switzerland.

Smeets, P.W.M.H., Rietveld, L.C., van Dijk, J.C., and Medema, G.J. (2010). Practical applications of quantitative microbial risk assessment (QMRA) for water safety plans. Water Sci. Technol. *61*, 1561–1571.

Stadler, P., Farnleiter, A.H., and Zessner, M. (2013). Interaction of on-site and near real time measured turbidity and enzyme activity in stream water. Geophys. Res. Abstr. *15*, EGU2013-8086.

Staley, J.T., and Konopka, A. (1985). Measurement of *in situ* activities of nonphotosynthetic microorganisms in aquatic and terrestrial habitats. Ann. Rev. Microbiol. *39*, 321–346.

Storey, M.V., van der Gaag, B., and Burns, B.P. (2011). Advances in on-line drinking water quality monitoring and early warning systems. Water Res. *45*, 741–747.

Swiss Food Book (2012). Swiss Food Book: Analysis method 333.1, December 2012. Determining the total cell count and ratios of high and low nucleic acid content cells in freshwater using flow cytometry. Federal Office of Public Health, Bern, Switzerland.

Szewzyk, U., Szewzyk, R., Manz, W., and Schleifer, K.-H. (2000). Microbiological safety of drinking water. Ann. Rev. Microbiol. *54*, 81–127.

Thyssen, M., Tarran, G.A., Zubkov, M., Holland, R.J., Grégori, G., Burkill, P.H., and Denis, M. (2008). The emergence of automated high-frequency flow cytometry: revealing temporal and spatial phytoplankton variability. J. Plank. Res. *30*, 333–343.

Troussellier, M., Courties, C., and Vaquer, A. (1993). Recent applications of flow cytometry in aquatic microbial ecology. Biol. Cell *78*, 111–121.

Tyndall, R.L., Hand, R.E. Jr, Mann, R.C., Evans, C., and Jernigan, R. (1985). Application of flow cytometry to detection and characterization of *Legionella* spp. Appl. Environ. Microbiol. *49*, 852–857.

United States Environmental Protection Agency (2005). Method 1623: Cryptosopridium and Giardia in water by filtration/IMS/FA, EPA 815-R-05-002 (Office of Water).

van der Kooij, D. (1992). Assimilable organic carbon as an indicator of bacterial regrowth. J. Am. Water Works Assoc. *84*, 57–65.

van der Kooij, D. (2000). Biological stability: a multidimensional quality aspect of treated water. Water, Air, and Soil Pollution *123*, 25–34.

van der Kooij, D., Visser, A., and Hijnen, W.A.M. (1982). Determination of easily assimilable organic carbon in drinking water. J. Am. Water Works Assoc. *74*, 540–545.

van der Kooij, D., van Lieverloo, J.H.M., Schellart, J.A., and Hiemstra, P. (1999). Distributing drinking water without disinfectant: highest achievement of height of folly? Aqua *48*, 31–37.

van Nevel, S., Kötzsch, S., Weilenmann, H.-U., Boon, N., and Hammes, F. (2013). Routine bacterial analysis with automated flow cytometry. J. Microbiol. Meth. *94*, 73–76.

Vaz-Moreira, I., Egas, C., Nunes, O.C., and Manaia, C.M. (2012). Bacterial diversity from the source to the tap: a comparative study based on 16S rRNA gene-DGGE and culture-dependent methods. FEMS Microbiol. Ecol. *83*, 361–374.

Vila-Costa, M., Gasol, J.M., Sharma, S., and Moran, M.A. (2012). Community analysis of high- and low-nucleic acid-containing bacteria in NW Mediterranean coastal waters using 16S rDNA pyrosequencing. Environ. Microbiol. *14*, 1390–1402.

Vital, M. (2010). Growth of pathogenic bacteria in freshwater and their competition with autochtonous bacterial flora. PhD thesis. Diss. ETH Nr. 19111, Swiss Federal Institute of Technology, Zurich, Switzerland.

Vital, M., Füchslin, H.P., Hammes, F., and Egli, T. (2007). Growth of *Vibrio cholera* O1 Ogawa Eltor in freshwater. Microbiology *153*, 1993–2001.

Vital, M., Hammes, F., and Egli, T. (2008). *E. coli* O157 can grow in natural freshwater at low carbon concentrations. Environ. Microbiol. *10*, 2387–2396.

Vital, M., Stucki, D., Egli, T., and Hammes, F. (2010). Evaluating the growth potential of pathogenic bacteria in water. Appl. Environ. Microbiol. *76*, 6477–6484.

Vital, M., Dignum, M., Magic-Knezev, A., Ross, P., Rietveld, L., and Hammes, F. (2012a). Flow cytometry and adenosine tri-phosphate analysis: alternative possibilities to evaluate major bacteriological changes in drinking water treatment and distribution systems. Water Res. *46*, 4665–4676.

Vital, M., Hammes, F., and Egli, T. (2012b). Competition of *Escherichia coli* O157 with a drinking water bacterial community at low concentrations. Water Res. *46*, 6279–6290.

Vives-Rego, J., Lebaron, P., and Nebe-von Caron, G. (2000). Current and future applications of flow cytometry in aquatic microbiology. FEMS Microbiol. Rev. *24*, 429–228.

Wang, Y., Hammes, F., Boon, N., and Egli, T. (2007). Quantification of the filterability of freshwater bacteria through 0.45, 0.22, and 0.1 µm pore size filters and shape-dependent enrichment of filterable bacterial communities. Environ. Sci. Technol. *41*, 7080–7086.

Wang, Y., Hammes, F., and Egli, T. (2008a). The impact of industrial-scale cartridge filtration on the native microbial communities from groundwater. Water Res. *42*, 4319–4326.

Wang, Y., Hammes, F., Düggelin, M., and Egli, T. (2008b). Influence of size, shape and flexibility on bacterial passage through micropore membrane filters. Environ. Sci. Technol. *42*, 6749–6754.

Wang, Y., Hammes, F., Boon, N., Chami, M., and Egli, T. (2009). Isolation and characterization of low nucleic acid content bacteria. ISME J. *3*, 889–902.

Wang, Y., Hammes, F., de Roy, K., Verstraete, W., and Boon, N. (2010). Past, present and future applications of flow cytometry in aquatic microbiology. Trends Biotechnol. *28*, 416–424.

Weinbauer, M. (2004). Ecology of prokaryotic viruses. FEMS Microbiol. Rev. *28*, 127–181.

Wildeboer, D., Amirat, L., Price, R.G., and Abuknesha, R.A. (2010). Rapid detection of *Escherichia coli* in water using a hand-held fluorescence detector. Water Res. *44*, 2621–2628.

World Health Organization (WHO) (2003). Guidelines for Drinking-water Quality, Volume 1: recommendations, 3rd edn., incorporating the first and second addenda. (World Health Organization, Geneva, Switzerland).

World Health Organization (WHO) (2008). Guidelines for Drinking-water Quality. Third edition, incorporating the first and second addenda, Vol. 1, Recommendations, 3rd edn. (World Health Organization, Geneva, Switzerland).

Yamaguchi, N., and Nasu, M. (1997). Flow cytometric analysis of bacterial respiratory and enzymatic activity in the natural aquatic environment. J. Appl. Microbiol. *83*, 43–52.

Yamaguchi, N., Torii, M., Uebayashi, Y., and Nasu, M. (2011). Rapid, semiautomated quantification of bacterial cells in freshwater by using a microfluidic device for on-chip staining and counting. Appl. Environ. Microbiol. 77, 1736–1739.

Zimmermann, R. (1977). Estimation of bacterial number and biomass by epifluorescence microscopy and scanning electron microscopy. In Microbial Ecology of a Brackish Water Environment, Rheinheimer, G. ed. (Springer Verlag, New York, NY, USA), pp. 103–120.

Zhao, X., Dong, T., Yang, Z., Pires, N., and Hoivik, N. (2012). Compatible immuno-NASBA LOC device for quantitative detection of waterborne pathogens: design and validation. Lab Chip *12*, 602–612.

ZoBell, C.E., and Grant, C.W. (1943). Bacterial utilization of low concentrations of organic matter. J. Bacteriol. *45*, 555–564.

Index

www.ingramcontent.com/pod-product-compliance
Lightning Source LLC
LaVergne TN
LVHW071337210926
848486LV00003B/168

9781910190111